Starch-Based Sustainable Food Packaging

APPLICATIONS OF STARCHES IN FOOD AND PACKAGING

STARCH-BASED SUSTAINABLE FOOD PACKAGING

CURRENT PRACTICES AND FUTURE TRENDS

Edited by

BILGE NAZLI ALTAY

Department of Packaging and Graphic Media Science, Rochester Institute of Technology, United States

BURAK AKSOY

Auburn University, College of Agriculture, School of Fisheries, Aquaculture & Aquatic Sciences, Auburn, AL, United States

Series Editor

SNEH PUNIA BANGAR

ELSEVIER

ACADEMIC PRESS

An imprint of Elsevier

Academic Press is an imprint of Elsevier
125 London Wall, London EC2Y 5AS, United Kingdom
50 Hampshire Street, 5th Floor, Cambridge, MA 02139, United States

ISBN: 978-0-443-36346-7

For information on all Academic Press publications visit our website at https://www.elsevier.com/books-and-journals

Publisher: Mara Conner
Acquisitions Editor: Palak Gupta
Editorial Project Manager: Ritika Dagar
Production Project Manager: Jenela Priscy
Cover Designer: Greg Harris

Typeset by TNQ Tech

Contents

Contributors

Özlem Kübra Akdoğan
Rutgers, The State University of New Jersey, Global Food Technology and Innovation, New Brunswick, NJ, United States; Eastern Michigan University, College of Technology, Polymers and Coatings Technology, Ypsilanti, MI, United States

Burak Aksoy
Auburn University, College of Agriculture, School of Fisheries, Aquaculture & Aquatic Sciences, Auburn AL, United States

Bilge Nazli Altay
Rochester Institute of Technology, College of Engineering Technology, Department of Packaging and Graphic Media Science, Rochester, NY, United States

Emine Arman Kandirmaz
Faculty of Applied Sciences, Marmara University, Istanbul, Turkiye

Aslı Beyler Cigil
Department of Chemistry and Chemical Processing Technologies, Technical Sciences Vocational School, Gazi University, Ankara, Turkiye

Hatice Birtane
Department of Chemistry, Faculty of Science, Marmara University, Istanbul, Turkiye

Fayçal Dergal
Scientific and Technical Research Center in Physico-Chemical Analysis (CRAPC), Tipaza, Algeria; Laboratory of Catalysis and Synthesis in Organic Chemistry (LCSCO), Faculty of Sciences, University of Tlemcen, Tlemcen, Algeria

Md Rabbi Hasan
Department of Paper and Packaging Technology, Indian Institute of Technology, Roorkee (IIT ROORKEE) Roorkee, Haridwar, Uttarakhand, India

Md Belal Hossain
Department of Paper and Packaging Technology, Indian Institute of Technology, Roorkee (IIT ROORKEE) Roorkee, Haridwar, Uttarakhand, India

Bekir Keskin
Ankara Haci Bayram Veli University, Fine Arts Faculty, Visual Communication Department, Ankara, Turkiye

Ramzi Khiari
University of Grenoble Alpes, CNRS, Grenoble INP, LGP2, Grenoble, France; Department of Textile, Higher Institute of Technological Studies of Ksar Hellal, Ksar Hellal, Tunisia

Engin Kocatürk
Department of Nanotechnology Engineering, Zonguldak Bülent Ecevit University, Zonguldak, Turkiye

Md Raju Mia
Department of Paper and Packaging Technology, Indian Institute of Technology, Roorkee (IIT ROORKEE) Roorkee, Haridwar, Uttarakhand, India

Tomasz Tadeusz Murawski
Faculty of Mechanical and Industrial Engineering, Warsaw University of Technology, Warsaw, Poland

Arif Ozcan
Faculty of Applied Sciences, Department of Printing Technologies, Marmara University, Istanbul, Turkey; Rochester Institute of Technology, College of Engineering Technology, Department of Packaging and Graphic Media Science, Rochester, NY, United States

Zafer Özomay
Marmara University, Faculty of Applied Sciences, Department of Printing Technologies, Istanbul, Turkiye

Magdolna Pal
Faculty of Technical Sciences, Department of Graphic Engineering and Design, University of Novi Sad, Novi Sad, Serbia

Cengiz Şahin
Ankara Yıldırım Beyazit University, Faculty of Architecture and Fine Arts, Department of Visual Communication Design, Ankara, Turkiye

M. Özgür Seydibeyoğlu
Department of Metallurgical and Materials Engineering, Izmir Katip Celebi University, Izmir, Turkiye

Afreen Sultana
Department of Food, Nutrition and Packaging Sciences, Clemson University, Clemson, SC, United States

Jerzy Szałapak
Faculty of Mechanical and Industrial Engineering, Warsaw University of Technology, Warsaw, Poland

Zuzanna Żołek-Tryznowska
Faculty of Mechanical and Industrial Engineering, Warsaw University of Technology, Warsaw, Poland

Acknowledgments

Chapter 6 was carried out within the framework of the DurInnPack project, part of the PRIMA program supported by the European Union. This project has received funding from the French National Research Agency (ANR) under grant number ANR-24-P013-0002 and from the Algerian Ministry of Higher Education and Scientific Research (MESRS) through the General Directorate for Scientific Research and Technological Development (DGRSDT).

CHAPTER ONE

Introduction to starch: Next-generation sustainable packaging material

Bekir Keskin[1], Zafer Özomay[2], Cengiz Şahin[3], Burak Aksoy[4]
[1]Ankara Haci Bayram Veli University, Fine Arts Faculty, Visual Communication Department, Ankara, Turkiye
[2]Marmara University, Faculty of Applied Sciences, Department of Printing Technologies, Istanbul, Turkiye
[3]Ankara Yıldırım Beyazit University, Faculty of Architecture and Fine Arts, Department of Visual Communication Design, Ankara, Turkiye
[4]Auburn University, College of Agriculture, School of Fisheries, Aquaculture & Aquatic Sciences, Auburn AL, United States

1.1 Introduction

The interest in biodegradable and renewable alternatives is increasing because of the global concerns over plastic pollution and the ecological implications of fossil fuel-based packaging materials. The long-term persistence of single-use plastics in the environment, microplastic formation, and carbon footprint have made it mandatory for policymakers and industry to seek sustainable solutions (Geyer et al., 2017; European Commission, 2019). The European Union's Directive 2019/904, for instance, outlines a strategy to phase out single-use plastic products and promote the utilization of biodegradable materials, such as starch (European Commission, 2019; Newburger, 2022). Single-use plastic policies in North America have also centered on downstream waste management with particular emphasis on expanding recycling/setting recycling targets, recycled-content mandates, and extended producer responsibility (EPR) for packaging (EPA, 2024). They stress targeted product bans and fees (bags, straws, foam, etc.) and litter prevention and clean-ups, along with limited substitution with compostable and biobased alternatives (SOR, 2022). Within these evolving paradigms, starch has emerged as a promising biopolymer for food packaging applications. Derived from renewable agricultural sources such as corn, potato, cassava, wheat, and rice, starch has attracted significant research interest because of its inherent film-forming capability, widespread availability, biodegradability, and low cost. In addition, its natural origin and compatibility with existing food systems position starch as an attractive

Starch-Based Sustainable Food Packaging
ISBN: 978-0-443-36346-7
https://doi.org/10.1016/B978-0-443-36346-7.00006-0

alternative to conventional petroleum-based packaging materials in the transition toward more sustainable and circular packaging solutions (Tharanathan, 2003; KARAKUŞ et al., 2021).

Starch extraction encompasses a continuum of processing strategies that reflect increasing levels of technological intervention, material specificity, and process control. At one end of this continuum, traditional aqueous-based methods exploit the intrinsic physicochemical properties of starch granules, namely their insolubility in cold water and higher density relative to most cellular components, to achieve separation through mechanical disintegration, washing, and gravitational settling (Dorantes-Fuertes et al., 2024). These approaches are particularly well suited to starch-rich roots, tubers, and cereals, where the cellular matrix can be readily disrupted and starch granules released with minimal chemical or thermal input. At the other end, modern extraction methodologies integrate mechanical intensification, phase separation technologies, and chemical or enzymatic aids to address limitations associated with yield variability, impurity carryover, and source-dependent structural complexity (Dorantes-Fuertes et al., 2024). Techniques such as controlled dry processing, pressing, centrifugation, solvent-assisted fractionation, and enzyme-mediated cell wall and protein degradation enable more selective liberation and recovery of starch, often producing materials with higher purity and more reproducible functional characteristics (Dorantes-Fuertes et al., 2024). The selection of an extraction route is therefore not merely operational but strategic, governed by the botanical origin of the starch, the morphology and composition of the plant tissue, processing scale, and the functional performance required in downstream food, industrial, or biobased applications. These diverse extraction paradigms illustrate how starch processing has evolved from empirically driven practices to increasingly tailored, science-based systems that directly influence starch structure–property relationships.

As an emerging material, thermoplastic starch (TPS) has gained increasing importance due to its compatibility with conventional polymer processing techniques. In recent years, TPS has been extensively investigated as a sustainable alternative to fossil-based polymers because of its ability to form transparent and flexible films while exhibiting moderate mechanical performance. Typical tensile strengths in the range of approximately 4.5–8 MPa and elongation at break values reaching up to about 100% have been reported, with these properties being strongly dependent on starch composition, plasticizer content, and processing conditions

(Domene-López et al., 2019). TPS also exhibits good film-forming ability and relatively homogeneous amorphous structures, which facilitate processing. However, TPS has important limitations, particularly its high sensitivity to moisture and relatively low thermal and mechanical stability compared with conventional plastics. These drawbacks can partially be mitigated through formulation strategies such as blending with additives (e.g., agar or waxes) or through advanced processing techniques that improve plasticization and intermolecular interactions. As a result, TPS is generally considered suitable for biodegradable packaging and other short-lifetime applications where full durability is not required. Chemical modifications such as esterification (the addition of ester bonds), oxidation (structural modification) and cross-linking (the linking together of polymer chains) alongside nanocomposite additives can significantly mitigate these disadvantages (Avérous and Halley, 2009; Lauer and Smith, 2020).

In recent years, active and smart packaging concepts have been integrated into starch-based packaging, imparting functional properties through plant extracts and antimicrobial agents (Cazón, P., Velazquez, G., Ramírez, J. A., & Vázquez, M, 2020). These developments offer major benefits in terms of food safety and optimizing shelf life. However, fundamental challenges remain, including production costs, water sensitivity, deficiencies in recycling infrastructure, and potential toxicity risks (Zimmermann et al., 2020; Rosenboom et al., 2022).

Chemical modification of starch via crosslinking with citric acid or sodium trimetaphosphate and sodium tripolyphosphate (STMP/STPP) or hydrophobic modification through silanization and esterification improves technical properties such as water resistance, tensile strength, and thermal stability for application in food packaging. However, these modifications may decrease biodegradability by limiting water penetration and access of enzymes as well as may compromise the recyclability of the materials they are applied to and limit their reprocessibility (N et al., 2024; Salgado-Delgado et al., 2022).

The following section provides an overview of starch-based packaging materials, with an emphasis on their fundamental characteristics, production routes, and application potential. It first examines starch sources and intrinsic properties, along with modification and improvement strategies aimed at enhancing performance. The discussion then addresses the advantages and limitations of starch-based packaging, current and emerging industrial applications, recycling and end-of-life considerations, and the role of these materials within a circular economy framework.

1.2 Starch sources

Starch, the main storage carbohydrate synthesized by plants through photosynthesis, accumulates in granular structures within kernels, endosperms, tubers, and root tissues. The botanical origin of starch obtained on an industrial scale determines its performance across a wide range of characteristics. Key source-dependent characteristics include granule morphology, the amylose and amylopectin ratio, phosphorus content, gelatinization behavior, and ultimately, the film and barrier properties obtained (Hoover, 2001; Tester et al., 2004). Therefore knowing which plant the starch is derived from is critical when designing sustainable and biodegradable packaging materials.

While the global starch industry relies heavily on cereal sources such as corn, wheat, and rice, as well as tuber and root crops such as potatoes, cassava, tapioca, sweet potatoes, and arrowroot; legumes such as peas, chickpeas, and lentils; and, pseudocereals such as quinoa and amaranth have increasingly become alternative sources (N et al., 2024; Moorthy, 2002; García-Guzmán et al., 2022; Ratnayake et al., 2002; Fatima et al., 2024). This botanical diversity enables tailoring of functional properties for packaging applications such as targeted mechanical strength, flexibility, transparency, gas, water vapor, oil and grease barrier properties (Chi et al., 2020). In addition, starch-based packaging has the potential to reduce the carbon footprint and resource consumption when linked to local or regional agricultural production patterns (FAOSTAT, 2023; SOR, 2022; Rosenboom et al., 2022). Fig. 1.1 illustrates starch sources across several categories, including cereal, tuber, and root starches; legume and pseudocereal starches; and starch derived from industrial by-products, together with their modifications and starch-based applications. The discussion also examines the functional differences among these starch sources, along with their environmental and economic implications.

1.2.1 Cereal starches

Cereals are the foundation of the global industrial starch supply. The main raw materials for starch are corn (also called maize), wheat, and rice. Corn-based starch dominates the global commercial starch supply due to its high starch yield, existing wet milling infrastructure, and agricultural production volume (FAOSTAT, 2023; Fatima et al., 2024; Tester et al., 2004). According to FAOSTAT data, global corn production exceeded 1.1 billion

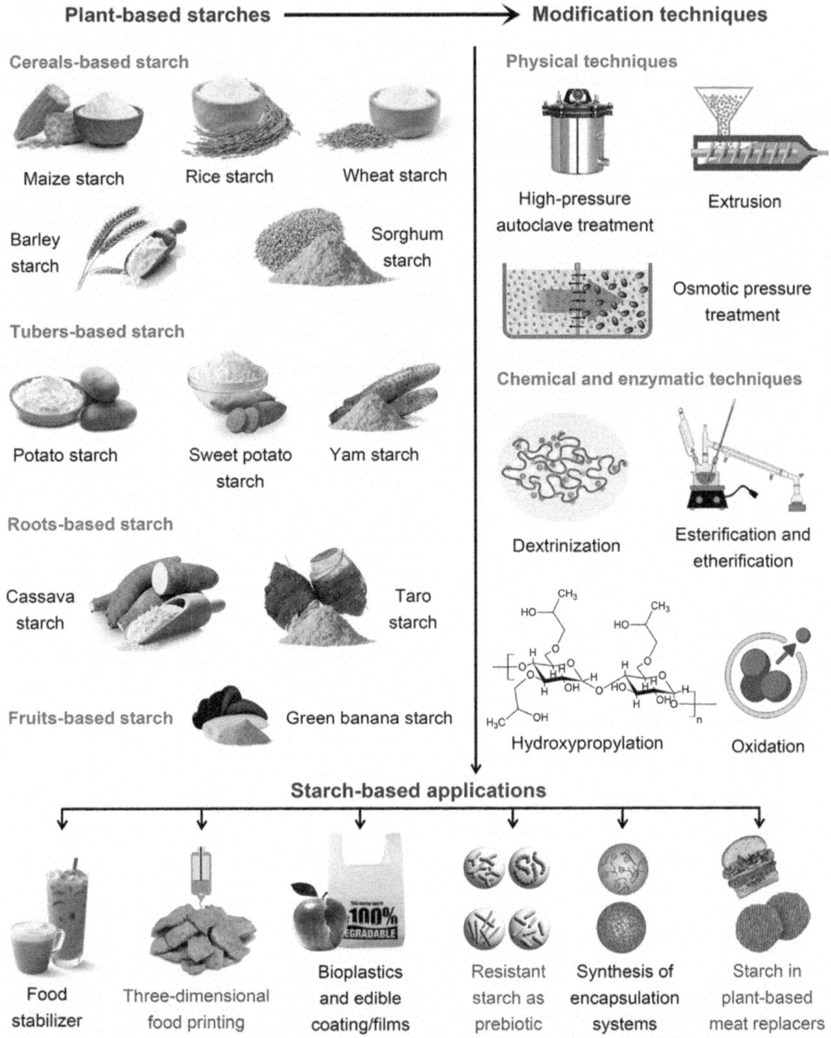

Figure 1.1 Overview of plant-derived starches, modification strategies, and their applications (Rashwan et al., 2024).

tons in 2022. This volume is distributed among the food, feed, and industrial channels (starches, sweeteners, and bioplastics), providing a robust supply of raw materials for industrial processing (FAOSTAT, 2023).

Production-wise, corn is produced on a large scale all over the globe, with top producers being the United States, China, and Brazil, and it is processed into a wide range of food and industrial items such as starch, sweeteners, oil, glue, industrial alcohol, and fuel ethanol. Particularly in the United States, a large portion of the corn crop has been used for ethanol, leading to increased competition among uses and rising prices and availability for food and industrial starch applications (Ranum et al., 2014).

Corn starch is the starch derived from the endosperm of the corn kernel, and its availability is one of the main drivers that make corn a great global source. Corn has approximately 72% starch (with around 10% protein and 4% fat), so it offers a huge, cheap carbohydrate source. At the molecular level, it is mainly amylose (mostly linear glucose chains with $\alpha(1 \rightarrow 4)$ links) and amylopectin (branched chains with $\alpha(1 \rightarrow 4)$ backbones and $\alpha(1 \rightarrow 6)$ branch points), which explains why corn starch thickens when heated in water (gelatinization) and can form gels and films on cooling (often stronger with higher-amylose starches). Corn starch typically contains 25%–30% amylose and 70%–75% amylopectin. High amylose specialty lines form more rigid, water-resistant film structures due to intense molecular interactions, making them ideal for bioplastic and packaging applications (Shogren, 1996; Avérous and Halley, 2009; Ranum et al., 2014).

Wheat starch is particularly important in Europe for the grain processing and baking industries. When starch is separated from wheat kernels, the gluten part is considered a very valuable by-product. This integrated processing model supports economic sustainability (Lindeboom et al., 2004). Wheat starch exhibits a bimodal granule size distribution consisting of A-type and B-type granules. This difference in granule size affects how quickly the material gelatinizes and how thick the resulting liquid is. This, in turn, affects how well the material can be processed into a film (Tester et al., 2004; Lindeboom et al., 2004). Incorporating gluten into wheat starch has been shown to enhance film strength and improve gas barrier properties (Fatima et al., 2024).

Rice starch is known for being small (about 2–8 μm), white, and not very allergenic (Fitzgerald et al., 2009). Its small, granular structure allows the production of thin films with a smooth surface, offering advantages in packaging applications where opacity is not desired (Tester et al., 2004). Rice starch is a good choice for making colorless or transparent films because it has low levels of protein and lipids. It has been tested in edible or biodegradable coating systems that are used on food products that are sensitive to the environment (Fitzgerald et al., 2009a, 2009b; Fatima et al., 2024).

Cereal-based starches continue to be the starting material of choice for starch-based packaging development studies because of their processability, availability, and cost. However, differences in the amylose and amylopectin ratio, granule morphology, and cocomponent content of each grain type must be carefully considered at the formulation level to optimize final packaging performance (Avérous and Halley, 2009; Tester et al., 2004).

1.2.2 Tuber and root starches

Starches obtained from tuber and root plants are widely used in the bioplastics and packaging industries due to their high purity, transparency, and viscosity. Potato (*Solanum tuberosum*) starch is notable for its water-holding capacity and gelatinization properties, which are due to its high phosphate content (0.08%–0.09%). Potato starch granules range in size from 15 to 100 μm, providing a smooth, flexible structure during film formation (Hoover, 2001; Moorthy, 2002). It is also preferred for producing composite bioplastics because of its high viscosity, thickening, and binding properties.

Cassava (*Manihot esculenta*) is a plant grown in tropical climates that produces large amounts of starch. Cassava starch is particularly suitable for the production of pure biopolymers because of its low protein and fat content (Moorthy, 2002). In addition, cassava has an environmentally friendly profile and reduces the carbon footprint because it requires few agricultural inputs (Rosenboom et al., 2022).

Starch granules from tuber and root sources are typically larger and more irregular in morphology, and the temperature at which they change to gelatin is between 58 and 65°C. This lower gelatinization temperature may reduce processing energy demand. Potato and cassava starches are widely used in biodegradable packaging because of their good film-forming ability. However, these starches can absorb a lot of water. This may require modification through blending or plasticization (Jane, 2006; Fatima et al., 2024). The composition of tuber and root starches is influenced by both genetics and environmental factors, resulting in a wide range of properties. This diversity is a significant advantage in achieving the desired mechanical and barrier properties for sustainable packaging applications.

1.2.3 Legume and pseudocereal starches

Cereal and tuber starches have a similar composition, but legume and pseudocereal starches differ substantially in composition and function. Legumes such as peas (*Pisum sativum*), chickpeas (*Cicer arietinum*), and lentils (Lens culinaris) usually have a very high amylose content at around the 35%–40% level. This content enhances the mechanical strength of the resulting films (Ratnayake et al., 2002). Pea starch granules of about 5–50 μm in size have a moderate gelatinization temperature. Chickpea starch increases water-binding capacity as well as the moisture barrier property of this type, owing to low lipid content (Mounir et al., 2024).

Pseudocereals, especially quinoa and amaranth, are known for their gluten-free formulation and high nutritional content. Their starches contain antioxidant components as well as essential amino acids, which creates niche markets in functional packaging (García-Guzmán et al., 2022). Due to smaller granule sizes and different chain lengths of these secondary starches, pseudocereals can differ slightly from grain starches in gelatinization and the behavior exhibited by them in films.

1.2.4 By-product source in industry and circular economy

Starch by-products derived from food processing industries offer great environmental and economic potential in the context of the circular economy. For example, barley husks from beer production and potato processing waste starch can be reformed into biomaterials that are biodegradable (Moorthy, 2002; Rosenboom et al., 2022). Starch recovery (starch recovery and by-product valorization) for the use of starch by-products of agricultural processing products can reduce food waste disposal costs while opening new biopolymer resource streams. This has connected resource efficiency, carbon reduction, and sustainable packaging goals (Rosenboom et al., 2022).

1.2.5 Functional property variation

The gelling properties of the structure of the packing material, as well as the ratio of amylose and amylopectin, gelatinization temperature, and water-binding capacity of starch sources, directly affect packaging properties. For instance, potato starch provides high transparency as well as flexibility, whereas corn starch offers the possibility to produce harder, more brittle films. The high content of amylose in legume starches can also support mechanical strength and water vapor barrier properties (Jane, 2006; Ratnayake et al., 2002). Conversely, cassava and sweet potato starches may reduce processing energy requirements due to lower gelatinization temperatures (Hoover, 2001). In formulation-based design, a general approach of obtaining satisfactory barrier and mechanical properties is to blend starches from various botanical materials or blend them with plasticizers and other bio-based components (Avérous and Halley, 2009; Fatima et al., 2024).

Starch from different botanical sources exhibits substantial variation in amylose content, granule morphology, crystallinity, and gelatinization behavior. These structural differences directly influence film-forming ability, mechanical performance, optical properties, moisture sensitivity, and barrier behavior in packaging applications. Table 1.1 is adapted from(Tester et al., 2004; Lindeboom et al., 2004), and related literature. It summarizes the major structure–property relationships relevant to source selection in starch-based packaging applications.

1.2.6 Environmental and economic dimensions

Starch production has been affected by environmental inputs such as fertilizers, water (considered irrigation water use), and pesticides, by energy use, and by process efficiency. Snack food crops, such as corn and potatoes, are associated with greater emissions, though in some cases the same plant-based food with little to no environmental impact can be used with crops such as cassava (very little in tropical regions can produce cassava) (Rosenboom et al., 2022). Dissemination to localized supply chains can minimize emissions related to transportation. Furthermore, by-products from industry can be recovered, and starch can also be reclaimed for waste reduction and cost-efficient activities (Moorthy, 2002; Rosenboom et al., 2022).

The "source-sensitive starch selection" strategy is emerging as a mainstay in sustainable packaging policies toward environmental and cost-saving and other environmental aspects (Avérous and Halley, 2009). Chemically modified starch films, such as those crosslinked with citric acid or STMP/STPP, could offer substantial environmental benefits for packaging, as they enable the use of biobased and biodegradable materials with performance somewhat comparable to that of conventional plastics (N et al., 2024).

1.3 Properties of starch

Starch is a type of polysaccharide found in nature, which has great potential for application to produce biodegradable materials in the packaging sector. The chemical composition, physical characteristics, and functional versatility of starch make it a promising candidate for environmentally friendly alternatives to fossil-based materials (Avérous and Halley, 2009). In the area of packaging engineering, starch is used for the development

Table 1.1 Botanical sources of starch and structure–property relationships relevant to packaging applications.

Botanical source	Widely reported structural characteristics	Packaging-relevant implications generally associated with these characteristics
Corn (maize)	Typically contains moderate amylose content and A-type crystallinity; granules are commonly polygonal and medium-sized	Widely used in starch-based films and TPS because of broad availability and balanced processing behavior
Wheat	Cereal starch with bimodal granule-size distribution (A- and B-type granules)	Granule-size distribution can influence gelatinization and film-forming behavior; often relevant in integrated starch–gluten processing systems
Rice	Characterized by very small granules and relatively low levels of nonstarch components	Small granule size is often associated with smooth film surfaces and favorable optical properties
Potato	Large granules, B-type crystallinity, and relatively high phosphate monoester content	High swelling and viscosity are useful for film formation, but moisture sensitivity remains an important limitation
Cassava	Generally low protein and lipid contents and relatively high starch purity	Frequently used in biodegradable film systems because of good film-forming ability and relatively easy processing
Pea	Legume starch often reported to have higher amylose content than many common cereal starches	Higher amylose content may support increased stiffness and strength in some film systems
Quinoa	Pseudocereal starch with very small granules and distinctive molecular features	Of interest for specialty applications where fine microstructure is desirable
Amaranth	Very small granules and pseudocereal starch characteristics distinct from major cereal starches	Relevant mainly to niche and specialty applications rather than large-volume industrial use

of flexible materials with suitable mechanical strength either as an isolated material or a composition as well as in coating formulations (Altay et al., 2024). An understanding of the physicochemical characteristics of starch is vital for developing high-performance, environmentally friendly packaging materials.

1.3.1 Molecular structure and packaging performance

Starch contains the macromolecules amylose and amylopectin. The linear structure of amylose promotes chain alignment and crystallinity, which contributes to improved mechanical strength and reduced gas permeability in starch-based films. On the other hand, amylopectin shows remarkable adaptability to processing and end-use conditions thanks to its branched organization (Tester et al., 2004). High amylose content is essential to increase the water vapor barrier and mechanical strength of packaging films (Shogren, 1996). As a result, high amylose starches are widely used in food packaging films.

The relative amylose and amylopectin ratio has a great impact on the observed gloss, transparency, and texture of the end film. Films rich in amylose have better mattness and stiffness, whereas amylopectin-rich films show higher gloss and flexibility. Such differences affect the choice of the starch source of the product to be processed according to the properties of the product that shall be packed onto the material (Jiménez et al., 2012).

Natural starch, also referred to as native, pearl, or unmodified starch, is an ultra-high-molecular-weight polymer, with molecular weights exceeding 300 million g/mol (Daltons), which presents significant challenges during cooking and processing (Altay et al., 2017). To function effectively as a binder in aqueous systems, the molecular weight of starch must be substantially reduced during cooking to obtain solutions with adequate solids content and manageable rheological properties. To function effectively in aqueous systems, the molecular weight of starch must be substantially reduced during cooking to obtain solutions with adequate solids content and manageable rheological properties (Altay et al., 2017). Inadequate cooking during paper coating operations can lead to several processing issues, including partial gelatinization, which causes gelling because of incomplete granule swelling, excessive viscosity increases during cooling, microbial growth in cooked starch systems, and poor viscosity stability during storage. These problems can ultimately result in compromised product quality and unplanned production interruptions.

1.3.2 Granule morphology and barrier effect

Starch granule morphology, their shape, size, and surface properties differ with the botanical origin. The particle features of corn starch were usually composed of 15–20 μm in diameter polygonal granules, whereas the film aspect of rice starch was found to be more transparent and smoother with 2–8 μm in diameter granules (Fitzgerald et al., 2009a, 2009b). On the other hand, potato starch gives flexible but relatively more permeable films having oval granules with dimensions of 15–100 μm (See Fig. 1.2) (Lindeboom et al., 2004).

Both starch granule size and granule size distribution directly influence film homogeneity and optical properties. Starches with smaller granules are generally better suited for producing thin, transparent films due to their more uniform dispersion within the film matrix. On the other hand, larger granules tend to produce thicker films with increased opacity. From a standpoint of packaging engineering, granule size distribution is a critical parameter for optimizing film barrier properties and achieving a smooth,

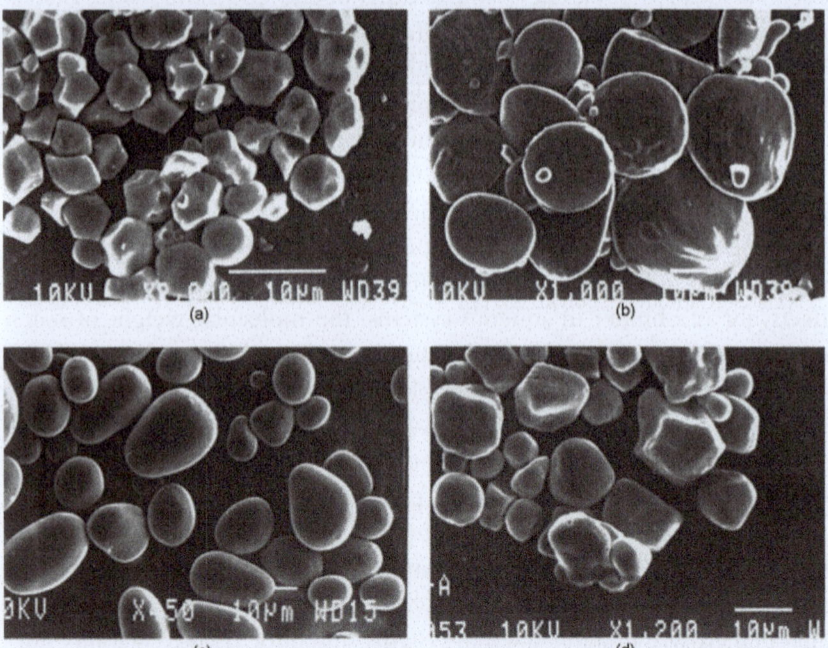

Figure 1.2 Different starch sources (SEM): (a) Rice, (b) wheat, (c) potato, (d) corn (bar = 10 μm) (Singh et al., 2003).

homogeneous surface structure (Fitzgerald et al., 2009a, 2009b; Lindeboom et al., 2004).

A significant advancement in this area was the development of starch nanoparticle biopolymer for the paper and packaging industry by EcoSynthetix Corporation, designed to replace conventional petroleum-derived synthetic latex binders (Altay et al., 2017). The company has patented a process that employs a twin-screw extruder operating under controlled shear conditions, upstream pressure, and the presence of a crosslinking agent to produce starch nanoparticles with an average size of about 100 nm, typically ranging from 20 to 150 nm (Fig. 1.3).

Fig. 1.4 illustrates a comparative distribution of particle sizes for starch biopolymer nanoparticle emulsions and native corn starch. The x-axis represents particle size on a logarithmic scale, ranging from the nanometer to micrometer scale, while the curves depict the dominant particle size ranges for each binder system. Starch biopolymer nanoparticle emulsions exhibit particle sizes predominantly in the nanometer range, typically below 200 nm. In contrast, native corn starch shows a substantially larger particle size, extending into the micrometer range, reflecting the intact starch granule structure. This large size difference explains the processing limitations of native starch, such as poor penetration, higher viscosity, and reduced uniformity in coating applications.

1.3.3 Crystallinity and barrier performance

The content of starch crystallinity directly impacts the performance of packaging materials in terms of moisture and gas barrier. The A-type crystallinity typical of cereal starches provides relatively good water vapor barrier properties. This property can be attributed to their high density as well as the compactness of their chains. By comparison, potato starch of B-type crystallinity is more susceptible to water owing to its comparatively loose chain structure (Jane, 2006). Amylose complexed with lipids leads to V-type crystallinity, which is responsible for hydrophobicity and enhanced water barrier properties.

In the packaging industries, the management of crystallinity is essential to maximize the shelf life of food products. Starch crystallinity is controlled in processing by destroying or altering its semicrystalline structure via thermal, physical, and chemical methods to design the functional properties such as solubility, viscosity, and digestibility. Among these, gelatinization (heating in water), heat-moisture treatment (HMT), annealing, and

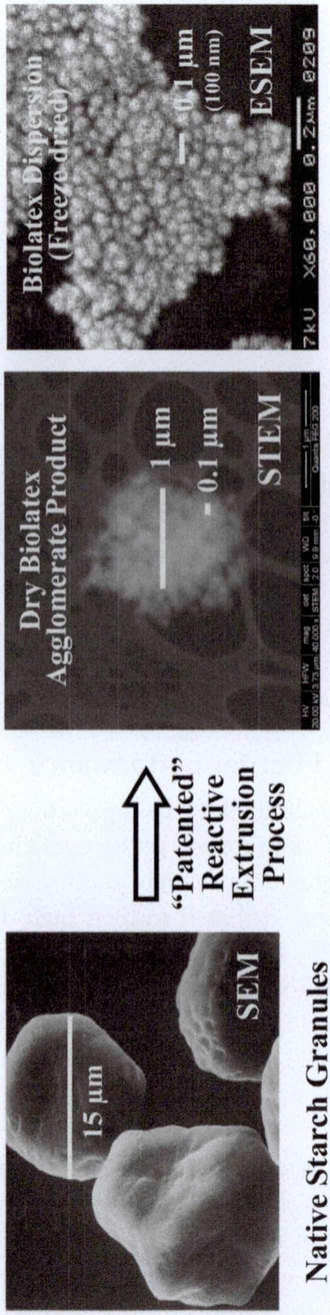

Figure 1.3 Native starch conversion to biobased latex using a twin-screw extruder.

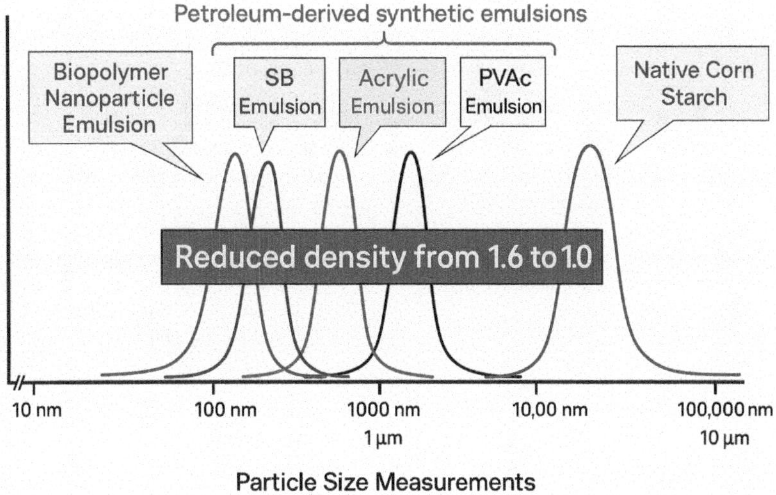

Figure 1.4 Particle size comparison of native corn starch versus starch biopolymer nanoparticles.

high-pressure processing (HPP) are important ones. These processes modify crystalline structure (A, B, or C type) and the degree of polymerization to meet industrial requirements, for example, production of resistant starch or thermoplastic materials (Yang et al., 2024).

The relative proportions of crystalline and amorphous domains in starch are not fixed and may be significantly altered by processing treatments such as annealing, extrusion, and enzymatic hydrolysis (Arruda et al., 2025). When starch develops highly ordered, compact crystalline regions, these structures can impede enzymatic access, thereby reducing the rate of starch breakdown during digestion. Starch exhibiting this digestion-resistant behavior is commonly classified as resistant starch (Arruda et al., 2025). Alterations in starch molecular organization and supramolecular structure can be systematically evaluated using a range of advanced characterization methods. Techniques including X-ray diffraction, nuclear magnetic resonance spectroscopy, infrared spectroscopy, differential scanning calorimetry, and atomic force microscopy provide complementary insights into changes in crystallinity, molecular mobility, thermal behavior, and surface morphology (Fig. 1.5).

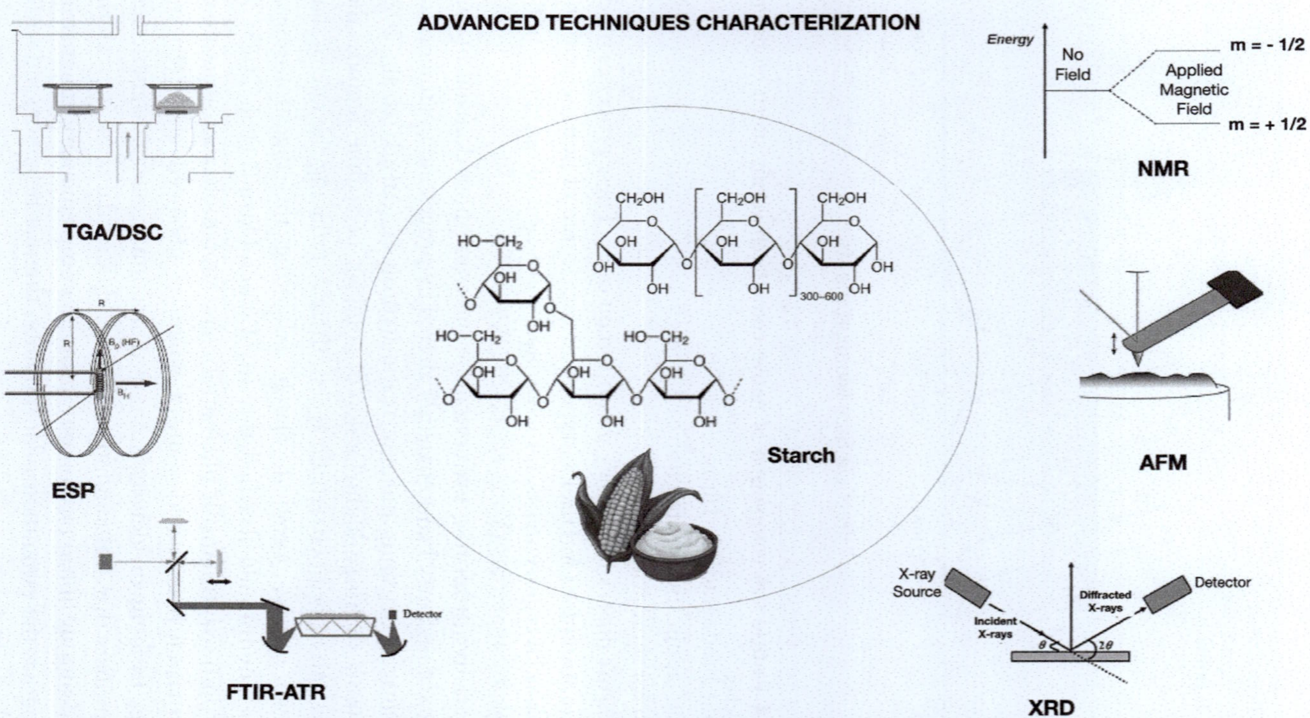

Figure 1.5 Illustration of the analytical techniques applicable for screening starch-based materials (Arruda et al., 2025).

1.3.4 The gelatinization, thermoplastic starch, and their processing

Native starch is not thermoplastic because its melting temperature is higher than its thermal degradation temperature; therefore, starch granules are plasticized with water and/or low-molecular-weight plasticizers (e.g., glycerol, sorbitol, urea, formamide, ethylene bisformamide, and citric acid) under heat and shear to form a continuous, viscous melt that can be shaped using conventional thermoplastic processing methods such as extrusion and injection molding, producing TPS (Kaseem et al., 2012). Starch granules absorb water and swell during gelatinization. Simultaneously, crystalline portions disintegrate, and polymer chains achieve mobility. This process enables the thermoplastic processing of starch using industrial techniques such as extrusion and film casting (Avérous and Halley, 2009).

Temperature, relative humidity, and the amount of plasticizer are critical factors in the production of TPS. Higher processing temperatures and energy input are generally needed for amylose-rich starch; however, more uniform and structured films can be obtained in such a case. Higher levels of plasticization are generally associated with better film flexibility; however, an excessively high amount of plasticizer may disintegrate the film matrix and/or impair barrier integrity (e.g., by increasing the permeability) (Kaseem et al., 2012; Arruda et al., 2025).

Fig. 1.6, adapted from work conducted at the Center for Industrial Rheology, illustrates the extensive rheological insights that can be extracted

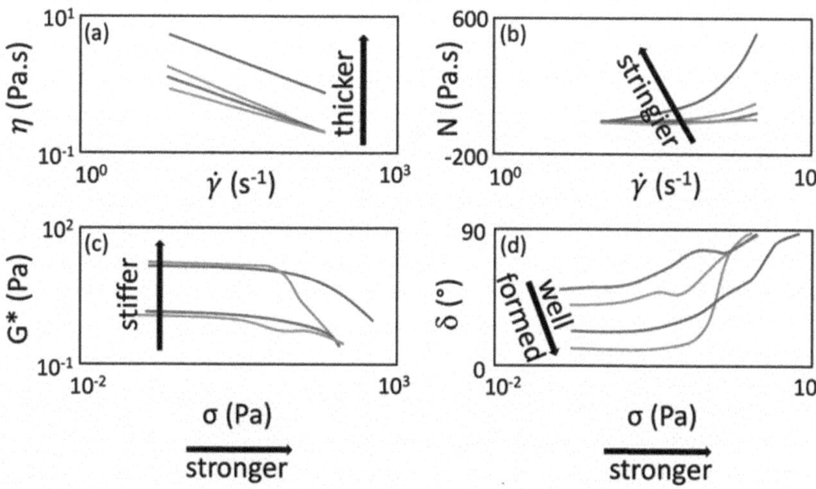

Figure 1.6 5% w/w gelatinized starch (potato, dent maize, waxy maize, and tapioca).

from a single oscillatory stress sweep experiment performed on starch systems gelatinized at 60°C. The viscosity–shear rate relationship, shown over a broad shear-rate spectrum (Fig. 1.6a), reveals that potato starch consistently exhibits higher viscosity than the other starch samples. At elevated shear rates, the remaining starches display comparable viscosity values; however, as shear rates decrease, clear distinctions emerge among the samples. These divergences suggest that differences in flow behavior would become increasingly evident under even lower shear conditions. Analysis of the normal stress response (Fig. 1.6b) further emphasizes the unique behavior of potato starch, while also differentiating samples that appeared similar based on viscosity alone. Additional interpretation of oscillatory stress data (Fig. 1.6c and d), particularly the complex modulus and phase angle, highlights variations in material stiffness and internal structural organization across the starch systems. Notably, the data reveal that samples grouped according to similar rigidity levels do not necessarily exhibit comparable yielding responses. For example, dent maize and potato starch reach yielding at approximately the same stress level, yet the manner in which yielding occurs differs markedly between them. This observation underscores the multifaceted and highly dynamic nature of the yielding behavior in gelatinized starch systems (Altay and Lewis, 2024).

1.3.5 Retrogradation and shelf life

Retrogradation in starch-based films describes the gradual recrystallization of starch molecules during storage. The rearrangement of starch chains during storage (retrogradation) can cause unwanted consequences (e.g., hardening and opacification in films) (Fig. 1.7). This process can deteriorate the film's performance by increasing brittleness, decreasing flexibility, and ultimately shortening shelf life. Thus the shelf life of the material products can be reduced, and mechanical properties may suffer. During thermal/mechanical processing (extrusion, compression molding, and film blowing), starch can undergo oxidative and thermal degradation, which can contribute to loss of whiteness and yellowing/browning (Van Soest and Borger, 1997). Incorporating additives such as glycerol into plasticized starch films has been confirmed to retard retrogradation processes and keep the films flexible. It is also known that retrogradation increases water permeability, leading to disturbance of moisture equilibrium in the food systems. Therefore the incorporation of antiretrogradation agents and the determination of suitable storage conditions are important factors in the development of packaging (Bangar et al., 2021; Onyeaka et al., 2022).

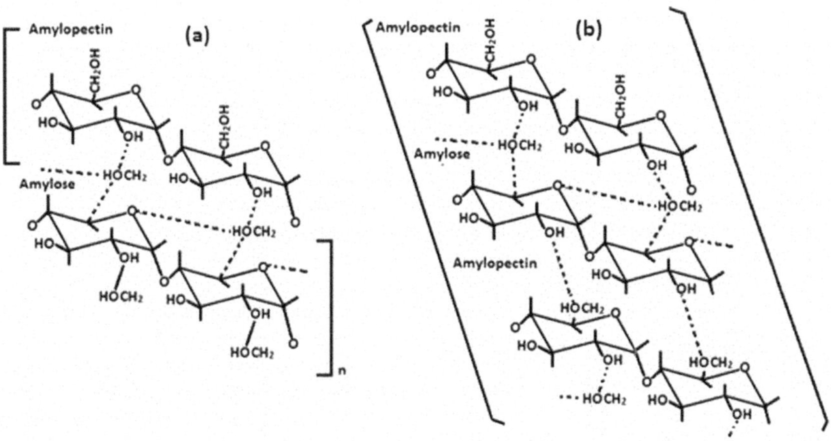

Figure 1.7 Hydrogen bonding between amylose and amylopectin molecules. (b) Retrogradation mechanism of starch. AY—amylose, AP—amylopectin, dotted lines—hydrogen bonding (Thakur et al., 2019).

1.3.6 Moisture sorption and barrier properties

Starch-based films have limited water barrier properties against water vapor since they are hydrophilic. In food packaging, this may speed the absorption and release of moisture in the packaging system. To enhance the moisture barrier ability of the starch-based materials, chemical modification, such as the acetylation, esterification, or succinylation of starch, is carried out (Tharanathan, 2003). Moreover, it has been shown that starch films can reduce water vapor permeability between the polymeric layers in multilayer structures.

Starch-based films can offer good oxygen barrier properties, but they are generally sensitive to moisture and water vapor, which limits their standalone use in packaging (Siracusa et al., 2008; Tang and Alavi, 2011). Therefore starch is frequently combined with other biodegradable polymers (e.g., PLA and PHA) in blends or multilayer structures to obtain a better balance of barrier and mechanical performance for food packaging applications (Siracusa et al., 2008; Sorrentino et al., 2007).

1.3.7 Chemical and physical changes

Starch is prepared physically and chemically to improve its properties for packaging. Starch resistance to water and mechanical properties can be improved through active chemical transformations such as acetylation, oxidation, crosslinking, and cationization. Nanofiller-based composites (e.g., nanoclay and cellulose nanofibers) improve gas barrier characteristics and thermal properties of films (Rosenboom et al., 2022; Courtene-Jones et al., 2022).

Physical modifications are done with extrusion, dry heating, ultrasonic processing, and pressure treatments (Altay et al., 2017). These methods are shown to change the particle size of starch (Fig. 1.1), and/or improve the strength of the packaging film by inducing a restructuring of the starch chains (He et al., 2023; Altay, 2023).

1.3.8 Rheology and mechanics

Starch pastes and TPS normally present shear-thinning (pseudoplastic) behavior, meaning the viscosity decreases as shear rate increases (Altay and Lewis, 2024). Thixotropy, recovering viscosity as a function of time, can be observed because of the alignment of polymer molecules and the disintegration of the granule structure. This flow behavior, with shear-thickening observed only in very concentrated, particular systems, has very important consequences for the production processes since flow, leveling, and setting behavior during extrusion or coating processing is largely determined by it and eventually influences the product thickness and defect rate (Barnes, 1989; Kistler & Schweizer, 1997; Liu et al., 2009; Rao, M. A., 2014). At industrial scales, plasticizer migration and rheological instability lead to greater thickness variation and reduced heat-seal consistency, thereby compromising field performance and manufacturing yield (Fang et al., 2024). To quantitatively describe the rheological flow behavior of starch-based systems and other complex fluids, experimental rheological data are commonly fitted using constitutive models that relate shear stress, shear rate, and viscosity. These models enable the prediction of material behavior under different processing conditions and facilitate comparison between formulations. Among the most widely used models is the Power Law model, which characterizes shear-dependent viscosity using the consistency coefficient and flow behavior index to indicate shear-thinning (pseudoplastic) behavior, which is commonly observed in starch dispersions and coatings (Altay and Lewis, 2024). However, because the Power Law model cannot accurately represent the low- and high-shear viscosity plateaus, more comprehensive models such as the Cross, Carreau, and Herschel–Bulkley models are often applied (Altay and Lewis, 2024). These models incorporate additional parameters to describe yield stress behavior or transitions between Newtonian and shear-thinning regions. The Herschel–Bulkley model, for instance, combines yield stress with shear-dependent viscosity and is particularly useful for structured fluids that require a minimum stress to initiate flow. Selecting an appropriate rheological model, therefore allows a more accurate interpretation of flow curves and improves the ability to predict processing behavior in coating, printing, and extrusion applications (Altay and Lewis, 2024).

The mechanical properties of starch-based packaging materials, the tensile strength, elongation, and elastic modulus, depend on the molecular structure of the starch and its plasticizers to a suitable level. High amylose content has been shown to improve tensile strength with less elongation. Plasticizers, such as glycerol, have also been shown to build flexibility for the material. However, overuse of these plasticizers may affect the film strength negatively (Avérous and Halley, 2009). Both rheology and mechanical properties are of fundamental importance within process design, especially in film extrusion and casting. In TPS systems, melt viscosity is an important parameter as it has a direct influence on the processability, for example, the selection of appropriate temperature, pressure, and screw speed during processing (Avérous, 2004; Liu et al., 2009).

1.4 Production processes and techniques for starch-based packaging

Starch-based packaging materials have been suggested as an alternative to traditional plastic packaging made from fossil fuels owing to environmental concerns. The processing of TPS-based materials is performed analogously to conventional plastic processing. However, because of the nature of starch, the production process must be modified.

1.4.1 Selection and preparation of raw materials

Starch is composed of two polysaccharides, amylose and amylopectin; the ratio of these two determines much of the usability of the starch. For example, high amylose content promotes the formation of strong, flexible films, or potato starch with a high viscosity, which is used for foam production (Zhang et al., 2014). The engineering property of starch-based materials, for instance, for bioplastics, is manifested in the processing as a determinant of the mechanical, barrier, and appearance properties of the material. Inputs may consist of starch as well as plasticizers (e.g., glycerol, sorbitol, and glucose), fillers such as cellulose, talc, clay, and biobased additive polymers like polylactic acid (PLA) and polyhydroxyalkanoates (PHA).

To prepare the raw materials, a series of steps is necessary, including drying and grinding the starch, reaching the required purity level, and premixing for uniformity of additive distribution. The exact amount/content of plasticizer is crucial for securing an optimal tradeoff between plasticity and biodegradability. In the right amount, it is said, too much of plasticizers makes the compound more water-susceptible while too little increases brittleness (Petersen et al., 1999). Starches obtained from lignocellulosic waste or industrial food waste have also gained interest in recent years (Dong et al., 2023). Alternative sources could potentially minimize environmental harm and provide tangible economic gains concurrently.

1.4.2 Plasticization process

Starch is a polymer that has both crystalline and semicrystalline regions in the natural state, and so is not directly molded because it has no thermoplastic behavior. This process involves gelatinization and plasticization of starch to give the TPS form. Gelatinization is the swelling of starch granules under heat and water, resulting in the dissolving of their crystalline component (Lourdin et al., 1997). In the plasticization step, glycerol, sorbitol, urea, and polyols are used to plasticize starch. These additives act to overcome the hydrogen bonds between the starch chains using a heat-modulated mechanical shear to increase chain mobility and develop an amorphous structure. TPS is a biopolymer that allows the process of melting, shaping, and reprocessing (Krochta & De Mulder-Johnston, 1997). TPS is generally processed within a temperature range of 120–180°C, where mechanical shear during blending and applied pressure are critical factors influencing starch plasticization, molecular mobility, and final material properties (Jiang et al., 2020; do Val Siqueira et al., 2021). The ratio of the additives in the plasticization is used to choose the hardness, flexibility, and barrier properties of the output.

Plasticization can be carried out during extrusion or before as a pretreatment in a mixer or kneader. The ratio of plasticizer is between 15% and 30%, depending on starch content. In the plasticizing process, molecular chains in the process are split into different lengths and hydrogen bonds are broken up (the hydrogen bonds of the chains in the matrix), yielding the new structure as a more flexible, heat-formable configuration. However, this process also has important limitations. Excessive plasticizer increases the water sensitivity of TPS, which may threaten TPS's dimensional stability because of its high plasticizer concentration (Paluch et al., 2022). So, the balance between plasticizer and moisture is very important for the product's mechanical strength and long-term durability.

1.4.3 Processing techniques

1.4.3.1 Extrusion and film casting

Extrusion is the most widely used approach to the large-scale production of TPS. This procedure involves the continuous addition of starch and plasticizers (e.g., glycerol and sorbitol) and additives (e.g., biopolymer reinforcements and natural fibers) at predetermined temperatures and pressure for a melt. The form of the mixture is essentially formed along the die at the end of the extruder with film/sheet/profile being produced (Müller et al., 2008). In general, extrusion has its advantages, such as high-speed production, low cost, and a diverse range of products. Nevertheless, specific process parameters must be controlled to improve the water vapor barrier property. In starch film casting, water is typically used as the main solvent, while plasticizers such as glycerol or sorbitol are added to improve film flexibility (Mali et al., 2005; Bourtoom, 2008). Starch solutions are ready until reaching a specific viscosity and spread uniformly over the glass, metal, or Teflon surfaces. The solvent evaporates carefully in a controlled manner, yielding homogeneous films, very fine grains, and a flexible shape.

Film casting is one of the most widely used laboratory methods for producing transparent food packaging films, edible coatings, and biodegradable films (Mali et al., 2005). In this process, starch is typically dispersed in water, gelatinized through heating, and then cast onto a surface where the solvent evaporates to form a film. Although solution casting is simple and effective for research-scale film preparation, its industrial application is limited. At an industrial scale, thermoprocessing methods such as extrusion and thermocompression are generally preferred to produce starch-based films. These techniques offer continuous processing and improved scalability compared with solution casting, making them more suitable for large-scale manufacturing (Avérous, 2004; Jiménez et al., 2012).

1.4.3.2 Injection molding

Injection molding is a widely used method for producing packaging in three dimensions from starch-based materials. This process is performed by heating the TPS at a temperature ranging from 140 to 170°C until melting and then injecting it into molds under high pressure. The cooling rate in the mold has a great influence on the shape and dimensional stability of the product (Liu et al., 2009). The process is especially well suited for the manufacture of food packaging such as disposable plates, cups, trays, and food service articles. Furthermore, the mechanical properties of the product can be improved by injection molding with filler materials and fiber reinforcements, which have the ability to deliver a good mechanical response (Yacob et al., 2019).

1.4.3.3 Foam production

Starch-based foams are a sustainable substitute for petroleum-based foams. In the production of foams, the starch and water mixture can usually be gelatinized by heat (150–200°C) and is processed under pressure in an extruder. As a result, the pressure drops dramatically at the extruder nozzle. This causes starch to expand, and a cellular porous structure is formed (Soykeabkaew et al., 2015; Figueir et al., 2024). The produced foam is utilized in energy-absorbing packaging, food transport containers, thermal insulating materials, and disposable packaging materials. The foam density, the cell size, and mechanical properties also depend on the starch type, plasticizer content, and process conditions. In recent studies, it has been demonstrated that additives such as biopolymer blends (PLA, polycaprolactone (PCL)) and natural oils (soybean, palm oil) can enhance the stability of starch foams (Fatima et al., 2024).

1.4.4 Improvement strategies

1.4.4.1 Enhancing mechanical strength

Starch is highly brittle, so improving the mechanical properties of its composites is highly desirable in applications. Fiber reinforcement, such as cellulose, bamboo, or flax fibers, is the most prevalent method in this sector of the industry. Uniform distribution of fiber load in the TPS matrix was reported to improve the load-bearing capacity and to prevent crack formation under stress (Reddy and Yang, 2005). Nanofillers, such as nano clay and nanocellulose fibers, may have positive effects on tensile strength and elasticity.

1.4.4.2 Improved water resistance

TPS is naturally hydrophilic and thus moisture exposure causes degradation of its mechanical properties. To overcome this limitation, crosslinking agents (e.g., citric acid and glutaraldehyde) or surface modifications (hydrophobic coating, biobased polyester blends) can be applied. Also, composites made from PLA have been demonstrated to greatly enhance the water resistance of TPS (Liu et al., 2009; Moreno-Bohorquez et al., 2026).

1.4.4.3 Improvement of barrier properties

Certain properties are required in the food packaging material, including good barriers to oxygen, carbon dioxide, aroma compounds, and other gases. Native starch exhibits low functionality and hence is limited in its use, but the characteristics of native starch can be dramatically modified by one or more modification techniques. For instance, polymeric composite films derived from modified starch, for example, acetylated starch, or starch blended with flexible polymers, have been reported to offer considerable enhancement in the mechanical and barrier properties of the attained materials. Also, wax-, chitosan-, or oil-based formulations applied during processing or as coatings have similar enhancements in performance (Tharanathan, 2003).

1.4.4.4 Chemical modifications

Chemical modification is often employed to improve the performance of starch-based packaging materials. Common methods such as acetylation, oxidation, carboxymethylation, and phosphorylation introduce functional groups into the starch backbone, altering its hydrophilicity and improving properties such as water resistance, mechanical strength, and processability (Tharanathan, 2003). Acetylation is a substitution reaction in which the hydroxyl group is replaced by an acetyl group, resulting in starch films with improved water resistance and flexibility. Oxidation of starch occurs when oxidizing agents convert hydroxyl groups in the starch structure into carbonyl and carboxyl groups. The introduction of these functional groups improves film structure and transparency. Carboxymethylation has been reported to increase starch solubility and biodegradability, whereas phosphorylation improves thermal stability and mechanical strength (Jiang et al., 2020; do Val Siqueira et al., 2021).

1.4.4.5 Composite and hybrid materials

Starch itself possesses poor mechanical and barrier performance. Therefore composite structures made from biopolymers and natural fibers are a common choice. When biopolymers, including PLA, PCL, and PHA, are mixed with starch, they provide both elastic, barrier, and strength properties not possible when used in the polymer industry, where they consume an amount of starch to manufacture it. In addition, the incorporation of natural fibers, including cellulose, flax, bamboo fiber, and chitosan into the starch matrix improves the mechanical strength and thermal resistance. It is well known that nanotechnology applications (e.g., nanoclay and graphene oxide) have improved the barrier properties of several materials (Kargarzadeh et al., 2017).

1.4.4.6 Coatings and surface treatments

Coating methodologies have been proven to be important for them to reduce water permeability in starch-structure films. The use of biobased waxes, cellulose derivatives, chitosan, and protein-coated films on the film surface led to a remarkable enhancement of barrier properties. In addition, surface treatments such as plasma treatment and UV crosslinking have been reported to improve film structure homogeneity on the surface through modification of surface energy (Arayaphan et al., 2020; de Lima et al., 2025).

1.4.4.7 Multilayer structures and laminating

Multilayer structure is one of the favored approaches to enhance the strength and barrier characteristics of the starch-based packaging. Starch layers are then merged with additional biopolymer layers, such as PLA or polyethylene (PE), in a process referred to as lamination. This leads to a remarkable rise in the barrier properties of oxygen, carbon dioxide, and water vapor. Lamination also has better resistance to heat and mechanical impacts (Avérous and Halley, 2009; Avila et al., 2023).

1.5 Advantages of starch-based packaging

Starch-based packaging materials, and in particular TPS, have seen a surge of interest in recent years, and they are now considered a very promising candidate among sustainable packaging solutions with the nature of renewable, biodegradable, and functionally tailorable materials. Starch films and their composites have potential multidimensional advantages such as low cost, availability, processability, renewability, biodegradability, and food-contact application compatibility; and for some uses, they can achieve a reasonable competitiveness in critical features such as oxygen barrier properties (Karnwal et al., 2025; Garavito et al., 2024).

1.5.1 Environmental benefits

Starch-based plastics have the environmental benefits of being biodegradable and compostable. Polysaccharides, such as starch-based polymers, can be hydrolyzed and it can be enzymatically catalyzed and degraded by microorganisms, and can be hydrolyzed effectively under the industrial composting conditions. Recent works report that, in case food is composted with control conditions, mass reduction is considered significant and is short, and the decay rate is also considered fast at higher temperature and moisture (Gastaldi et al., 2024; Gioia et al., 2021). Several commercial starch-containing products disintegrate visibly within a few months, but the quality of the blend, the film thickness, and many other design elements may significantly affect the speed of deterioration (Afshar et al., 2025).

From a sustainability point of view, starch-based systems contribute to the reduction of plastic waste and pollution. While starch-based polymers are channeled to a suitable disposal site (such as industrial composting and organic recycling), fossil-based polymers tend to take more years to degrade, allowing the gradual accumulation of microplastics. Life-cycle evaluations emphasize that proper end-of-life management and infrastructure are essential to reduce plastic leakage and persistence in the environment (N et al., 2024). Standards, such as ASTM D6400 and EN 13,432, define the compostability criteria in a way that is measurable and verifiable, which fosters the credibility of environmental claims (Nizamuddin et al., 2024).

1.5.2 Economic and industrial advantages

Starch is predominantly obtained from various kinds of crops, such as corn, potato, cassava, and wheat with a huge variety of growing areas throughout the globe. Such availability contributes to enable economies of scale based on lower feedstock unit cost and establishes a reliable supply. It has been recently demonstrated that starch can be a fraction of the cost of other biopolymers, such as PLA, for certain applications and that local and national agricultural production can contribute to mitigating supply-chain risks (Garavito et al., 2024; Yudhistira et al., 2024). TPS can be processed by conventional polymer techniques, such as extrusion, film blowing, casting, foaming, and thermoforming. This compatibility also allows lower switching costs and capital outlays during implementation in an industrial setting (Tapia-Blácido et al., 2022; Garavito et al., 2024). In the industrial-scale process, the plasticizers, polymer blends, and additives methods broaden the processing window, which leads to high efficiency and product uniformity (Figueiró et al., 2024). Also, starch can be sourced as a by-product of the agricultural processing industry (milling of agricultural products), providing both environmental/economic benefits. It also reduces emissions from imported raw materials and generates multiple agricultural revenue streams for farmers and supports local employment (Karnwal et al., 2025).

1.5.3 Carbon footprint and sustainability advantages

Starch has significant potential for leveraging sustainability and decreased carbon footprint implications because of its biogenic nature and renewable carbon cycle. As a biopolymer and a photosynthesis carbon-neutral solution, it contributes to decreasing fossil carbon dependency. The associated literature indicates that starch-based plastics could reduce the greenhouse gas (GHG) emissions associated with their production, compared to the production of oil-based ones, considering the assumptions of realistic energy mixes and sustainable agricultural practices (Yadav and Nikalje, 2024; Forfora et al., 2024). This means that from a human perspective, such a conversion might in principle lead to a huge reduction in greenhouse gas emissions and fossil energy use. But the capacity for mitigation is quite sensitive to boundary conditions such as land use and agricultural inputs. This also means packaging and food systems should be modeled holistically, including the environmental impact of food waste in the equation (He et al., 2024; N et al., 2024; Bremenkamp and Sousa Gallagher, 2024; Hemachandra et al., 2024). As such, comparative assessments need to be performed on an equivalent basis. In addition, carbon footprint alone is not sufficient, and other environmental impact indicators such as acidification, eutrophication, and photochemical ozone formation may also be assessed along the life cycle. It also highlights the contextual and place-specific benefits of starch-based systems, thereby providing a fact-based baseline to accelerate policymakers' support for sustainable product and investment paths (Forfora et al., 2024).

1.5.4 Technical and application advantages

Starch is a multifunctional and tailorable material that can be converted into films and coatings by casting or extrusion blowing for packaging or protection, containers, and edible coatings (Tapia-Blácido et al., 2022; Liu et al., 2022). This forms the basis for design for use, and consequently optimization by application. For example, a greater oxygen barrier may be preferred for fresh foods, but a greater mechanical strength may be necessary for dry foods. These desired properties can be obtained by semirational tailoring of the formulation and processing conditions.

Sufficiently safe starch-based films can be prepared, fulfilling the requirements for food-contact and safety regulations with appropriate additives under suitable processing conditions. It is also anticipated that they will be utilized as carrier matrices in active, smart packaging systems when delivering antimicrobial and antioxidant agents, freshness indices, and farm-to-fork trackability to end users (Karnwal et al., 2025; Dutta and Sit, 2024).

At some defined thickness, certain starch-based films with various amylose-to-amylopectin ratios and crystallinities can possess comparable oxygen barrier values to those of a conventional oxygen barrier material for their functional performance against oxygen. This is particularly crucial for such products, as they distort with oxidation (Karnwal et al., 2025). Hydrophilicity may restrict water vapor flow resistance; however, nanofillers, hydrophobic coatings, and multilayers have been proven to improve water vapor transmission rates. Recently, it has been reported that a starch-based nanocomposite film possesses good water vapor and oxygen barrier properties. Moreover, the mechanical properties can be further adjusted for particular application requirements by the degree of plasticization and with a certain method of efficiently achieving crosslinking and by the inclusion of novel polymer types in the film, among other methods (Dong et al., 2024).

Water vapor barrier and mechanical properties of starch-based systems are poorer than those of fossil-based polyethylene and polypropylene, even using conventional plastic packaging as the baseline. The customized performance profile of starch-based materials in polymer blends with polymers such as PLA and poly(butylene adipate-co-terephthalate) (PBAT), and by coating or lamination processes, can be tailored for specific end-use requirements (Cheng et al., 2021, 2024; Sikora, Majewski, and Puszka, 2021). This type of flexibility is what can enable a function and context-oriented material-architecture design to be more suitable, rather than trying to find one material that solves all problems.

1.6 Challenges in starch-based packaging

Starch-based packaging materials have a particularly strong value proposition, in particular TPS, as both a renewable resource and a biodegradable material with large design space. However, a number of coupled barriers currently limit large-scale industrial application: moisture sensitivity, mechanical strength, processing window sizes, cost and upscaling, as well as standardization and compatibility with infrastructure. Based on the literature and analysis results, it has been identified that these barriers are by far not exclusively material problems, and they demand an integrated engineering solution for overcoming them. The hydrophilic nature of starch and crystallization and retrogradation kinetics are to be considered at the level of structure, while at the processing stage, the precise control of plasticizer type and amount, moisture content, and temperature profile is required and at the system level (supply chain, industrial composting infrastructure, regulatory compliance synchronization) (Garavito et al., 2024; Karnwal et al., 2025; Pei et al., 2024).

1.6.1 Moisture sensitivity and water vapor issues

Starch is hydrophilic and thus suffers a loss of performance in films and molded items due to water vapor permeability (WVTR) and moisture sorption. Higher relative humidity acts as a plasticizer by decreasing the glass transition temperature with an increase in polymer chain mobility and a decrease in elastic modulus, tensile strength, and optical and dimensional stabilities (Pei et al., 2024). This problem is especially accentuated in high-moisture foods and at processing stages featuring high vapor loads. In addition, moisture sensitivity also speeds retrogradation, which is a process whereby gelatinized starch chains reform into crystalline gels, which are susceptible to breaking (Chang et al., 2021).

There are two dominant approaches to mitigate water sensitivity. One involves chemical or physical alterations to molecules in the system (e.g., crosslinking), which eliminates free –OH groups and weakens starch's interactions with water. Techniques such as citric acid esterification, acetalization, and hydrophobic substitution can be considered as practical and scalable process options (Stelescu et al., 2024). The second method is to fabricate structural barriers that increase the diffusion path and reduce the permeability. This involves nano clays, cellulose nanofibers, or zinc oxide (ZnO) nanoparticles to create tortuosity effects, multilayer structures, and hydrophobic surface coatings to reduce WVTR (Dong et al., 2024; Dejene et al., 2024).

Recently suggested evidence indicated that these enabled efforts can be tuned to enhance both water- and oxygen barrier "boosted performance" of starch-based nanocomposite films (Dong et al., 2024). Hence, an engineering strategy that combines material chemistry with structural design to control moisture is a strategic imperative.

1.6.2 Mechanical strength and physical properties

TPS films and molded products exhibit lower tensile strength, elongation, and resistance to crack propagation compared with conventional polyolefins. This constraint is due to the vitreous nature of the starch matrices and their strong sensitivity against plasticization and recrystallization (Karnwal et al., 2025). Therefore excellent mechanical strength can be obtained via a triaxial scheme. First, a combined approach (TPS with polymers such as PLA, PBAT, and polybutylene succinate (PBS) along with interfacial compatibilizers) leads to improved toughness–strength performance. The maleic-anhydride-grafted or epoxy-functionalized copolymers are proven to be good compatibilizers (Nam et al., 2024; Gong et al., 2025; Farshbaf Taghinezhad et al., 2025). Second, reinforcement with fibers and nanofillers (such as cellulose nanofibers, clays, or metal-oxide nanoparticles) enhances the load transfer and triggers crack-deflection and microcrack-bridging mechanisms, which lead to an increase in structural toughness (Karnwal et al., 2025; Dejene et al., 2024). Finally, crosslinking based on citric acid, network formation, and controlled esterification prevent creep and physical aging, thus minimizing time-dependency in the biodegradation of mechanical properties and offering dimensional stability (Stelescu et al., 2024). However, higher amounts of TPS may lead to a reduction in tensile strength in some combinations of biopolymers, particularly those based on PBAT. This indicates a critical TPS level in formulation above which plasticizer type (glycerol, acetyl tributyl citrate (ATBC), etc.) has a macroscopic effect on the mechanics through water uptake and phase morphology (He et al., 2024; Zhang et al., 2025). To achieve good properties in TPS-based systems, a materials-systems view that enables blend architecture, multi-scale reinforcement, and chemical-network design in concert with processing control and moisture management will be needed.

1.6.3 Processing and technological limitations

TPS production and processing need to be carried out within the limited processing conditions. Small variations in temperature, moisture, plasticizer content, and shear rate can have a significant influence on the extent of gelatinization, crystallization kinetics, and ultimate morphology (Garavito et al., 2024). Moisture and temperature insufficiently controlled during film blowing lead to an unstable formation of bubbles and result in blowing-out or bubble breakage, causing the bubble-formation stability to be affected and film thickness to be uneven during film blowing; meanwhile, they also cause the extrusion pressure to fluctuate more seriously, and then line efficiency is weakened. At the industrial scale, plasticizer migration and rheological instability cause an increase in thickness variation and a decrease in heat-seal uniformity, which leads to inferior field performance as well as reduced manufacturing yield (Fang et al., 2024).

The measurements need to be correct and proven so that the properties of the material comply with the applicable standards. Mechanical, gas, and water vapor barrier properties can be determined according to certain ASTM standards, including ASTM D882 for tensile properties, ASTM D3985 for oxygen transmission rate, and ASTM F1249 for water vapor transmission rate. Claims of compostability are validated according to standards such as EN 13,432 or ASTM D6400 and may also be verified via industrial composting facilities (Nizamuddin et al., 2024; Garavito et al., 2024). To meet these requirements, stringent quality assurance and quality control (QA/QC) procedures, including batch traceability, process capability analysis, and periodic revalidation, must be implemented and maintained throughout the R&D to commercial scale production.

1.6.4 Economic and industrial constraints

While raw materials such as corn, potatoes, and cassava are easy to obtain and inexpensive, the total cost of ownership (TCO) may be directly proportional to processing energy demands, compatibilizer and nanofiller usage, and coating or lamination processes. Advanced polymeric or nanocomposite structures with higher performance add to formulation and processing complexity and to greater capital and operating expenses. In the environment of limited fabrication time and strict quality specifications, this limits cost flexibility (Pei et al., 2024). In addition, starch-based polymers are still nowhere near the scale and size of fossil-based plastics. More costly cycle times and throughput restraints, such as bottlenecks, raise the unit cost and compete for customers with products (Garavito et al., 2024).

Market adoption is also challenged by uncertain disposal pathways, certification requirements, and the associated costs. Participation in certification programs such as OK Compost (OK-Compost) and the Biodegradable Products Institute certification (BPI) requires expenses related to testing, validation, documentation, and compliance monitoring. In addition, most compostability standards are designed for industrial composting conditions rather than home composting, which further complicates end-of-life management. As a result, it becomes difficult to provide clear value to end users and to bridge the gap between consumer expectations and the existing waste management infrastructure(Yadav and Nikalje, 2024; Beiras et al., 2025). In addition, commercialization of TPS-based solutions must be pursued with technical scalability and regulatory/infrastructural adoptability in mind, and with a focus on life-cycle cost at the system level.

1.6.5 Mitigation strategies and future perspectives

Reinforcements based on composite and nanotechnologies can act as bases to improve barrier and mechanical properties at the same time. Additives such as cellulose nanofibers, montmorillonite, and ZnO can increase tortuosity, lengthen the diffusion path, and lead to load transfer mechanism, which promotes crack deflection and then increases toughness. These approaches are in line with significant improvements in the WVTR and oxygen barrier performance (Dong et al., 2024). Polymeric processing and compatibilization also adjust the TPS property window for the application-specific requirements. For PLA/PBAT or PLA/PBS blends, compatibilizers such as maleic-anhydride (MA) derivatives, epoxy-functional reactive agent epoxy-glycidyl-methacrylate (EGMA) are applied to control the extent of phase separation to improve the toughness–strength tradeoff (Nam et al., 2024; Gong et al., 2025; Farshbaf Taghinezhad et al., 2025).

The effect of the type of plasticizer, particularly between glycerol and ATBC, on the water sorption and phase morphology is so pronounced that the multiobjective best-performing optimization has to be carried out at the formulation level (He et al., 2024).

Chemical modifications and crosslinking modify chain dynamics to improve mechanical and thermal stability and water affinity. Esterification-based networked architectures formed on multifunctional acids (such as citric acid) exhibit reduced creep and physical aging, and increased dimensional stability (Stelescu et al., 2024).

Processing and design features contribute to increased manufacturing rates and uniformity. Controlled-Rheology extrusion profiles, moisture-controlled feed systems, and in-line thickness, oxygen transmission rate (OTR), and WVTR monitoring contribute to line stability. Multilayer coextrusion architectures are designed to provide superior barrier and mechanical performance in a single step, allowing scalability (Garavito et al., 2024).

The standardized compliance and infrastructure alignment act as a pillar of confidence to assure success in the market. Protocols to verify compostability as per EN 13,432 and ASTM D6400 lend credibility to product claims, and the compatibility with waste management infrastructure allows for better alignment with industrial and home-composting pathways (Yadav and Nikalje, 2024; Nizamuddin et al., 2024).

1.7 Applications of starch in sustainable packaging

Starch-based packaging materials such as TPS, starch composites, and starch foams have also established themselves in the sustainable packaging sector because of the renewable nature of the raw materials, biodegradability, and various shape possibilities. Their application includes a variety of items from food service products (plates, cups, and cutlery) to edible coatings, cushion-foams for fragile-goods logistics, smart and active packaging systems, and 3D printed protective structures (Karnwal et al., 2025; Ghasemlou et al., 2024).

1.7.1 Applications in single-use packaging

In the single-use application sector, starch-based films and containers formed by heat have reached functionality as a barrier and optical properties for the needs of snacks, confectionery, and ready-to-eat (RTE) food products. Starch films display a high oxygen barrier at low or intermediate relative humidity. However, their limitations in water vapor barrier and thermal stability may be overcome by employing multilayer structures or coatings (Karnwal et al., 2025; Surendren et al., 2022). In disposable tableware, life-cycle assessments demonstrate that starch-filled or starch-based plates and cutlery can significantly lower environmental impact when handled through appropriate end-of-life infrastructure, especially when composted alongside food waste (UNEP, 2021). In parallel systematic reviews, it is found that increasing starch content increases photodegradation and compostability potential (Pantović et al., 2025). Taken together, these results emphasize starch-based products as viable alternatives to single-use petrochemical plastics in circular packaging systems.

1.7.2 Applications in foam packaging

Both extruded starch and thermoformed starch foam have been used successfully to provide lightweight protective packaging for delicate or easily scratched items during shipping, and could be the best options as renewable replacements for polystyrene or polyethylene foams. Many of these investigate the effect of the formulation and processing variables on the parameters' density, energy absorption, and specific energy absorption. Compressive strength, thermal stiffness, and further processing, such as chemical modification, can be utilized to improve the compressive strength and thermal resistance of fibers (Tapia-Blácido et al., 2022; Kossalbayev et al., 2025). In addition, dual-modified cassava-starch foams have been reported to biodegrade rapidly without compromising good mechanical properties, confirming their potential for commercial production (Amaraweera et al., 2022).

Starch-based foam packaging design is controlled by cell morphology and impact absorption performance as well as degradation rates. Cell size distribution and the closed cell to open cell ratio affect the energy adsorption, and the foam's stability and heat resistance can be enhanced by different starch matrices and fiber reinforcements, that is, cellulose and other natural fibers as reinforcing agents. In the packaging industry, starch-based foams are gaining ground in e-commerce and electronics packaging as they are compatible with composting end-of-life solutions and can, in many cases, be manufactured on existing production lines with minimum modifications, which makes them feasible drop-in substitutes (Tapia-Blácido et al., 2022; Kossalbayev et al., 2025).

1.7.3 Food packaging and edible coatings

Starch serves as the main polysaccharide matrix in edible films and coatings. These coatings minimize moisture loss, regulate respiration rate, and retard oxidative and microbial decay, thus extending the shelf life of fruits, vegetables, and bakery products (Blancas-Benitez et al., 2022; Nunes et al., 2023). Recent reports describe that in combination with suitable formulations and cold-chain management, coatings can extend shelf life by 1 week at room temperature and by 1 month at refrigerated storage (Krishnan et al., 2025).

The range of applications is broad, including thin starch films that alter gas exchange for fresh-cut produce, barrier layers that inhibit staling in bakery products, and antioxidant-containing coatings that reduce lipid oxidation in confectionery and snack products. Active packaging interest in the availability of controlled-release active packaging systems with natural products or with ZnO nanoparticles included into a starch matrix, as the incorporation of the extracts or of organic acids is increasing. These systems attenuate microbial load and maintain sensory quality as a function of time (Dejene et al., 2024; Mariño-Cortegoso et al., 2025). Starch can be used as a clean-label material in edible and active food coatings, providing barrier properties and extending shelf life. Edible coatings are biodegradable and help reduce plastic waste, making them a sustainable alternative to conventional packaging.

1.7.4 Industrial and commercial applications

TPS and starch composites are being used more widely in a wide range of products, including plates, cups, food trays, shopping bags, and transport dividers. TPS and starch composites can be manufactured at an industrial scale through established film blowing, casting, thermoforming, molding, coating, and lamination lines (Garavito et al., 2024). In the food service domain, the starch-based solution is compatible with corporate sustainability goals and ongoing procurement needs, which emphasize compostability, biodegradability, and renewable content. Yet, clear differentiation between industrial and home-composting criteria, including transparent labeling and sorting, can be a key factor affecting a product's chances of success in practical scenarios (UNEP, 2021; Ghasemlou et al., 2024).

The starch-based coating is also used in low-cost modules to enhance oil and water resistance and print quality in both paper and cellulose-based packaging. Biodegradable composite coatings create a fluorine-free grease-repellent system, which is a key requirement for per- and polyfluoroalkyl substances (PFAS) regulatory compliance, as these substances are being phased out (Karnwal et al., 2025; Surendren et al., 2022).

1.7.5 Emerging applications and future perspectives

In starch matrices, the incorporation of ZnO, clay, or cellulose nanofiber can enhance the barrier and mechanical properties and impart antimicrobial and antioxidant properties in nanocomposite smart/active systems (Dejene et al., 2024). Recently, there has been an explosive growth of smart packaging applications, mainly freshness indicators and pH-sensitive labeling with pigments and antioxidant capsules incorporated in starch films. They can track the product quality and potential spoilage in real time (Mkhari et al., 2025; Yudhistira et al., 2024).

Tuning the rheological properties (shear-thinning and gelation relaxation) in the additive manufacturing process, TPS-based inks and pastes are promising materials for the fabrication of tailored protective systems. This approach is consistent with supply-chain policies and uses features of small inventory, high variability, and small waste and storing costs (Ghasemlou et al., 2024).

TPS provides a way to tailor the microstructure of materials for applications when used in combination with other polymers, including PLA, PBAT, and PBS and reactive compatibilizers. This strategy overcomes the mechanical and moisture-sensitivity hindrances, enabling stability during processing and ultimately performance. Industrial solutions are available for monolayer to multilayer coextrusion systems with functional requirements (Karnwal et al., 2025; Surendren et al., 2022).

At the policy and standardization level, success is contingent upon the testing of compostability according to well-known standards (including EN 13,432 and ASTM D6400, among others) and the confirmation of compliance with current waste management and composting practices (UNEP, 2021; Ghasemlou et al., 2024).

1.8 Starch in recycling and the circular economy

Starch derived from renewable agro-feedstock (e.g., corn, potato, or wheat) is regarded as favorable in circular systems owing to its broad availability, ability to be modified, and, in most instances, because of being compostable/biodegradable contingent on formulation and certification (El Halal et al., 2019; Di Bartolo, Infurna and Dintcheva, 2021). Circular outcomes, however, are a matter of system fit and determination, material design, collection and sorting, and standards. The local and national infrastructure will ultimately determine the success and/or failure (Forfora et al., 2024).

1.8.1 The role of starch in recycling

Recycling systems are associated with waste diversion from landfills, conservation of energy/resource, and better waste management execution that is beneficial to both environmental and economic aspects (Tilton, 1999; Kharola et al., 2022). Starch plays a practical role at several points within the recycling-associated value chains. First, starch is a key feedstock for bioplastics used in packaging and certain single-use products. Those products are sometimes formulated to be biobased and degradable under suitable circumstances, which may worsen long-term lasting results compared to traditional polymers (Bashir and Aggarwal, 2019; Kim et al., 2023; Di Bartolo, Infurna and Dintcheva, 2021; La Fuente, Maniglia and Tadini, 2023a, 2023b). Second, starch is an essential ingredient in adhesives/coatings for the paper and board industry, for use in fiber-based packaging solutions, and for the possibility of better adequacy with paper recycling process (Hubbe et al., 2008; Gadhave and Gadhave, 2022). Third, in areas where industrial composting is available, registered compostable starch-based products can be sent for their organic recycling with food waste to yield compost and reduce disposal in landfill (Di Bartolo, Infurna and Dintcheva, 2021; Rujnić-Sokele and Pilipović, 2017).

Starch is biogenic and displaces fossil feedstocks, so entailing a potential reduction in fossil fuel use and associated emissions, yet the results are sensitive to the definition of agricultural inputs, energy mix, and processing scenarios (Guo et al., 2020; Gökçe, 2018). In general, starch has the potential to contribute to recycling/circularity goals when products are designed and managed to align with functional recovery options (Bastioli, 2005; Dobrucka, 2019; Rosenboom et al., 2022).

1.8.2 The role of starch in the circular economy

Circular economy principles aim to minimize waste and keep materials in use using improved design, environmentally and economically efficient resource use, and recovery pathways (reuse, recycling, and biological cycling) (Geissdoerfer et al., 2017; Ghisellini et al., 2016). Starch-based materials can support these objectives through renewable feedstock replacement and, in many instances, compostable and biodegradable end-of-life solutions that contribute to biological cycling where the necessary infrastructure is in place (Mestre and Cooper, 2017; Rosenboom et al., 2022; Di Bartolo, Infurna, and Dintcheva, 2021).

In practice, circular performance depends on: (i) tentative products to the intended end-of-life solution; (ii) clear labeling and standardization; and (iii) services providing collection, sorting, and treatment. Without this coordination, the claims of compostability may not lead to meaningful reduction of impact (Forfora et al., 2024; Suleman, A., Bora, J., Kalita, A. J., & Chetia, B., 2025).

1.8.3 Sustainable management of starch waste

Regeneration of starchy waste and by-products may be achieved via enhancing resource utilization while minimizing disposal problems. Typical pathways are composting of plate-processing residues into a soil amendment, application of starch-rich by-products as a raw material in the production of starch-based materials, bioenergy production through anaerobic digestion (biogas) with land application of digestate, and application in industry, such as paper and textile sizing and adhesive use, in which it may replace petrochemical-derived materials partially (Rosenboom et al., 2022; Singh et al., 2024; Ren et al., 2021). The optimum pathway choice will depend on context and take into account of contamination, logistics, local infrastructure, and quality needs for the recovered streams. When properly adapted to fitting local systems, starch-based materials and starch waste valorization have great potential for advancing circular economy and sustainability goals.

1.9 Conclusion

In summary, starch stands out as a key factor in the sustainable packaging challenge, and its use brings varied environmental and economic advantages. As a biopolymer that is both renewable and biodegradable, starch is not only a viable option to help solve the urgent issue of plastic pollution, but it also complements the tenets of the circular economy. It is also a multiuse commodity, as its applications do not end with food packaging but extend to many other areas. The development of starch-based packaging materials, also with active and smart functions, is an encouraging application that may bring benefits to both food protection and safety. Nevertheless, the road to mass adoption is paved with challenges. Problems such as moisture sensitivity, mechanical strength, and production cost must be addressed by technology innovation and/or supply chain collaborations. In the future, it will also be important that further R&D is conducted to improve the starch-modification and processing techniques to fulfill the needs of environment-conscious packaging. By switching to starch-based material, a company can dramatically lessen its carbon impact and help build a more sustainable future for us all. In the end, moving to starch-based packaging is not only a material change, but a wholesale rethinking of consumption and waste and a pathway to a more sustainable global economy.

References

Afshar, S.V., Boldrin, A., Christensen, T.H., Corami, F., Daugaard, A.E., Rosso, B., Hartmann, N.B., 2025. Disintegration of commercial biodegradable plastic products under simulated industrial composting conditions. Sci. Rep. 15 (1). https://doi.org/10.1038/s41598-025-91647-z.

Altay, B.N., 2023. Investigating the Effects of Using Biopolymer in Paper Coatings on Printability and Oil Absorption, PhD Dissertation. Marmara University, Istanbul, Turkiye.

Altay, B.N., Bloembergen, S., Aydemir, C., Karademir, A., Fleming, P.D., 2017. Use of nanoparticle binders for paper coatings: a review. J. Graph. Eng. Des. 8 (1), 39–43. https://doi.org/10.24867/jged-2017-1-039.

Altay, B.N., Kubik, Al C., Williams, S., Huq, A., Sugiyama, M., Dhote, Y., Zunjarrao, A., Aydemir, C., Karademir, A., 2024. Starch-based nanoparticles as a replacement for synthetic latex: a comprehensive assessment of printability and colorimetric characteristics. Color Res. Appl. 49 (6), 538–552. https://doi.org/10.1002/col.22941.

Altay, B.N., Lewis, C.L., 2024. Rheological characterization. In: Punia Bangar, S. (Ed.), Standardized Procedures and Protocols for Starch. Methods and Protocols in Food Science. Humana, New York, NY. https://doi.org/10.1007/978-1-0716-3866-8_6.

Amaraweera, S.M., Gunathilake, C., Gunawardene, O.H.P., Dassanayake, R.S., Fernando, N.M.L., Wanninayaka, D.B., Rajapaksha, S.M., Manamperi, A., Gangoda, M., Manchanda, A., Fernando, C.A.N., Kulatunga, A.K., Manipura, A., 2022. Preparation and characterization of dual-modified cassava starch-based biodegradable foams for sustainable packaging applications. ACS Omega 7 (23), 19579–19590. https://doi.org/10.1021/acsomega.2c01292.

Arayaphan, J., Boonsuk, P., Chantarak, S., 2020. Enhancement of water barrier properties of cassava starch-based biodegradable films using silica particles. Iran. Polym. J. 29 (9), 749–757. https://doi.org/10.1007/s13726-020-00837-1.

Arruda, T.R., Machado, G.d.O., Marques, C.S., Souza, A.L.d., Pelissari, F.M., Oliveira, T. V.d., Silva, R.R.A., 2025. An overview of starch-based materials for sustainable food packaging: recent advances, limitations, and perspectives. Macromolecules 5 (2). https://doi.org/10.3390/macromol5020019.

Avérous, L., 2004. Biodegradable multiphase systems based on plasticized starch: a review. J. Macromol. Sci. Polym. Rev. 44 (3), 231–274. https://doi.org/10.1081/mc-200029326.

Avérous, L., Halley, P.J., 2009. Biocomposites based on plasticized starch. Biofuel Bioprod. Biorefining 3 (3), 329–343. https://doi.org/10.1002/bbb.135France.

Avila, L.B., Schnorr, C., Silva, L.F.O., Morais, M.M., Moraes, C.C., da Rosa, G.S., Dotto, G.L., Lima, É.C., Naushad, M., 2023. Trends in bioactive multilayer films: perspectives in the use of polysaccharides, proteins, and carbohydrates with natural additives

for application in food packaging. Foods 12 (8). https://doi.org/10.3390/foods12081692.
Bangar, S.P., Singh Purewal, S., Trif, M., Maqsood, S., Kumar, M., Manjunatha, V., Rusu, A.V., 2021. Functionality and applicability of starch-based films: an eco-friendly approach. Foods 10 (9), 2181. https://doi.org/10.3390/foods10092181.
Barnes, H.A., 1989. Shear-thickening ("dilatancy") in suspensions of nonaggregating solid particles dispersed in Newtonian liquids. J. Rheol. 33 (2), 329–366. https://doi.org/10.1122/1.550017.
Bashir, K., Aggarwal, M., 2019. Physicochemical, structural and functional properties of native and irradiated starch: a review. J. Food Sci. Technol. 56 (2), 513–523. https://doi.org/10.1007/s13197-018-3530-2.
Bastioli, C., 2005. Starch-Based Technology. Handbook of Biodegradable Polymers. Rapra Technology Limited, London (United Kingdom), pp. 257–286.
Beiras, R., Concha-Graña, E., Laranjeiro, F.M.G., Fernández-González, V., López-Ibáñez, S., Moscoso-Pérez, C., Vilas, A., Muniategui-Lorenzo, S., 2025. Compostable and recycled plastics do not improve environmental chemical safety compared to conventional single-use polyethylene. Sci. Total Environ. 996, 180116. https://doi.org/10.1016/j.scitotenv.2025.180116.
Blancas-Benitez, F.J., Montaño-Leyva, B., Aguirre-Güitrón, L., Moreno-Hernández, C.L., Fonseca-Cantabrana, Á., Romero-Islas, L.d.C., González-Estrada, R.R., 2022. Impact of edible coatings on quality of fruits: a review. Food Control 139. https://doi.org/10.1016/j.foodcont.2022.109063.
Bourtoom, T., 2008. Edible films and coatings: characteristics and properties. Int. Food Res. J. 15 (3), 237–248.
Bremenkamp, I., Sousa Gallagher, M.J., 2024. Life cycle assessment methods for investigating novel food packaging systems. Environ. Sci. Adv. 3 (10), 1337–1350. https://doi.org/10.1039/d3va00380a.
Cazón, P., Velazquez, G., Ramírez, J.A., Vázquez, M., 2020. Polysaccharide-based films and coatings for food packaging: a review. Food Hydrocoll. 103, 105630.
Chang, Q., Zheng, B., Zhang, Y., Zeng, H., 2021. A comprehensive review of the factors influencing the formation of retrograded starch. Int. J. Biol. Macromol. 186, 163–173. https://doi.org/10.1016/j.ijbiomac.2021.07.050.
Cheng, H., Chen, L., McClements, D.J., Yang, T., Zhang, Z., Ren, F., Miao, M., Tian, Y., Jin, Z., 2021. Starch-based biodegradable packaging materials: a review of their preparation, characterization and diverse applications in the food industry. Trends Food Sci. Technol. 114, 70–82. https://doi.org/10.1016/j.tifs.2021.05.017.
Cheng, J., Gao, R., Zhu, Y., Lin, Q., 2024. Applications of biodegradable materials in food packaging: a review. Alex. Eng. J. 91, 70–83. https://doi.org/10.1016/j.aej.2024.01.080.
Chi, K., Wang, H., Catchmark, J.M., 2020. Sustainable starch-based barrier coatings for packaging applications. Food Hydrocoll. 103, 105696. https://doi.org/10.1016/j.foodhyd.2020.105696.
Courtene-Jones, W., De Falco, F., Napper, I.E., 2022. A Review of Biodegradable Plastics from Multidisciplinary Perspectives Plastic Pollution in the Global Ocean. World Scientific Publishing Co., United Kingdom, pp. 339–374. https://doi.org/10.1142/9789811259111_0012.
de Lima, G.G., Zakaluk, I.C.B., Artner, M.A., Pedro, A.C., Gonzalez de Cademartori, P. H., Muniz, G.I.B.d., Magalhães, W.L.E., 2025. Enhancing barrier and antioxidant properties of nanocellulose films for coatings and active packaging: a review. ACS Appl. Nano Mater. 8 (9), 4397–4421. https://doi.org/10.1021/acsanm.4c04805.
Dejene, B.K., Birilie, A.A., Yizengaw, M.A., Getahun, S.A., 2024. Thermoplastic starch-ZnO nanocomposites: a comprehensive review of their applications in functional

food packaging. Int. J. Biol. Macromol. 282. https://doi.org/10.1016/j.ijbiomac.2024.137099.

Di Bartolo, A., Infurna, G., Tzankova Dintcheva, N., 2021. A review of bioplastics and their adoption in the circular economy. Polymers 13 (8), 1229. https://doi.org/10.3390/polym13081229.

do Val Siqueira, L., Arias, C.I.L.F., Maniglia, B.C., Tadini, C.C., 2021. Starch-based biodegradable plastics: methods of production, challenges and future perspectives. Curr. Opin. Food Sci. 38, 122–130. https://doi.org/10.1016/j.cofs.2020.10.020.

Dobrucka, R., 2019. Bioplastic packaging materials in circular economy. Poznan School of Logistics, Poland Logforum 15 (1), 129–137. https://doi.org/10.17270/J.LOG.2019.322.

Domene-López, D., García-Quesada, J.C., Martin-Gullon, I., Montalbán, M.G., 2019. Influence of starch composition and molecular weight on physicochemical properties of biodegradable films. Polymers 11 (7). https://doi.org/10.3390/polym11071084.

Dong, M., Mastroianni, G., Bilotti, E., Zhang, H., Papageorgiou, D.,G., 2024. Biodegradable starch-based nanocomposite films with enhanced barrier properties. ACS Sustain. Chem. Eng. 12 (29), 11056–11066.

Dong, C.D., Tsai, M.L., Nargotra, P., Kour, B., Chen, C.W., Sun, P.P., Sharma, V., 2023. Bioprocess development for the production of xylooligosaccharide prebiotics from agro-industrial lignocellulosic waste. Heliyon 9 (7). https://doi.org/10.1016/j.heliyon.2023.e18316.

Dorantes-Fuertes, M.G., López-Méndez, M.C., Martínez-Castellanos, G., Meléndez-Armenta, R.Á., Jiménez-Martínez, H.E., 2024. Starch extraction methods in tubers and roots: a systematic review. Agronomy 14 (4). https://doi.org/10.3390/agronomy14040865.

Dutta, D., Sit, N., 2024. Comprehensive review on developments in starch-based films along with active ingredients for sustainable food packaging. Sustain. Chem. Pharm. 39, 101534. https://doi.org/10.1016/j.scp.2024.101534.

El Halal, S.L.M., Kringel, D.H., Zavareze, E.d.R., Dias, A.R.G., 2019. Methods for extracting cereal starches from different sources: a review. Starch/Staerke 71 (11–12). https://doi.org/10.1002/star.201900128.

EPA (2024). National strategy to prevent plastic pollution part three of a series on building a circular economy for all November, 2024 www.epa.gov/system/files/documents/2024-11/final_national_strategy_to_ prevent_plastic_pollution.pdf.

European Commission, 2019. Directive (EU) 2019/904 on the reduction of the impact of certain plastic products on the environment. Off. J. Eur. Union L155, 1–19.

Fang, K., Wang, H., Wang, X., Xia, F., Su, X., Jiang, L., Li, J., Sun, Z., Li, J., Lin, Q., 2024. An intensive exploration of plasticization, tensile, flexural and ecotoxicity of styrene-grafted thermoplastic starch (TPS) composite material. Ind. Crop. Prod. 220, 119285. https://doi.org/10.1016/j.indcrop.2024.119285.

FAOSTAT. (2023). FAO statistical database. http://www.fao.org/faostat/A.D: 01.25.2026.

Farshbaf Taghinezhad, S., Mansourieh, M., Abbasi, A., Major, I., Pezzoli, R., 2025. Improved compatibilized TPS/PLA blends: effects of singular and binary compatibilization systems. Carbohydr. Polym. Technol. Appl. 10, 100819. https://doi.org/10.1016/j.carpta.2025.100819.

Fatima, S., Khan, M.R., Ahmad, I., Sadiq, M.B., 2024. Recent advances in modified starch based biodegradable food packaging: a review. Heliyon 10 (6), e27453. https://doi.org/10.1016/j.heliyon.2024.e27453.

Figueiró, C.d.S., Calcagno, C.I.W., Santana, R.M.C., 2024. Starch foams and their additives: a brief review. Starch/Staerke 76 (3–4). https://doi.org/10.1002/star.202300012.

Fitzgerald, M.A., Bergman, C.J., Resurreccion, A.P., Möller, J., Jimenez, R., Reinke, R.F., Martin, M., Blanco, P., Molina, F., Chen, M.H., Kuri, V., Romero, M.V., Habibi, F., Umemoto, T., Jongdee, S., Graterol, E., Reddy, K.R., Bassinello, P.Z., Sivakami, R., Rani, N.S., Das, S., Wang, Y.J., Indrasari, S.D., Ramli, A., Ahmad, R., Dipti, S.S., Xie, L., Lang, N.T., Singh, P., Toro, D.C., Tavasoli, F., Mestres, C., 2009a. Addressing the dilemmas of measuring amylose in rice. Cereal Chem. 86 (5), 492–498. https://doi.org/10.1094/CCHEM-86-5-0492.

Fitzgerald, M.A., McCouch, S.R., Hall, R.D., 2009b. Not just a grain of rice: the quest for quality. Trends Plant Sci. 14 (3), 133–139. https://doi.org/10.1016/j.tplants.2008.12.004.

Forfora, N., Azuaje, I., Kanipe, T., Gonzalez, J.A., Lendewig, M., Urdaneta, I., Venditti, R., Gonzalez, R., Argyropoulos, D., 2024. Are starch-based materials more eco-friendly than fossil-based? a critical assessment. Clean. Environ. Syst. 13, 100177. https://doi.org/10.1016/j.cesys.2024.100177.

Gadhave, R.V.I., Gadhave, C.R., 2022. Adhesives for the paper packaging industry: an overview. Open J. Polym. Chem. 12 (02), 55–79. https://doi.org/10.4236/ojpchem.2022.122004.

Garavito, J., Peña-Venegas, C.P., Castellanos, D.A., 2024. Production of starch-based flexible food packaging in developing countries: analysis of the processes, challenges, and requirements. Foods 13 (24). https://doi.org/10.3390/foods13244096.

García-Guzmán, L., Cabrera-Barjas, G., Soria-Hernández, C.G., Castaño, J., Guadarrama-Lezama, A.Y., Rodríguez Llamazares, S., 2022. Progress in starch-based materials for food packaging applications. Polysaccharides 3 (1), 136–177. https://doi.org/10.3390/polysaccharides3010007.

Gastaldi, E., Buendia, F., Greuet, P., Benbrahim Bouchou, Z., Benihya, A., Cesar, G., Domenek, S., 2024. Degradation and environmental assessment of compostable packaging mixed with biowaste in full-scale industrial composting conditions. Bioresour. Technol. 400, 130670. https://doi.org/10.1016/j.biortech.2024.130670.

Geissdoerfer, M., Savaget, P., Bocken, N.M.P., Hultink, E.J., 2017. The circular economy – a new sustainability paradigm? J. Clean. Prod. 143, 757–768. https://doi.org/10.1016/j.jclepro.2016.12.048.

Geyer, R., Jambeck, J.R., Law, K.L., 2017. Production, use, and fate of all plastics ever made. Sci. Adv. 3 (7). https://doi.org/10.1126/sciadv.1700782.

Ghasemlou, M., Barrow, C.J., Adhikari, B., 2024. The future of bioplastics in food packaging: an industrial perspective. Food Packag. Shelf Life 43, 101279. https://doi.org/10.1016/j.fpsl.2024.101279.

Ghisellini, P., Cialani, C., Ulgiati, S., 2016. A review on circular economy: the expected transition to a balanced interplay of environmental and economic systems. J. Clean. Prod. 114, 11–32. https://doi.org/10.1016/j.jclepro.2015.09.007.

Gioia, C., Giacobazzi, G., Vannini, M., Totaro, G., Sisti, L., Colonna, M., Marchese, P., Celli, A., 2021. End of life of biodegradable plastics: composting versus re/upcycling. ChemSusChem 14 (19), 4167–4175. https://doi.org/10.1002/cssc.202101226.

Gökçe, E., 2018. Rethinking sustainability: a research on starch based bioplastic. Journal of Sustainable Construction Materials and Technologies 3 (3), 249–260. https://doi.org/10.29187/jscmt.2018.28.

Gong, K., Chen, Y., Lu, Y., Zhao, Z., Portela, A., Xu, H., Hu, M., Liu, H., Collins, M.N., 2025. Study of Thermoplastic starch/poly (butylene succinate) blends: the effect of reactive compatibilizers. Macromolecules 5 (3). https://doi.org/10.3390/macromol5030042.

Guo, L., Jin, Y., Xiao, Y., Tan, L., Tian, X., Ding, Y., He, K., Du, A., Li, J., Yi, Z., Wang, S., Fang, Y., Zhao, H., 2020. Energy-efficient and environmentally friendly production of starch-rich duckweed biomass using nitrogen-limited cultivation. J. Clean. Prod. 251, 119726. https://doi.org/10.1016/j.jclepro.2019.119726.

He, R., Li, S., Zhao, G., Zhai, L., Qin, P., Yang, L., 2023. Starch modification with molecular transformation, physicochemical characteristics, and industrial usability: a state-of-the-art review. Polymers 15 (13). https://doi.org/10.3390/polym15132935.

He, X., Zhang, F., Li, C., Ding, W., Jin, Y., Tang, L., Huang, R., 2024. Effect of starch plasticization on morphological, mechanical, crystalline, thermal, and optical behavior of poly(butylene adipate-co-terephthalate)/thermoplastic starch composite films. Polymers 16 (3), 326. https://doi.org/10.3390/polym16030326.

Hemachandra, S., Hadjikakou, M., Pettigrew, S., 2024. A scoping review of food packaging life cycle assessments that account for packaging-related food waste. Int. J. Life Cycle Assess. 29 (10), 1899–1915. https://doi.org/10.1007/s11367-024-02349-z.

Hoover, R., 2001. Composition, molecular structure, and physicochemical properties of tuber and root starches: a review. Carbohydr. Polym. 45 (3), 253–267. https://doi.org/10.1016/s0144-8617(00)00260-5.

Hubbe, M.A., Rojas, O.J., Lucia, L.A., Sain, M., 2008. Cellulosic nanocomposites: a review. Bioresources 3 (3), 929–980. https://doi.org/10.15376/BIORES.3.3.929-980.

Jane, J.-lin, 2006. Current understanding on starch granule structures. J. Appl. Glycosci. 53 (3), 205–213. https://doi.org/10.5458/jag.53.205.

Jiang, T., Duan, Q., Zhu, J., Liu, H., Yu, L., 2020. Starch-based biodegradable materials: challenges and opportunities. Adv. Ind. Eng. Polym. Res. 3 (1), 8–18. https://doi.org/10.1016/j.aiepr.2019.11.003.

Jiménez, A., Fabra, M.J., Talens, P., Chiralt, A., 2012. Edible and biodegradable starch films: a review. Food Bioprocess Technol. 5 (6), 2058–2076. https://doi.org/10.1007/s11947-012-0835-4.

Karakuş, E., Kibar Balballi, E., Ara, İ., Ayhan, Z., 2021. Polisakkarit ve protein bazli aktif biyokompozit malzemelerin gida ambalajlama açisindan değerlendirilmesi. Akademik Gıda 19 (1), 74–88. https://doi.org/10.24323/akademik-gida.927700.

Kargarzadeh, H., Mariano, M., Huang, J., Lin, N., Ahmad, I., Dufresne, A., Thomas, S., 2017. Recent developments on nanocellulose reinforced polymer nanocomposites: a review. Polymer 132, 368–393. https://doi.org/10.1016/j.polymer.2017.09.043.

Karnwal, A., Rauf, A., Jassim, A.Y., Selvaraj, M., Al-Tawaha, A.R.M.S., Kashyap, P., Kumar, D., Malik, T., 2025. Advanced starch-based films for food packaging: innovations in sustainability and functional properties. Food Chem. X 29. https://doi.org/10.1016/j.fochx.2025.102662.

Kaseem, M., Hamad, K., Deri, F., 2012. Thermoplastic starch blends: a review of recent works. Polym. Sci. 54 (2), 165–176. https://doi.org/10.1134/s0965545x1202006x.

Kharola, S., Ram, M., Goyal, N., Mangla, S.K., Nautiyal, O.P., Rawat, A., Kazancoglu, Y., Pant, D., 2022. Barriers to organic waste management in a circular economy. J. Clean. Prod. 362. https://doi.org/10.1016/j.jclepro.2022.132282.

Kim, M.S., Chang, H., Zheng, L., Yan, Q., Pfleger, B.F., Klier, J., Nelson, K., Majumder, E.L.W., Huber, G.W., 2023. A review of biodegradable plastics: chemistry, applications, properties, and future research needs. Chem. Rev. 123 (16), 9915–9939. https://doi.org/10.1021/acs.chemrev.2c00876.

Kistler, S.F., Schweizer, P.M., 1997. In: Liquid Film Coating. Springer.

Kossalbayev, B.D., Belkozhayev, A.M., Abaildayev, A., Kadirshe, D.K., Tastambek, K.T., Kurmanbek, A., Toleutay, G., 2025. Biodegradable packaging from agricultural wastes: a comprehensive review of processing techniques, material properties, and future prospects. Polymers 17 (16). https://doi.org/10.3390/polym17162224.

Krishnan, R., Misra, M., Subramanian, J., Mohanty, A., 2025. Emerging trends and application of edible coating as a sustainable solution for postharvest management in stone fruits: a comprehensive review. Compr. Rev. Food Sci. Food Saf. 24 (3). https://doi.org/10.1111/1541-4337.70179.

Krochta, J.M., De Mulder-Johnston, C., 1997. Edible and biodegradable polymer films: challenges and opportunities. Food Technol. 51 (2), 61–74.

La Fuente, C.I.A., Maniglia, B.C., Tadini, C.C., 2023a. Biodegradable polymers: a review about biodegradation and its implications and applications. Packag. Technol. Sci. 36 (2), 81–95.

La Fuente, C.I.A., Maniglia, B.C., Tadini, C.C., 2023b. Biodegradable polymers: a review about biodegradation and its implications and applications. Packag. Technol. Sci. 36 (2), 81–95. https://doi.org/10.1002/pts.2699.

Lauer, M.K., Smith, R.C., 2020. Recent advances in starch-based films toward food packaging applications: physicochemical, mechanical, and functional properties. Compr. Rev. Food Sci. Food Saf. 19 (6), 3031–3083. https://doi.org/10.1111/1541-4337.12627.

Lindeboom, N., Chang, P.R., Tyler, R.T., 2004. Analytical, biochemical and physicochemical aspects of starch granule size, with emphasis on small granule starches: a review. Starch - Stärke 56 (3–4), 89–99. https://doi.org/10.1002/star.200300218.

Liu, H., Xie, F., Yu, L., Chen, L., Lin, L., 2009. Thermal processing of starch-based polymers. Prog. Polym. Sci. 34 (12), 1348–1368. https://doi.org/10.1016/j.progpolymsci.2009.07.001.

Liu, D., Zhao, P., Chen, J., Yan, Y., Wu, Z., 2022. Recent advances and applications in starch for intelligent active food packaging: a review. Foods 11 (18), 2879. https://doi.org/10.3390/foods11182879.

Lourdin, D., Coignard, L., Bizot, H., Colonna, P., 1997. Influence of equilibrium relative humidity and plasticizer concentration on the water content and glass transition of starch materials. Polymer 38 (21), 5401–5406. https://doi.org/10.1016/s0032-3861(97)00082-7.

Mali, S., Sakanaka, L.S., Yamashita, F., Grossmann, M.V.E., 2005. Water sorption and mechanical properties of cassava starch films and their relation to plasticizing effect. Carbohydr. Polym. 60 (3), 283–289. https://doi.org/10.1016/j.carbpol.2005.01.003.

Mariño-Cortegoso, S., Lestido-Cardama, A., Sendón, R., Ana Rodríguez, B.de Q., Barbosa-Pereira, L., 2025. The state of the art and innovations in active and edible coatings and films for functional food applications. Polymers 17 (18), 2472. https://doi.org/10.3390/polym17182472.

Mestre, A., Cooper, T., 2017. Circular product design. A multiple loops life cycle design approach for the circular economy. Des. J. 20 (Suppl. 1), S1620. https://doi.org/10.1080/14606925.2017.1352686.

Mkhari, T., Adeyemi, J.O., Fawole, O.A., 2025. Recent advances in the fabrication of intelligent packaging for food preservation: a review. Processes 13 (2). https://doi.org/10.3390/pr13020539.

Moorthy, S.N., 2002. Physicochemical and functional properties of tropical tuber starches: a review. Starch - Stärke 54 (12), 559–592. https://doi.org/10.1002/1521-379x(200212)54:12<559::aid-star2222559>3.0.co;2-f.

Moreno-Bohorquez, E., Arias-Tapia, M.J., Jaramillo, A.F., 2026. Recent advances in thermoplastic starch (TPS) and biodegradable polyester blends: a review of compatibilization strategies and bioactive functionalities. Polymers 18 (2). https://doi.org/10.3390/polym18020289.

Mounir, S., Ghandour, A., Shatta, A., Farid, E., 2024. Starch: Properties and Functionality Starch: Structure, Properties, and Modifications for Food Applications. CRC Press, Egypt, pp. 1–46. https://doi.org/10.1201/9781032655598-1.

Müller, C.M.O., Yamashita, F., Laurindo, J.B., 2008. Evaluation of the effects of glycerol and sorbitol concentration and water activity on the water barrier properties of cassava starch films through a solubility approach. Carbohydr. Polym. 72 (1), 82–87. https://doi.org/10.1016/j.carbpol.2007.07.026.

Naa, P., Shashank, S., Singh, B.N., Mazumder, N., 2024. Advancing sustainable bioplastics: chemical and physical modification of starch films for enhanced thermal and barrier properties. RSC Adv. 14 (33), 23943–23951. https://doi.org/10.1039/d4ra04263h.

Nam, K., Kim, S.G., Kim, D.Y., Lee, D.Y., 2024. Enhanced mechanical properties of polylactic acid/poly(butylene adipate-co-terephthalate) modified with maleic anhydride. Polymers 16 (4). https://doi.org/10.3390/polym16040518.

Newburger, E., 2022. U.S. To Ban Sale of Single-Use Plastic on Public Lands. CNBC national parks by 2032. https://www.cnbc.com/2022/06/08/us-to-ban-sale-of-single-use-plasticon-public-lands-national-parks-by-2032.html.

Nizamuddin, S., Baloch, A.J., Chen, C., Arif, M., Mubarak, N.M., 2024. Bio-based plastics, biodegradable plastics, and compostable plastics: biodegradation mechanism, biodegradability standards and environmental stratagem. Int. Biodeterior. Biodegrad. 195. https://doi.org/10.1016/j.ibiod.2024.105887.

Nunes, C., Silva, M., Farinha, D., Sales, H., Pontes, R., Nunes, J., 2023. Edible coatings and future trends in active food packaging–fruits' and traditional sausages' shelf life increasing. Foods 12 (17), 3308. https://doi.org/10.3390/foods12173308.

Onyeaka, H., Obileke, K.C., Makaka, G., Nwokolo, N., 2022. Current research and applications of starch-based biodegradable films for food packaging. Polymers 14 (6), 1126. https://doi.org/10.3390/polym14061126.

Paluch, M., Ostrowska, J., Tyński, P., Sadurski, W., Konkol, M., 2022. Structural and thermal properties of starch plasticized with glycerol/urea mixture. J. Polym. Environ. 30 (2), 728–740. https://doi.org/10.1007/s10924-021-02235-x.

Pantović, A., Tomić, N., Petrović, T., Djekić, I., 2025. Sustainable perspectives of disposable tableware: a systematic review. Sustainability 17 (4), 1434. https://doi.org/10.3390/su17041434.

Pei, J., Palanisamy, C.P., Srinivasan, G.P., Panagal, M., Kumar, S.S.D., Mironescu, M., 2024. A comprehensive review on starch-based sustainable edible films loaded with bioactive components for food packaging. Int. J. Biol. Macromol. 274. https://doi.org/10.1016/j.ijbiomac.2024.133332.

Petersen, K., Væggemose Nielsen, P., Bertelsen, G., Lawther, M., Olsen, M.B., Nilsson, N.H., Mortensen, G., 1999. Potential of biobased materials for food packaging. Trends Food Sci. Technol. 10 (2), 52–68. https://doi.org/10.1016/S0924-2244(99)00019-9.

Ranum, P., Peña-Rosas, J.P., Garcia-Casal, M.N., 2014. Global maize production, utilization, and consumption. Ann. N. Y. Acad. Sci. 1312 (1), 105–112. https://doi.org/10.1111/nyas.12396.

Rao, M.A., 2014. Rheology of Fluid and Semisolid Foods: Principles and Applications, third ed. Springer.

Rashwan, A.K., Younis, H.A., Abdelshafy, A.M., Osman, A.I., Eletmany, M.R., Hafouda, M.A., Chen, W., 2024. Plant starch extraction, modification, and green applications: a review. Environ. Chem. Lett. 22 (5), 2483–2530. https://doi.org/10.1007/s10311-024-01753-z.

Ratnayake, W.S., Hoover, R., Warkentin, T., Starch, P., 2002. Composition, structure and properties — a review. Starch - Stärke 54 (6), 217–234. https://doi.org/10.1002/1521-379x(200206)54:6<217::aid-star217>3.0.co;2-r.

Reddy, N., Yang, Y., 2005. Biofibers from agricultural byproducts for industrial applications. Trends Biotechnol. 23 (1), 22–27. https://doi.org/10.1016/j.tibtech.2004.11.002.

Ren, Y., Yuan, T.Z., Chigwedere, C.M., Ai, Y., 2021. A current review of structure, functional properties, and industrial applications of pulse starches for value-added utilization. Compr. Rev. Food Sci. Food Saf. 20 (3), 3061–3092. https://doi.org/10.1111/1541-4337.12735.

Rosenboom, J.G., Langer, R., Traverso, G., 2022. Bioplastics for a circular economy. Nat. Rev. Mater. 7 (2), 117–137. https://doi.org/10.1038/s41578-021-00407-8.

Rujnić-Sokele, M., Pilipović, A., 2017. Challenges and opportunities of biodegradable plastics: a mini review. Waste Manag. Res.: The Journal for a Sustainable Circular Economy 35 (2), 132–140. https://doi.org/10.1177/0734242x16683272.

Salgado-Delgado, A.M., Lozano-Pineda, E., Salgado-Delgado, R., Hernández-Uribe, J.P., Olarte-Paredes, A., Granados-Baeza, M.J., 2022. Chemical modification of rice (Oryza sativa) and potato (Solanum tuberosum) starches by silanization with trimethoxy (methyl)silane. Rev. Mex. Ing. Quim. 21 (3), 1–16. https://doi.org/10.24275/rmiq/alim2802.

Shogren, R., 1996. Preparation, thermal properties, and extrusion of high-amylose starch acetates. Carbohydr. Polym. 29 (1), 57–62. https://doi.org/10.1016/0144-8617(95)00143-3.

Sikora, M., 2021. Puszka, modern biodegradable plastics-processing and properties part II. Materials 14 (10).

Singh, P., Pandey, V.K., Singh, R., Singh, K., Dash, K.K., Malik, S., 2024. Unveiling the potential of starch-blended biodegradable polymers for substantializing the eco-friendly innovations. J. Agric. Food Res. 15. https://doi.org/10.1016/j.jafr.2024.101065.

Singh, N., Singh, J., Kaur, L., Sodhi, N.S., Gill, B.S., 2003. Morphological, thermal and rheological properties of starches from different botanical sources. Food Chem. 81 (2), 219–231. https://doi.org/10.1016/S0308-8146(02)00416-8.

Siracusa, V., Rocculi, P., Romani, S., Rosa, M.D., 2008. Biodegradable polymers for food packaging: a review. Trends Food Sci. Technol. 19 (12), 634–643. https://doi.org/10.1016/j.tifs.2008.07.003.

SOR (2022). Single-use plastics prohibition regulations (SOR/2022-138) https://laws-lois.justice.gc.ca/PDF/SOR-2022-138.pdf.

Sorrentino, A., Gorrasi, G., Vittoria, V., 2007. Potential perspectives of bio-nanocomposites for food packaging applications. Trends Food Sci. Technol. 18 (2), 84–95. https://doi.org/10.1016/j.tifs.2006.09.004.

Soykeabkaew, N., Thanomsilp, C., Suwantong, O., 2015. A review: starch-based composite foams. Compos. Appl. Sci. Manuf. 78, 246–263. https://doi.org/10.1016/j.compositesa.2015.08.014.

Stelescu, M.D., Oprea, O.C., Sonmez, M., Ficai, A., Motelica, L., Ficai, D., Georgescu, M., Gurau, D.F., 2024. Structural and thermal characterization of some thermoplastic starch mixtures. Polysaccharides 5 (4), 504–522. https://doi.org/10.3390/polysaccharides5040032.

Suleman, A., Bora, J., Kalita, A.J., Chetia, B., 2025. Starch-based nanocomposite for food packaging applications. In: Engineering Applications of Polymer-Based Nanocomposites. Springer Nature Switzerland, Cham, pp. 47–68.

Surendren, A., Mohanty, A.K., Liu, Q., Misra, M., 2022. A review of biodegradable thermoplastic starches, their blends and composites: recent developments and opportunities for single-use plastic packaging alternatives. Green Chem. 24 (22), 8606–8636. https://doi.org/10.1039/d2gc02169b.

Tang, X., Alavi, S., 2011. Recent advances in starch, polyvinyl alcohol based polymer blends, nanocomposites and their biodegradability. Carbohydr. Polym. 85 (1), 7–16. https://doi.org/10.1016/j.carbpol.2011.01.030.

Tapia-Blácido, D.R., Aguilar, G.J., de Andrade, M.T., Rodrigues-Júnior, M.F., Guareschi-Martins, F.C., 2022. Trends and challenges of starch-based foams for use as food packaging and food container. Trends Food Sci. Technol. 119, 257–271. https://doi.org/10.1016/j.tifs.2021.12.005.

Tester, R.F., Karkalas, J., Qi, X., 2004. Starch - composition, fine structure and architecture. J. Cereal. Sci. 39 (2), 151–165. https://doi.org/10.1016/j.jcs.2003.12.001.

Thakur, R., Pristijono, P., Scarlett, C.J., Bowyer, M., Singh, S.P., Vuong, Q.V., 2019. Starch-based films: major factors affecting their properties. Int. J. Biol. Macromol. 132, 1079–1089. https://doi.org/10.1016/j.ijbiomac.2019.03.190.

Tharanathan, R.N., 2003. Biodegradable films and composite coatings: past, present and future. Trends Food Sci. Technol. 14 (3), 71–78. https://doi.org/10.1016/s0924-2244(02)00280-7.

Tilton, J.E., 1999. The future of recycling. Resour. Policy 25 (3), 197–204. https://doi.org/10.1016/S0301-4207(99)00027-6.

UNEP. (2021). Single-use plastic tableware and its alternatives: recommendations from life cycle assessments. (United Nations Environment Programme. Life Cycle Initiative/).

Van Soest, J.J.G., Borger, D.B., 1997. Structure and properties of compression-molded thermoplastic starch materials from normal and high-amylose maize starches. J. Appl. Polym. Sci. 64 (4), 631–644. https://doi.org/10.1002/(sici)1097-4628(19970425)64:4<631::aid-app2>3.0.co;2-o.

Yacob, N., Yusof, M.R., Mohamed, A.Z., Badri, K.H., 2019. Effect of cellulose fiber loading on the properties of starch-based films. In: AIP Conference Proceedings. American Institute of Physics Inc., Malaysia https://doi.org/10.1063/1.5111257.

Yadav, K., Nikalje, G.C., 2024. Comprehensive analysis of bioplastics: life cycle assessment, waste management, biodiversity impact, and sustainable mitigation strategies. PeerJ 12 (9). https://doi.org/10.7717/peerj.18013.

Yang, Y., Fu, J., Duan, Q., Xie, H., Dong, X., Yu, L., 2024. Strategies and methodologies for improving toughness of starch films. Foods 13 (24), 4036. https://doi.org/10.3390/foods13244036.

Yudhistira, B., Husnayain, N., Punthi, F., Gavahian, M., Chang, C.K., Hsieh, C.W., 2024. Progress in the application of emerging technology for the improvement of starch-based active packaging properties: a review. ACS Food Sci. Technol. 4 (9), 1997–2012. https://doi.org/10.1021/acsfoodscitech.4c00260.

Zhang, Y., Rempel, C., Liu, Q., 2014. Thermoplastic starch processing and characteristics—a review. Crit. Rev. Food Sci. Nutr. 54 (10), 1353–1370. https://doi.org/10.1080/10408398.2011.636156.

Zhang, C., Sun, Z., Yang, X., Duan, G., Wang, M., Qiao, Y., 2025. Effect of thermoplastic starch content on the properties of poly(butylene adipate-co-terephthalate) (PBAT) composites. Bioresources 20 (2), 3910–3922. https://doi.org/10.15376/biores.20.2.3910-3922.

Zimmermann, L., Dombrowski, A., Völker, C., Wagner, M., 2020. Are bioplastics and plant-based materials safer than conventional plastics? In vitro toxicity and chemical composition. Environ. Int. 145, 106066. https://doi.org/10.1016/j.envint.2020.106066.

CHAPTER TWO

Starch based food packaging: Advantages and disadvantages

Md Raju Mia, Md Belal Hossain and Md Rabbi Hasan
Department of Paper and Packaging Technology, Indian Institute of Technology, Roorkee (IIT ROORKEE) Roorkee, Haridwar, Uttarakhand, India

2.1 Introduction

The packaging industry plays a critical role in protecting products, facilitating transportation, preserving product quality, and communicating essential information to consumers. Market analysis indicate that the global packaging industry is expected to grow at an annual rate of about 3%, surpassing $1.2 trillion between 2018 and 2028, while contributing to notable environmental impacts (Trends in Packaging Sector). It is estimated that around 8.3 billion metric tons of plastic have been produced worldwide, yet only about 9% has been recycled (Geyer et al., 2017), with nearly 79% accumulating in natural environments. Increasing concerns over plastic waste and its environmental consequences are driving both manufacturers and consumers toward recyclable and biodegradable alternatives (Paper & Paperboard Packaging market Size, Share).

The rapid rise in single use packaging, particularly for food and consumer goods, has resulted in massive amounts of plastic waste filling landfills, waterways, and oceans. Recent estimates suggest that millions of tons of plastic packaging waste are created each year, but only a small portion is effectively recycled. Poor disposal practices result in long-lasting environmental pollution, pose dangers to wildlife, and negatively affect human health due to microplastics. Additionally, extracting and processing fossil fuels to make plastic adds to greenhouse gas emissions and depletes resources, making climate change issues worse.

As public awareness grows, along with stricter environmental rules and global sustainability goals, there is an increasing need for eco friendly, biodegradable, and renewable alternatives to traditional plastic packaging. In this context, biopolymers have become an important solution for the

Starch-Based Sustainable Food Packaging
ISBN: 978-0-443-36346-7
https://doi.org/10.1016/B978-0-443-36346-7.00002-3

environmental issues caused by traditional petroleum based packaging materials. They are derived from renewable biological sources such as plants, animals, and microorganisms, making them a sustainable option due to their biodegradability, compostability, and reduced carbon footprint. Unlike synthetic plastics, biopolymers break down naturally in the right environmental conditions. Some common biopolymers used in packaging are starch, cellulose, polylactic acid (PLA), polyhydroxyalkanoates (PHAs), and proteins like casein and soy protein. Starch is abundant, renewable, biodegradable, and affordable. Starch is among the most studied and used biopolymers that has great potential to lower the environmental impact of packaging materials due to its availability, low cost, and ease of modification (Mironescu et al., 2021). However, to effectively replace conventional plastics, challenges regarding performance, processing, and cost must be addressed (Lauer and Smith, 2020).

Biopolymer based food packaging materials can be produced in various forms, including films, coatings, foams, trays, and edible wraps. These materials effectively serve as alternatives to conventional single use plastics in numerous applications, particularly within the food packaging sector, where biodegradability and safety are of utmost importance. Nonetheless, biopolymers encounter certain limitations, such as poor moisture resistance, mechanical fragility, and sensitivity to environmental conditions. To address these challenges, ongoing research and innovation are focused on blending biopolymers with other materials, implementing chemical or physical modifications, and developing composites that exhibit enhanced performance. The adoption of biopolymer based packaging directly supports the principles of a circular economy by encouraging the use of renewable resources, decreasing reliance on fossil fuels, and enabling end of life biodegradation. As industries and policymakers increasingly prioritize sustainability, biopolymers are expected to play a pivotal role in advancing greener, more responsible packaging solutions.

Starch is one of the most renewable, and cost effective natural polymers available worldwide. It is primarily obtained from crops such as corn, potatoes, wheat, rice, and cassava. Its widespread availability, biodegradability, and versatility make starch an attractive raw material for developing sustainable packaging solutions. Native starch consists mainly of two polysaccharides: amylose (a mostly linear molecule) and amylopectin (a highly branched molecule). This unique structural composition imparts useful film forming and thermoplastic properties, making starch suitable for producing biodegradable films, coatings, foams, and composites. In food packaging applications, starch based materials have gained considerable attention for their potential to replace petroleum based plastics in disposable packaging products such as bags, trays, and wrapping films. They are particularly promising for food packaging due to their non toxic nature and ability to form good oxygen barriers, which help maintain food quality and extend shelf life.

Despite its advantages, native starch has limitations such as high water sensitivity, brittleness, and poor mechanical strength, which restrict its direct use as a packaging material. To overcome these drawbacks, various modification techniques, including blending with other biodegradable polymers, plasticization, crosslinking, and reinforcement with fibers or nanoparticles, have been developed to enhance their performance. Growing environmental awareness, stricter regulations on plastic waste, and the push for sustainable alternatives have created significant opportunities for starch based packaging materials in the global market. Research continues to focus on developing cost effective, high performance starch based products that meet industrial standards and consumer expectations. By addressing technical challenges and scaling up production, starch based packaging has the potential to contribute meaningfully to a circular economy, reducing reliance on fossil resources and mitigating the environmental impact of conventional plastic waste (Arruda et al., 2025).

2.2 The composition and characteristics of starch

The two polysaccharides that make up starch, a naturally occurring, biodegradable carbohydrate polymer, are amylose, a linear molecule, and amylopectin, a highly branched molecule. It is widely accessible from a variety of plant sources, such as cassava, corn, potatoes, wheat, and rice. The low cost, renewable nature, and environmental friendliness of starch make it a promising raw material for the creation of environmentally friendly and sustainable products. A variety of natural polymeric matrices, such as protein, chitosan, gelatin, pectin, and starch, have been used to create antimicrobial food packaging solutions. Starch is the most promising biodegradable natural polymer for use as a matrix, as it is inexpensive, widely available, and compostable, leaving no harmful residue. There are several applications for starch based films in both food and non food applications (Fig. 2.1).

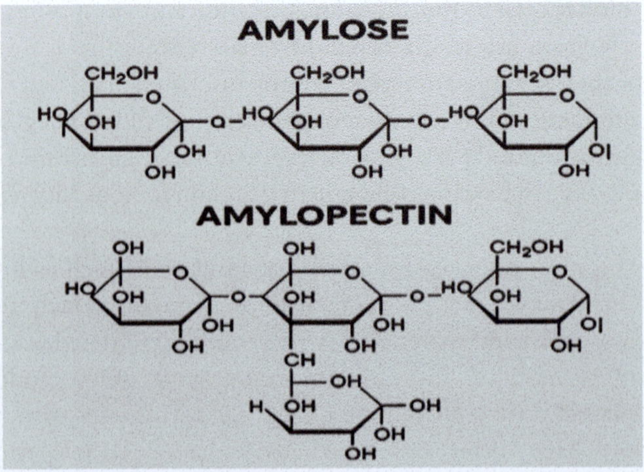

Figure 2.1 Chemical structure of starch displaying branching amylopectin (containing both α-1,4- and α-1,6-glycosidic links) and linear amylose (linked by α-1,4-glycosidic bonds) (Cornejo-Ramírez et al., 2018).

2.2.1 Physicochemical characteristics that are important for packaging

Due in large part to its distinct physicochemical properties, starch is a suitable raw material for environmentally friendly packaging. The way starch behaves during processing and in the finished packaging product is determined by these characteristics. Optimizing starch based materials for real world packaging applications requires an understanding of these properties (Luchese et al., 2017).

2.2.2 Water sensitivity and moisture content

Due to the large number of hydroxyl (-OH) groups on its glucose units, starch is naturally hydrophilic. High moisture absorption follows, which may affect the mechanical strength, dimensional stability, and barrier properties of starch based films. Consequently, when designing starch packaging, moisture sensitivity is a key factor that is frequently addressed by coating or blending with hydrophobic materials (Bertuzzi et al., 2007).

2.2.3 Gelatinization behavior

For packaging applications, one of the key functional features of starch is its ability to gelatinize. Starch granules swell, absorb water, and rupture when heated in the presence of water, transforming the crystalline structure into an amorphous, viscous paste. Because of this trait, starch can be processed using common thermoplastic techniques like molding, extrusion, and film casting (Lauer and Smith, 2020).

2.2.4 Retrogradation

Retrogradation is the term for the tendency of starch chains, particularly amylose, to reassociate into more ordered structures as they cool after gelatinization. Over time, excessive retrogradation can cause starch films to become brittle and less flexible, whereas moderate retrogradation can enhance the film's strength and barrier qualities. Therefore, maintaining consistent product quality requires control of retrogradation (Ning et al., 2025).

2.2.5 Film-forming ability

Because starch can create hydrogen bonds between polymer chains, it has a strong natural capacity to form films. Starch based films are appropriate for applications involving food contact because they are typically transparent, odorless, and non toxic. Nevertheless, plasticizers (such as sorbitol or glycerol) or blending agents are frequently added because the mechanical qualities and water resistance of pure starch films are frequently insufficient (Lin et al., 2017).

2.2.6 The mechanical characteristics

For packaging performance, starch based films' elasticity, elongation at break, and tensile strength are essential. Films of native starch frequently show low elongation and are brittle. These mechanical qualities can be enhanced to satisfy useful packaging needs by changing starch or adding additives (Lin et al., 2017).

2.2.7 Barrier characteristics

Due to the dense hydrogen bonding within the polymer matrix, starch based films exhibit excellent oxygen barrier properties, which is highly desirable for maintaining the freshness of food products. But their high water vapor permeability still presents a problem, particularly for packaging that needs to be moisture resistant.

2.2.8 Biodegradability

The full biodegradability of starch in natural settings is one of its most valuable properties. In contrast to traditional plastics, microorganisms can easily break starch into carbon dioxide, water, and biomass, thereby significantly reducing environmental pollution (Lin et al., 2017).

2.3 Modification techniques for starch

2.3.1 Physical alterations

Physical modification refers to processes that tailor the structural organization of starch while maintaining its original chemical composition, making these approaches environmentally friendly and well aligned with sustainable material development. These methods are widely used to improve processability, mechanical performance, and barrier functionality of starch materials. Heat moisture treatment (HMT) involves conditioning starch at low moisture levels and elevated temperatures, which enhances thermal stability, reduces excessive swelling, and improves dimensional stability during processing. Annealing, conducted in excess water at temperatures below gelatinization, promotes molecular reorganization, leading to improved crystallinity and more stable pasting behavior, which is beneficial for film formation. Pregelatinization, achieved by gelatinizing starch followed by drying, produces instant starch with enhanced cold water solubility, facilitating easier dispersion in coating and film forming formulations. In addition, extrusion and milling apply mechanical shear that disrupts granule integrity, enabling the production of thermoplastic starch with modified viscosity, improved processability, and tunable mechanical and biodegradation properties suitable for packaging systems (Nawaz et al., 2020).

2.3.2 Chemical alterations

Chemical modification involves introducing functional groups onto starch molecules to tailor their physicochemical and functional performance, enabling improved processability, stability, and end-use properties. These modifications are widely used to overcome the inherent limitations of native starch, such as high water sensitivity, poor mechanical strength, and limited thermal stability. Esterification, including acetylation, introduces ester groups that enhance flexibility, improve stability, and reduce retrogradation, which is beneficial for maintaining film integrity during storage. Etherification produces starch ethers with improved solubility and controlled viscosity, supporting more uniform coating and film formation. Oxidation reduces molecular weight while introducing carbonyl and carboxyl groups, leading to improved film forming ability, smoother surfaces, and enhanced adhesion. Cross linking creates covalent bonds between starch chains, significantly increasing resistance to shear, heat, and acidic conditions, which is critical for processing robustness and dimensional stability in packaging applications. Acid hydrolysis partially depolymerizes starch chains, lowering viscosity while increasing gel strength, which can be advantageous for tailoring rheology in coating and adhesive formulations (Nawaz et al., 2020).

2.3.3 Enzymatic and biological alterations

Enzymatic and biological modifications provide highly specific and controlled pathways to tailor starch structure under relatively mild processing conditions. These approaches leverage enzymes or microbial systems to selectively modify molecular architecture, enabling precise control over functionality while minimizing harsh processing conditions. Enzymatic hydrolysis, typically carried out using enzymes such as α-amylase and glucoamylase, breaks starch into smaller fractions such as dextrins, maltose, or glucose, which can be used to tailor viscosity, improve processability, and modify film-forming behavior. Debranching enzymes, including pullulanase and isoamylase, selectively alter the branched structure of amylopectin, promoting molecular reorganization that can enhance crystallinity and improve mechanical and barrier performance in films. Fermentation processes, driven by microbial activity, can modify molecular weight distribution and introduce functional metabolites, thereby influencing flexibility, biodegradability, and overall material performance. In addition, genetic modification involves altering plant biosynthetic pathways to produce starches with controlled amylose to amylopectin ratios, enabling the development of starch feedstocks with inherently improved film strength, reduced water sensitivity, and more predictable processing behavior for packaging applications (Choton et al., 2024).

2.3.4 Combining with additional polymers and additives

For starch based packaging materials, blending with complementary polymers and functional additives is a widely used strategy to enhance performance and expand application potential. Because native starch films often exhibit brittleness, moisture sensitivity, and limited barrier performance, combining starch with other materials enables the design of more robust and functional systems. Polymer blending involves incorporating starch with natural or synthetic polymers, such as cellulose, gelatin, or polyvinyl alcohol, to improve mechanical strength, flexibility, and gas and moisture barrier properties while maintaining biodegradability. The addition of plasticizers, including glycerol or sorbitol, reduces intermolecular forces within the starch matrix, enhancing flexibility, reducing brittleness, and improving processability during film formation or extrusion. Reinforcing agents, such as natural fibers, nanocellulose, or inorganic nanoparticles, can further increase tensile strength, stiffness, and dimensional stability while potentially improving barrier performance. Functional additives, including antioxidants, surfactants, and emulsifiers, are incorporated to tailor surface properties, improve compatibility between components, and provide additional functionalities such as enhanced stability or active packaging performance. Together, these formulation strategies enable the development of starch based materials with tunable mechanical, barrier, and functional properties suitable for diverse packaging applications (Ambrogi et al., 2017; Fearon et al., 2002).

2.4 Sources of starch

The natural carbohydrate starch is mostly found in plants, where it is used as a substance to store energy. The primary starch sources are categorized according to their botanical origin.

2.4.1 Cereal origins

Cereals represent the most significant and widely utilized sources of starch for industrial applications due to their global availability, consistent supply, and well established processing infrastructure. Among these, corn (maize) is the dominant commercial source, widely used because of its high starch yield and cost effectiveness. Other important cereal sources include wheat, rice, barley, and sorghum, each offering distinct functional properties depending on their amylose amylopectin composition and granule structure. Starches derived from these cereals are extensively applied across multiple industries, including paper manufacturing, adhesives, textiles, and food processing, as well as in the development of sustainable packaging materials where they serve as film forming agents, binders, and coating components (James et al., 2003; Hannah and James, 2008).

2.4.2 Sources of tuber

Tuber crops are important sources of starch, storing substantial amounts of reserve carbohydrates in their underground plant structures, which allows for relatively efficient extraction. Major examples include potato, cassava (tapioca), sweet potato, and yam, each providing starch with distinctive functional characteristics. Compared with cereal starches, tuber starches typically possess larger granule sizes, higher swelling power, and greater paste clarity, which contribute to their excellent thickening and film forming capabilities. These properties make tuber derived starches particularly suitable for applications in food and pharmaceutical formulations, as well as in biodegradable films and coatings for packaging, where high viscosity and smooth film formation are desirable (Reyniers et al., 2020; Moorthy, 2002).

2.4.3 Sources of the root

Roots represent another important source of starch, particularly in tropical and subtropical regions where these crops are widely cultivated and serve as major carbohydrate reserves. Common examples include manioc (cassava) and arrowroot, both of which yield starches known for their high purity, light color, and neutral flavor profile. These characteristics make root derived starches especially valuable in applications requiring minimal taste or odor interference, such as food products, pharmaceuticals, and certain coating formulations. In addition, their clean pasting behavior and smooth texture support their use in film formation and specialty packaging applications where clarity and uniformity are important performance attributes (Moorthy, 2002).

2.4.4 Sources of legume

Legumes constitute another important botanical source of starch and are characterized by their balanced composition of carbohydrates, protein, and dietary fiber. Common examples include peas, beans, lentils, and chickpeas, all of which provide starch with distinctive functional properties compared to cereal and tuber sources. Legume starches typically contain higher amylose content, which contributes to stronger gel formation, improved film strength, and reduced swelling behavior. These characteristics make them particularly valuable in specialty food formulations and functional ingredients, and they are increasingly explored in biodegradable packaging and coating systems where enhanced mechanical integrity and structural stability are desired (Tayade et al., 2019; Hoover and Zhou, 2003).

2.4.5 Sources of fruit and plant stems

Starch is also found in certain fruits and plant stems, particularly during immature developmental stages when it serves as a primary energy reserve before being converted into sugars during ripening. Notable examples include green bananas and plantains, as well as sago starch derived from the pith of the sago palm. These starch sources play an important role in traditional and regional diets due to their availability and functional properties. From a materials perspective, such starches are increasingly gaining attention for specialty applications because their unique granule morphology and composition can influence gelatinization behavior, viscosity, and film forming performance, making them of interest not only in food systems but also in niche biodegradable material and packaging applications (MacNeill et al., 2017; Nath and Dutta, 2024).

2.4.6 Industrial and unconventional sources

Emerging starch sources are increasingly being explored to support industrial innovation and sustainability goals. These include nontraditional feedstocks such as algae, which offer rapid growth rates and do not compete directly with food crops, as well as underutilized agricultural residues and alternative crops that provide opportunities for value-added utilization of biomass. Investigating these novel sources can help diversify raw material supply, reduce environmental impact, and enable the development of next generation bio based materials for packaging and other industrial applications (Phukan and Nongkhlaw).

2.5 Starch applications

Packaging materials based on starch have attracted a lot of scientific attention because of their renewable source, biodegradability, biocompatibility, and adaptable physicochemical characteristics. Starch is a vital biopolymer for developing sustainable material science and is mostly derived from agricultural sources such as corn, potatoes, wheat, and cassava. Starch based materials are essential for scientific study in the fields of polymer science, food technology, biomedical engineering, and environmental studies, in addition to commercial packaging (Onyeaka et al., 2022). The development of renewable and biodegradable packaging materials has been spurred by growing environmental concerns about petroleum based plastics. Because it is plentiful, inexpensive, biodegradable, and adaptable, starch based packaging has become a viable option among various substitutes. Sustainable packaging materials are often made using starch derived from corn, potatoes, wheat, and cassava, either by itself or in conjunction with other biopolymers. Applications for starch based packaging are found in a variety of industries, such as industrial logistics, food, agriculture, and healthcare.

2.5.1 Food packaging

The most important area for materials based on starch is food packaging. Biodegradable films, trays, containers, and coatings for food items are all made with starch. These materials reduce plastic waste while offering protection against environmental variables and physical damage (García-Guzmán et al., 2022). To prevent moisture loss, postpone oxidation, and extend shelf life, starch based films are used as edible coatings for fruits, vegetables, baked goods, and confections. Furthermore, to increase its mechanical strength and flexibility, starch is commonly combined with plasticizers or other biopolymers, which qualify for use in dry and semi moist food packaging (Onyeaka et al., 2022).

2.5.2 Functional packaging

In active packaging systems, where the packaging interacts with the packaged product to improve quality and safety, starch based packaging materials are being used more widely. Antimicrobials, antioxidants, and flavor preserving substances can be incorporated into starch matrices to enable controlled release during storage. Because starch based active packaging can prevent microbial development and delay the rate of spoiling, these applications are especially pertinent to perishable food products. With this feature, packaging can now serve as an active preservation strategy in addition to passive confinement (Lauer and Smith, 2020).

2.5.3 Cushioning and protective packaging

Starch based polymers are also used extensively in protective packaging, where they take the role of synthetic foams and expanded polystyrene. To safeguard delicate items while in transit, molded inserts, cushioning materials, and loose fill packing made of starch are frequently utilized. Because they are lightweight, biodegradable, and water soluble, these materials provide similar stress absorption qualities, making them better options for transportation and logistical applications (Kaur et al., 2026; Amir et al., 2018).

2.5.4 Agricultural packaging

The agriculture industry uses starch based packaging for biodegradable mulch films, fertilizer bags, and seed packing. Because these materials naturally break down in soil, there is no need for collection or disposal, and environmental contamination is reduced. Additionally, starch based films are being investigated as controlled release pesticide carriers and seed coatings, which could enhance crop productivity and promote sustainable farming methods (Glenn et al., 2014).

2.5.5 Medical and pharmaceutical packaging

Starch based materials are being researched for use in pharmaceutical and medical packaging due to their non toxic and biocompatible properties. These consist of sachets, blister packs, and disposable parts for medical packaging. Starch based packaging is a promising option for environmentally friendly healthcare solutions because research in this field focuses on preserving product sterility, guaranteeing chemical stability, and regulating degradation rates (Mironescu et al., 2021).

2.5.6 Retail packaging and consumer products

The packaging of consumer goods, such as cosmetics, personal hygiene products, and domestic goods, is increasingly using materials based on starch. To cut down on the usage of single use plastic, biodegradable shopping bags and wrapping films manufactured from starch blends are frequently employed in retail environments. The printability, visual attractiveness, and favorable environmental perception of starch among consumers are advantageous for such applications (Deng et al., 2026).

2.5.7 E commerce and industrial packaging

Starch based packaging is utilized for transport packaging, protective mailers, and void filling in the industrial and e commerce industries. By decreasing trash production and dependency on non renewable plastics, these materials promote sustainable supply chains. Because they are easily disposed of through biodegradation or composting, they are especially appealing for extensive logistical operations.

2.5.8 Environment and sustainability applications

To promote sustainable packaging solutions and the circular economy, starch based packaging is essential. By using it, trash reduction, compostability, and lower greenhouse gas emissions are all supported. Consequently, pilot projects, environmental impact studies, and life cycle assessments that evaluate sustainable packaging alternatives frequently use materials based on starch (Deng et al., 2026).

2.5.9 Composite packaging applications

Starch is frequently combined with polymers like polylactic acid (PLA), polyhydroxyalkanoates (PHA), or reinforced with natural fibers to get around drawbacks like low mechanical strength and moisture sensitivity. By enhancing durability, thermal stability, and barrier qualities, these starch based composites increase the variety of packaging applications (Mironescu et al., 2021; Deng et al., 2026). Numerous uses for starch based packaging can be found in the consumer goods, food, healthcare, and industrial sectors. Because of its functional versatility, biodegradability, and renewability, it is a promising alternative to traditional plastics. Its position in environmentally friendly packaging systems is anticipated to be strengthened by additional research and material optimization, which will increase its applications and improve its performance.

2.6 Advantages of starch based food packaging materials

Extensive use of plastics is a global concern that is now the biggest threat to the environment. Minimizing environmental threats by reducing plastic use as people become more interested in recycling, reusing, and promoting eco friendly alternatives. For this reason, alternative packaging materials are gaining priority. Biomass generates the most prevalent biodegradable polymers, which are mostly polypeptides and polysaccharides. Starch is a leading bio based polymer for future food packaging materials, being the second most abundant natural polysaccharide after cellulose. It offers advantages like good compatibility with other materials, excellent processability, and effective barriers against oxygen and carbon dioxide. As a safe and low cost renewable resource, starch is readily available from various agricultural and industrial by products (Arruda et al., 2025). One of the most prevalent sources of carbohydrates is known to be starch. Its manufacturing procedure is reasonably easy and economical. Unconventional sources of this biopolymer also present creative possibilities for producing biodegradable films with conventional qualities (Buenrostro-Figueroa and Tafolla-Arellano, 2025). Numerous starches, including cassava, corn, sago, and tamarind, have been investigated for their potential in active packaging applications, representing a significant progression in advocating starches as biodegradable materials for active packaging (Samsudin and Hani, 2017).

Packaging materials based on starch are particularly intriguing to produce thermoplastic starch since they are often low in toxicity and show high compatibility with natural plasticizers. Furthermore, films made from starch have advantageous mechanical, thermal, biodegradation, and barrier qualities for thermoplastic starch applications (Niranjana Prabhu and Prashantha, 2018). Compared to synthetic plastics, biopolymers are thought to be a great substitute. They have many of the same properties as conventional plastics, but they biodegrade far more quickly, are highly plentiful, and are biocompatible with other materials. The number of biopolymers produced worldwide rose to 2.1 million tons in 2018 over the previous 10 years (García-Guzmán et al., 2022). Approximately 36% of packaging materials produced in 2015 were biodegradable, whereas 64% were not. In 2019, the manufacturing of non-biodegradable plastics decreased while the percentage of packaging made of biodegradable materials rose to 56%. By 2025, the EU Commission hopes to have 55% of plastic packaging recycled, and by 2030, all plastics should be recyclable or reusable. This is part of their circular economic strategy. According to a recent estimate, biodegradable packaging accounts for roughly 53% of the market for bioplastics, or 1.14 million tons, especially in the food packaging sector (García-Guzmán et al., 2022). Starch based polymers, which are used extensively nowadays, are based on natural starch. Before being used as a packaging material, starch must first be extracted from plants. Therefore, a biodegradable bioplastic is produced. One of the most popular varieties of starch is thermoplastic, which can be enhanced by blending it with other biopolymers (Understanding Starch-Based Plastics in Sustainable Packaging).

Starch offers numerous advantages that make it an attractive material for food packaging applications, particularly in the context of sustainability and circular material design. As a naturally derived biopolymer, starch is environmentally compatible and exhibits inherent biodegradability, supporting the reduction of long-term plastic waste. It is sourced from abundant and renewable agricultural feedstocks, ensuring a sustainable and widely available raw material base. In packaging systems, starch based films and coatings can contribute to extended shelf life and improved food safety by providing barrier functions and, when appropriately formulated, enabling active or intelligent packaging features such as freshness indicators. Starch can also function as a foaming agent in the production of lightweight, biodegradable cushioning materials. Furthermore, it serves as a versatile matrix for film and coating development, offering tunable mechanical and barrier properties through modification and blending. Importantly, starch remains a cost effective alternative to many petroleum-based polymers, making it economically viable for large scale packaging applications.

2.6.1 Environmental suitability and phenomenological biodegradability

For food packaging, starch is a potentially biodegradable material that has the potential to significantly decrease environmental pollution (Cheng et al., 2021). Because of its total degradability, low cost, and wide availability, starch is an auspicious natural biodegradable material. Excellent biodegradable materials based on starch can be used in a variety of ways by improving their characteristics (Yu et al., 2021). Microorganisms view starch, a common polysaccharide that acts as an energy transporter in plants, as an easily available energy source (Polman et al., 2021). It is composed of linear amylose chains and branched amylopectin chains, both of which are derived from glucose molecules. These glucose molecules are joined by α glycosidic bonds, which are comparatively weak. A helical structure was perhaps first proposed for amylose, a biopolymer (Dufresne, 2014; Ai and Jane, 2017). Glycoside hydrolases, a broad class of enzymes that catalyze the hydrolysis of glycosidic bonds, are the main enzymes responsible for breaking down starch. The process is started by the enzyme α-amylase, which breaks down long starch polymers into smaller pieces. Various enzymes, including β-amylase, glucoamylase, and α glucosidase, which specifically target and hydrolyze α glycosidic bonds, subsequently further hydrolyze these fragments (German et al., 2011; Warren, 1996). A considerable amount of starch degrades within 3 days of entering the soil, according to Haslam's (2004) review. Starch degradation rates often follow an asymptotic curve. Starch-based materials are being successfully used for producing nanocomposites. Starch is also essential for cost reduction and improves the processability of other polymers (Jiang et al., 2020). Given its many benefits, using biodegradable materials based on starch in food packaging is a great idea. Because it is biodegradable, non toxic, and abundant in nature, starch helps lower the amount of fossil fuels. It is a desirable alternative due to its abundance and low cost. Moreover, starch has no impact on CO2 levels in the global ecosystem, assists in preserving energy, offering an environmentally beneficial way to dispose of waste, and is safe to eat when used in food packaging (Onyeaka et al., 2022). Starch based biodegradable materials require further development to enhance their environmental friendliness for sustainable packaging. Future work in this area is essential (Jiang et al., 2020).

2.6.2 Abundant and sustainable resource base

Environmental consciousness is growing quickly, and more individuals are actively searching for renewable, bio based, and sustainable materials instead of petroleum based ones (Onyeaka et al., 2022). A plentiful and sustainable resource, starch is a promising polysaccharide that is frequently utilized to substitute plastics made from fossil fuels (Gamage et al., 2022). Starch is abundant in rice, corn, wheat, and potatoes. Plant biomass is the source of renewable materials like starch, which may be produced rapidly and have low carbon emissions, assisting in maintaining the natural equilibrium of the ecosystem. A good alternative to synthetic polymers, starch is widely available, inexpensive, fully biodegradable, highly biocompatible, and has good film forming qualities (Cheng et al., 2021; Geyer et al., 2017; Gabor and Tita, 2012).

2.6.3 Extended shelf life and better food safety

To distribute food from the producer to the final consumer and maintain its safety, packaging is essential. The purpose of the packaging is to serve as a barrier between the food items inside and to prevent contamination and unintended spoiling (Adeyeye et al., 2019). Starch based films have good oxygen barrier qualities that improve food preservation and prolong their shelf life. They are also cheap, widely available, tasteless, colorless, and odorless (Sadeghizadeh-Yazdi et al., 2019).

2.6.4 Intelligent packaging based on starch

To use active and intelligent food packaging, various technologies have been developed in recent years. Intelligent instruments like active films, sensors, and indications are examples of this development. To increase the packaging system's performance, active packaging especially entails purposefully incorporating extra parts into the packaging material or the headspace of the package. This method, which is always changing, improves food quality and prolongs shelf life by allowing the food product to directly interact with the bioactive ingredients that are purposefully added to the packaging. Food preservation is better achieved with active packaging than with passive packaging (Rodrigues Arruda et al., 2022). However, a packaging system that is capable of several tasks, including detection, sensing, recording, tracing, communication, and the application of scientific reasoning, is referred to as intelligent packaging. These features aid in extending the shelf life, boosting quality, safety, and informational capacities, as well as alerting customers to possible issues (Medina-Jaramillo et al., 2017). When adding additives like antimicrobials and antioxidants, starch based films play an essential role in enhancing the functional qualities. By preventing moisture migration, these active packaging options can offer a semi-permeable barrier that helps prolong the shelf life of food items (García-Guzmán et al., 2022). When starch is exposed to varying pH levels, it changes color, making it a useful pH indicator. They can be found in green tea and basil extracts, which include carotenoids and chlorophyll (Medina-Jaramillo et al., 2017). Apart from performing practical research on active packaging, researchers have developed intelligent packaging films using starch. Indicator intelligent packaging materials are currently the primary emphasis. By combining conventional package technologies with intelligent characteristics, these materials enable consumers to acquire information through color changes on the outside (Liu et al., 2022). Using nitrite and cassava myoglobin extract, a biodegradable thermochromic sensor film has been developed as an alternative to conventional electronic sensors. Using the thermochromic qualities of these proteins, this natural and non toxic sensor uses color shifts to identify temperature changes. It has potential for use in marking and packaging applications (Kusuma, Ma'arif and Rahmadhia, 2024).

2.6.5 Starch as a material for films and coatings

The film forming, renewable, and biodegradable properties of starch make it a viable basis polymer for biodegradable films and coatings. In order to improve performance, advanced starch based films can be strengthened or altered while still utilizing the polymer's natural structure (Karnwal et al., 2025). The two main components of starch are amylose, which mostly forms linear chains, and amylopectin, which creates branched chains. Mechanical strength, flexibility, and barrier qualities are determined by the ratio of amylose to amylopectin. While higher amylopectin content enhances extensibility but may lessen water resistance, higher amylose content usually produces denser and stronger films (Karnwal et al., 2025; Onyeaka et al., 2022; Wittaya, 2012). The most common method for making films is gelatinization, which is boiling starch in water to break up the granules, then combining it with plasticizers like sorbitol or glycerol, casting, and finally drying it (Tharanathan, 2003). In comparison to untreated starch films, starch modified by citric acid cross linking shows decreased water vapor permeability (WVP), increased mechanical strength, and decreased swelling (Kumar et al., 2021). Due to their hydrophilic nature, starch films can absorb moisture and lose structural integrity in humid settings. To solve this, scientists incorporate hydrophobic elements, combine starch with other biopolymers, or add nanofillers to improve the mechanical and barrier qualities (Liu et al., 2009). In one case, a cassava starch film enhanced with essential oils and sodium bentonite nanoclay demonstrated antibacterial efficacy in addition to mechanical stability (Iamareerat et al., 2018). Five percent citric acid reduced swelling and solubility while increasing film compaction and water resistance, according to a different study on maize starch + citric acid cross linking (Kumar et al., 2021).

2.6.6 Using starch as a foaming agent

Due to its ability to gelatinize and trap air, starch has been extensively studied as a natural foaming agent, resulting in the creation of stable, porous structures (Soykeabkaew et al., 2015). However, under humid conditions, native starch frequently shows poor foam stability, which restricts its direct use in cushioning and packaging materials (Tapia-Blácido et al., 2022). Pre gelatinization, hydroxypropylation, and enzymatic hydrolysis are examples of physical, chemical, and enzymatic modifications of starch that can be used to get around these restrictions (Amaraweera et al., 2022). These changes increase starch's surface activity and stabilize its bubbles, enabling the creation of long lasting and biodegradable foams (Asghari et al., 2016). Therefore, for use in packaging, cushioning, and lightweight building materials, starch based foams have become sustainable substitutes for traditional synthetic foams (Kulkarni et al., 2024).

2.7 Disadvantages of strach based food packaging materials

Starch based packaging faces challenges in replacing traditional plastic in commercial applications due to its water sensitivity and poor mechanical properties. Their excessive water vapor permeability and subpar mechanical properties are particularly problematic. Despite their biodegradability and renewable sources, starch based polymers have little mechanical strength and are susceptible to moisture, compromising their longevity and protection. Their inability to withstand oxygen, water vapor, and oils also hinders their performance in food and medicinal applications (Ismail et al., 2017; Ribba et al., 2017). Manufacturing starch based materials is challenging due to the need for unique tools or chemical changes, and their higher production costs and reliance on agricultural feedstock raise questions about raw material stability and economic viability. Therefore, the primary drawbacks of materials based on starch are thoroughly explained in this chapter. Particularly those about their excessive water vapor permeability and subpar mechanical properties. To expand their industrial and commercial uses, several solutions to these problems are also presented in this context (Lim et al., 2008).

2.7.1 Mechanical and structural limitations

The mechanical and thermal performance of starch based biodegradable packaging is inferior to that of traditional plastics like PET, PP, and PE. Its employment in high stress, high temperature, or high barrier applications, like industrial packaging, hot fill bottling, or transportation grade film, is severely hampered by these restrictions (Ribba et al., 2017).

2.7.1.1 Durability and tensile strength

Starch based films generally exhibit limited load bearing capacity, low elongation at break, and relatively poor tensile strength, which restrict their use in demanding packaging applications. Upon drying, native starch forms a glassy and brittle matrix that is highly susceptible to mechanical fatigue, tearing, and cracking under stress. As a result, unmodified starch materials are typically unsuitable for load bearing or pressure sensitive uses such as bulk packaging, carrier bags, or stretch films (Arruda et al., 2025). This mechanical fragility is largely attributed to the extensive hydrogen bonding between starch chains, which restricts molecular mobility and leads to increased stiffness and brittleness. In addition, thermoplastic starch (TPS) often exhibits relatively low crystallinity, contributing to poor fracture resistance and reduced durability under mechanical stress (Ma et al., 2005).

From an industrial perspective, these limitations can translate into inadequate performance during shipping and handling, particularly for products requiring structural integrity such as vacuum sealed goods. Furthermore, the brittle nature of starch films may lead to higher rejection rates in high speed automated packaging operations due to tearing or failure during processing, highlighting the need for modification, plasticization, or reinforcement strategies to improve mechanical robustness.

2.7.1.2 Problems with brittleness and flexibility

Plasticization is commonly employed to overcome the inherent brittleness of starch by incorporating low molecular weight compounds such as glycerol, sorbitol, urea, or polyethylene glycol (PEG). These plasticizers disrupt intermolecular hydrogen bonding within the starch matrix, increasing chain mobility and thereby enhancing flexibility, ductility, and elongation of the resulting films. Plasticizers increase flexibility, but they also present new difficulties (Narancic et al., 2018). For example, thermoplastic starch (TPS) formulations containing approximately 15% to 20% glycerol can achieve moderate flexibility suitable for certain general purpose packaging applications (Ribba et al., 2017; Diah et al., 2024). However, the use of plasticizers introduces important trade offs that must be carefully managed. Excessive plasticizer content can reduce tensile strength and cohesive integrity, leading to softer films that may become tacky or greasy over time (Chaudhary et al., 2011). Because many plasticizers are hygroscopic, they readily absorb moisture from the environment, which can further degrade mechanical performance and, in some cases, create conditions favorable for microbial growth (Diah et al., 2024). In addition, plasticizer migration is a well documented challenge, particularly under elevated temperature and humidity conditions, where gradual loss or redistribution of the plasticizer can cause shrinkage, embrittlement, and the development of off odors, ultimately affecting both material performance and product acceptability (Narancic et al., 2018).

2.7.1.3 Thermal consistency

Starch based materials exhibit limited thermal stability, which restricts their use in processes that require elevated temperatures. Typically, starch begins to degrade at relatively moderate temperatures in the range of approximately 160 to 180°C, making it challenging to process using conventional high temperature polymer processing techniques (Ismail et al., 2017). At temperatures above about 180°C, native starch undergoes thermochemical degradation, including depolymerization, browning, and eventual charring. This behavior creates significant processing challenges for operations that commonly run at higher temperatures, such as extrusion, injection molding, and thermoforming, which often operate in the range of 200 to 240°C (Arruda et al., 2025; Chaudhary et al., 2011). These thermal limitations also translate into performance constraints in end use packaging applications. In hot fill or microwave environments, many starch based films and structures tend to soften, deform, or lose structural integrity, making them unsuitable for packaging hot liquids such as soups, beverages, or ready to eat meals. Additionally, starch materials are prone to steam absorption, which can lead to swelling, dimensional instability, and delamination in multilayer trays or pouches, further limiting their applicability in high-temperature or high moisture conditions (Ismail et al., 2017; Averous and Boquillon, 2004).

2.7.2 Properties of the barrier and moisture sensitivity

Due to their intrinsic hydrophilicity, which impairs their capacity to withstand moisture, maintain their dimensional stability, and function as a gas barrier, starch based packaging materials are not widely used. Due to these properties, starch based films are not appropriate for a wide range of commercial packaging applications, particularly those that include liquids, wet goods, or food systems that must be kept for a long time (Ribba et al., 2017; Navigating the Pros and Cons of Starch-Based Food Packaging - NguyenStarch).

2.7.2.1 Swelling and water absorption

Amylose and amylopectin, the primary structural components of starch, contain numerous hydroxyl (–OH) groups that readily interact with water molecules through hydrogen bonding, giving starch a highly hydrophilic character (Ribba et al., 2017). As a result, starch based materials face significant challenges related to moisture sensitivity. Starch films readily absorb moisture when exposed to humid environments or direct water contact, leading to swelling and plasticization of the polymer matrix. This moisture uptake can substantially reduce mechanical strength, causing the material to become soft, tacky, or, in some cases, brittle depending on exposure conditions and environmental factors (Müller et al., 2011). Water penetration also disrupts intermolecular hydrogen bonds within the starch network, promoting chain separation and gelatinization like behavior. Over prolonged exposure to moisture, these effects can lead to warping, dimensional instability, and, in severe cases, structural collapse of the film. Consequently, the high moisture sensitivity of starch significantly limits its suitability for packaging applications involving high water activity products such as dairy, fresh produce, and beverages, as well as for environments with elevated humidity. These limitations are particularly critical in cold chain logistics, where temperature fluctuations and condensation can further compromise structural integrity and performance (Müller et al., 2011; Abe et al., 2021).

2.7.2.2 *Poor properties of gas barriers*

In addition to their moisture sensitivity, starch based films generally exhibit limited barrier performance against gases, particularly oxygen (O_2) and carbon dioxide (CO_2), which poses a major limitation for food packaging applications that depend on modified atmosphere packaging (MAP) or controlled atmosphere packaging (CAP) (Diah et al., 2024). The relatively loose and predominantly amorphous structure of native starch, coupled with its low crystallinity, allows gas molecules to diffuse more readily through the polymer matrix. As a result, starch based films typically show significantly higher O_2 and CO_2 permeability compared with conventional high barrier materials such as PET or EVOH (Jiang et al., 2020). The elevated gas permeability can negatively affect food quality and shelf life. Oxygen ingress promotes oxidative degradation of lipids and vitamins and supports microbial growth, leading to rancidity, spoilage, and off flavor development. Similarly, inadequate CO_2 retention in fresh cut fruits and vegetables can accelerate respiration rates and senescence, further reducing product shelf life (Tharanathan, 2003). These challenges are exacerbated under humid conditions, where moisture absorption causes swelling of the starch matrix and increases free volume, thereby facilitating even greater gas diffusion. This moisture gas coupling effect further compromises barrier performance in real world storage environments (Halley et al., 2009).

2.7.3 Economic viability and high production costs

Even though starch based biodegradable packaging comes from renewable and reasonably priced agricultural sources, several financial barriers are preventing its wider commercial use. The expense of processing, material change, and the wider ethical and economic ramifications of using food based crops for industrial purposes are the main causes of these difficulties.

2.7.3.1 *Costs of processing and modification*

Native starch must undergo modification to meet the performance requirements of modern packaging applications, as its natural form exhibits several limitations, including low mechanical strength, limited thermal stability, and poor water resistance. However, these necessary modifications often lead to increased production costs. One common approach is blending and plasticization, where plasticizers such as glycerol, sorbitol, or urea are incorporated to reduce intermolecular hydrogen bonding and convert starch into thermoplastic starch (TPS) (Ribba et al., 2017). While these additives improve flexibility and processability, they add to raw material costs and require precise control during compounding to avoid issues such as brittleness or tackiness (Diah et al., 2024). Another widely used strategy involves reinforcing starch through blending with biodegradable or synthetic polymers such as PLA (polylactic acid), PCL (polycaprolactone), or PBAT (polybutylene adipate terephthalate) to enhance mechanical performance. Although effective, these polymers are generally more expensive than conventional plastics, and the need for compatibilizers and additional processing steps further increases overall production costs (Navigating the Pros and Cons of Starch-Based Food Packaging - NguyenStarch). In addition, post treatment processes are often required to improve water resistance and extend shelf life, including surface modifications such as crosslinking, lamination, or the application of hydrophobic coatings. These additional processing steps demand extra equipment, labor, and energy, contributing to higher operational expenses and limiting cost competitiveness in some applications (Jiang et al., 2020; Avérous, 2004).

2.7.3.2 Conflict with the food supply

The large scale use of starch derived from major food crops such as maize, cassava, potatoes, and wheat raise important ethical, economic, and geopolitical considerations. One of the central concerns is the food versus materials debate, as diverting edible crops toward packaging and industrial uses may reduce the availability of resources for human and animal consumption. This issue is particularly sensitive in regions experiencing food insecurity, where agricultural land could otherwise be dedicated to staple food production.

In addition, expanding demand for starch based bioplastics can contribute to volatility in agricultural commodity markets. Increased industrial consumption may place upward pressure on food prices, disproportionately affecting low income populations, especially in developing countries (Jiang et al., 2020). The expansion of starch crop cultivation for industrial purposes may also intensify land use, potentially leading to soil degradation, biodiversity loss, and deforestation as more land is converted to agriculture.

Geopolitical risks further complicate the landscape. Countries that rely heavily on imported food grains may face supply instability if domestic production is redirected toward industrial starch demand. Moreover, competition for bio based raw materials can lead to international trade tensions, export restrictions, and policy conflicts, highlighting the need for careful resource management and sustainable sourcing strategies (Arruda et al., 2025; Jiang et al., 2020).

2.7.4 Industrial difficulties and limited scalability

Although starch based biodegradable polymers are increasingly recognized as promising sustainable alternatives to conventional plastics, their large scale industrial implementation remains constrained by a combination of technological, economic, and infrastructure related challenges. These limitations arise not only from the agricultural origin of starch but also from its variable behavior during polymer processing and manufacturing operations (Ismail et al., 2017; Krogars et al., 2002; Talja et al., 2007). As a naturally occurring biopolymer, starch is primarily derived from crops such as corn, wheat, cassava, and potatoes, yet its performance as a feedstock for bioplastics is inherently inconsistent due to biological diversity and environmental influences (Krogars et al., 2002; Talja et al., 2007, 2008). Differences among botanical sources lead to variations in physicochemical properties, including granule size, amylose to amylopectin ratio, swelling behavior, and gelatinization temperature, all of which directly affect processing behavior and final material performance (Lim et al., 2008; Halley et al., 2009). For instance, corn starch typically exhibits smaller granules and higher gelatinization temperatures, whereas potato starch is characterized by larger granules and greater swelling capacity, and cassava starch is valued for good film forming ability but often shows lower mechanical strength (Ribba et al., 2017; Arruda et al., 2025).

In addition, seasonal and environmental factors such as soil fertility, rainfall patterns, fertilization practices, and pest management can significantly influence starch composition, purity, and molecular structure. Even within the same crop type, seasonal variability can lead to fluctuations in starch content and functionality, which in turn affect plasticization efficiency, thermal stability, and mechanical performance during processing (Ismail et al., 2017; Ribba et al., 2017; Abe et al., 2021). This inherent variability in raw material quality presents a major challenge for industrial standardization and consistent large scale production, highlighting the need for improved feedstock control, material characterization, and process optimization to support scalable manufacturing.

2.7.4.1 Variability related to processing

Starch's moisture content and propensity for retrogradation are influenced by the extraction, purification, and drying techniques (such as spray vs. drum drying). Extrusion stability and melt viscosity may be impacted by irregular water content and leftover proteins or lipids. Because of this fluctuation, product quality varies from batch to batch, requiring producers to make constant formulation adjustments that make process control and quality assurance more difficult (Ribba et al., 2017; Navigating the Pros and Cons of Starch-Based Food Packaging - NguyenStarch; Müller, Laurindo and Yamashita, 2011; Abe et al., 2021).

2.7.4.2 Problems with manufacturing compatibility

Starch based polymers often face significant compatibility challenges when processed on conventional plastic production lines originally designed for synthetic thermoplastics such as polyethylene (PE), polypropylene (PP), and polyethylene terephthalate (PET). One of the primary limitations is their sensitivity to thermal degradation. Native starch begins to degrade at temperatures above approximately 180°C, which is lower than the typical processing temperatures used for most conventional polymers. Although starch can tolerate temperatures approaching 200°C for short durations, prolonged exposure leads to instability, depolymerization, and discoloration. Without proper plasticization, such as the addition of glycerol or sorbitol, starch does not exhibit smooth melt behavior and instead tends to burn or decompose during processing (Narancic et al., 2018; Effect of glycerol and sorbitol plasticizers on physical and thermal properties of sugar palm starch-based films; Garavito et al., 2024).

The inherently hydrophilic nature of starch further complicates processing and storage. Because starch readily absorbs moisture, excess water during operations such as extrusion, thermoforming, or injection molding can lead to poor melt flow, dimensional instability, and the formation of bubbles or voids within the material (Ribba et al., 2017; Müller et al., 2011). To address these issues, conventional equipment often requires significant modifications, including the implementation of moisture controlled feeding systems, screw designs optimized for lower shear sensitivity, and processing zones with reduced temperature profiles. Such retrofitting can be both costly and time intensive, posing a substantial barrier for small and medium sized manufacturers seeking to adopt starch based materials.

In addition, processing efficiency remains a concern due to the relatively low melt strength of starch based bioplastics. These materials frequently require reinforcement with fibers or nanofillers, or blending with other biodegradable polymers, to achieve acceptable processing performance.

Even with these modifications, starch based blends may still exhibit lower throughput, higher defect rates, and variability in film thickness compared to conventional plastics, ultimately reducing operational efficiency and increasing production complexity (Ismail et al., 2017; Averous and Boquillon, 2004; Avérous, 2004).

2.8 Conclusion

Starch based packaging materials represent a promising alternative to conventional single use plastics, which continue to contribute significantly to global environmental challenges (Cheng et al., 2021). The development of starch based systems is primarily driven by the goals of sustainability, biodegradability, and reliance on renewable resources, making them attractive candidates for next generation packaging solutions (Arruda et al., 2025). However, despite these advantages, starch based materials still face notable limitations. Their sensitivity to moisture and relatively weak mechanical performance remain key challenges that require continued material optimization and technological development (Pei et al., 2024).

As scientific understanding and processing technologies continue to advance, starch based materials are expected to play an increasingly significant role in the transition toward more sustainable and environmentally responsible packaging systems (Muñoz-Gimena et al., 2023; Liu et al., 2022). Ongoing research efforts are actively addressing these constraints through strategies such as nanoparticle incorporation and the design of composite and foam structures, which have shown considerable potential to enhance mechanical strength, barrier properties, and overall functionality (Muñoz-Gimena et al., 2023; Liu et al., 2022; Altay et al., 2017; Keskin, 2018; Keskin et al., 2020).

References

Abe, M.M., Martins, J.R., Sanvezzo, P.B., Macedo, J.V., Branciforti, M.C., Halley, P., Botaro, V.R., Brienzo, M., 2021. Advantages and disadvantages of bioplastics production from starch and lignocellulosic components. Polymers 13 (15). https://doi.org/10.3390/polym13152484.

Adeyeye, O.A., Sadiku, E.R., Babu Reddy, A., Ndamase, A.S., Makgatho, G., Sellamuthu, P.S., Perumal, A.B., Nambiar, R.B., Fasiku, V.O., Ibrahim, I.D., Agboola, O., Kupolati, W.K., Daramola, O.O., Machane, M.J., Jamiru, T., 2019. The Use of Biopolymers in Food Packaging Materials Horizons: From Nature to Nanomaterials. Springer Nature, South Africa, pp. 137–158. https://doi.org/10.1007/978-981-13-8063-1_6.

Ai, Y., Jane, J.l., 2017. Understanding Starch Structure and Functionality Starch in Food: Structure, Function and Applications. Elsevier, Canada, pp. 151–178. https://doi.org/10.1016/B978-0-08-100868-3.00003-2.

Altay, B.N., Bloembergen, S., Aydemir, C., Karademir, A., Fleming, P.D., 2017. Use of nanoparticle binders for paper coatings: a review. J. Graph. Eng. Des. 8 (1), 39–43. https://doi.org/10.24867/JGED-2017-1-039.

Amaraweera, S.M., Gunathilake, C., Gunawardene, O.H.P., Dassanayake, R.S., Fernando, N.M.L., Wanninayaka, D.B., Rajapaksha, S.M., Manamperi, A., Gangoda, M., Manchanda, A., Fernando, C.A.N., Kulatunga, A.K., Manipura, A., 2022. Preparation and characterization of dual-modified cassava starch-based biodegradable foams for sustainable packaging applications. ACS Omega 7 (23), 19579–19590. https://doi.org/10.1021/acsomega.2c01292.

Ambrogi, V., Carfagna, C., Cerruti, P., Marturano, V., 2017. Additives in Polymers Modification of Polymer Properties. Elsevier Inc., Italy, pp. 87–108. https://doi.org/10.1016/B978-0-323-44353-1.00004-X.

Amir, N., Hisham, M.S.M., Abidin, K.A.Z., 2018. EDP sciences Malaysia study of physical properties and shock absorption abilities of starch polymer foam as cushioning material for packaging. MATEC Web of Conferences 225. https://doi.org/10.1051/matecconf/201822506010.

Arruda, T.R., Machado, G.de O., Marques, C.S., Lelis de Souza, A., Pelissari, F.M., Oliveira, T.V.de, Silva, R.R.A., 2025. An Overview of starch-based materials for sustainable food packaging: recent advances, limitations, and perspectives. Macromol 5 (2), 19. https://doi.org/10.3390/macromol5020019.

Asghari, A.K., Norton, I., Mills, T., Sadd, P., Spyropoulos, F., 2016. Interfacial and foaming characterisation of mixed protein-starch particle systems for food-foam applications. Food Hydrocoll. 53, 311–319. https://doi.org/10.1016/j.foodhyd.2015.09.007.

Avérous, L., 2004. Biodegradable multiphase systems based on plasticized starch: a review. J. Macromol. Sci., Polym. Rev. 44 (3), 231–274. https://doi.org/10.1081/MC-200029326.

Averous, L., Boquillon, N., 2004. Biocomposites based on plasticized starch: thermal and mechanical behaviours. Carbohydr. Polym. 56 (2), 111–122. https://doi.org/10.1016/j.carbpol.2003.11.015.

Bertuzzi, M.A., Castro Vidaurre, E.F., Armada, M., Gottifredi, J.C., 2007. Water vapor permeability of edible starch based films. J. Food Eng. 80 (3), 972–978. https://doi.org/10.1016/j.jfoodeng.2006.07.016.

Buenrostro-Figueroa, J.J., Tafolla-Arellano, J.C., 2025. Food Security, Safety, and Sustainability. Apple Academic Press. https://doi.org/10.1201/9781003518440.

Chaudhary, D.S., Adhikari, B.P., Kasapis, S., 2011. Glass-transition behaviour of plasticized starch biopolymer system - a modified Gordon-Taylor approach. Food Hydrocoll. 25 (1), 114–121. https://doi.org/10.1016/j.foodhyd.2010.06.002.

Cheng, H., Chen, L., McClements, D.J., Yang, T., Zhang, Z., Ren, F., Miao, M., Tian, Y., Jin, Z., 2021. Starch-based biodegradable packaging materials: a review of their preparation, characterization and diverse applications in the food industry. Trends Food Sci. Technol. 114, 70–82. https://doi.org/10.1016/j.tifs.2021.05.017.

Choton, S., Bandral, J.D., Singh, J., Bhat, A., Sood, M., Gupta, N., Reshi, M., Kaur, D., 2024. Enzymatic modification of starch: a review. Saudi Journal of Medical and Pharmaceutical Sciences 10 (01), 1–8. https://doi.org/10.36348/sjmps.2024.v10i01.001.

Cornejo-Ramírez, Y.I., Martínez-Cruz, O., Del Toro-Sánchez, C.L., Wong-Corral, F.J., Borboa-Flores, J., Cinco-Moroyoqui, F.J., 2018. The structural characteristics of starches and their functional properties. CyTA - J. Food 16 (1), 1003–1017. https://doi.org/10.1080/19476337.2018.1518343.

Deng, H., Guo, X., Sun, C., Zheng, Y., Chen, M., Zhang, B., Deng, S., Chen, J., Chen, J., 2026. A novel pH-responsive corn starch/polylactic acid bilayer film: application for the freshness preservation of hairtail (Trichiurus haumela). Carbohydr. Polym. Technol. Appl. 13. https://doi.org/10.1016/j.carpta.2025.101075.

Diah, S.B.M., Abdullah, S.H.Y.S., Seok, Y.B., Fatah, I.Y.A., Rahman, N.I.A., Hafizulhaq, F., Alias, N.N., 2024. Towards sustainable food packaging: a review of thermoplastic starch (TPS) as a promising bioplastic material, its limitations, and improvement strategies with bio-fillers and essential oils. Journal of Advanced Research in Fluid Mechanics and Thermal Sciences 119 (1), 80–104. https://doi.org/10.37934/arfmts.119.1.80104.

Dufresne, A., 2014. Crystalline starch based nanoparticles. Curr. Opin. Colloid Interface Sci. 19 (5), 397–408. https://doi.org/10.1016/j.cocis.2014.06.001.

"(PDF) Effect of glycerol and sorbitol plasticizers on physical and thermal properties of sugar palm starch-based films." Accessed: Jul. 07, 2025. Online. Available: https://www.researchgate.net/publication/279861467_Effect_of_glycerol_and_sorbitol_plasticizers_on_physical_and_thermal_properties_of_sugar_palm_starch_based_films.

Fearon, P.K., Phease, T.L., Billingham, N.C., Bigger, S.W., 2002. A new approach to quantitatively assessing the effects of polymer additives. Polymer 43 (17), 4611–4618. https://doi.org/10.1016/S0032-3861(02)00295-1.

Gabor, D., Tita, O., 2012. Biopolymers used in food packaging: a review. Acta Univ. Cibiniensis. Ser. E Food Technol. 16 (2), 3–19.

Gamage, A., Liyanapathiranage, A., Manamperi, A., Gunathilake, C., Mani, S., Merah, O., Madhujith, T., 2022. Applications of starch biopolymers for a sustainable modern agriculture. Sustainability 14 (10). https://doi.org/10.3390/su14106085.

Garavito, J., Peña-Venegas, C.P., Castellanos, D.A., 2024. Production of starch-based flexible food packaging in developing countries: analysis of the processes, challenges, and requirements. Foods 13 (24), 4096. https://doi.org/10.3390/FOODS13244096, 2024, Vol. 13, Page 4096.

García-Guzmán, L., Cabrera-Barjas, G., Soria-Hernández, C.G., Castaño, J., Guadarrama-Lezama, A.Y., Rodríguez Llamazares, S., 2022. Progress in starch-based materials for food packaging applications. Polysaccharides 3 (1), 136–177. https://doi.org/10.3390/polysaccharides3010007.

German, D.P., Chacon, S.S., Allison, S.D., 2011. Substrate concentration and enzyme allocation can affect rates of microbial decomposition. Ecology 92 (7), 1471–1480. https://doi.org/10.1890/10-2028.1.

Geyer, R., Jambeck, J.R., Law, K.L., 2017. Production, use, and fate of all plastics ever made. Sci. Adv. 3 (7). https://doi.org/10.1126/sciadv.1700782.

Glenn, G.M., Orts, W., Imam, S., Chiou, B.S., Wood, D.F., 2014. Starch Plastic Packaging and Agriculture Applications Starch Polymers: From Genetic Engineering to Green Applications. Elsevier B.V., United States, pp. 421–452. https://doi.org/10.1016/B978-0-444-53730-0.00032-4.

Halley, P.J., Truss, R.W., Markotsis, M.G., Chaleat, C., Russo, M., Lisa Sargent, A., Tan, I., Sopade, P.A., 2009. A Review of Biodegradable Thermoplastic Starch Polymers. American Chemical Society (ACS), pp. 287–300. https://doi.org/10.1021/bk-2007-0978.ch024.

Hannah, L.C., James, M., 2008. The complexities of starch biosynthesis in cereal endosperms. Curr. Opin. Biotechnol. 19 (2), 160–165. https://doi.org/10.1016/j.copbio.2008.02.013.

Haslam, M., 2004. The decomposition of starch grains in soils: implications for archaeological residue analyses. J. Archaeol. Sci. 31, 1715–1734. https://doi.org/10.1016/j.jas.2004.05.006.

Hoover, R., Zhou, Y., 2003. In vitro and in vivo hydrolysis of legume starches by α-amylase and resistant starch formation in legumes - a review. Carbohydr. Polym. 54 (4), 401–417. https://doi.org/10.1016/S0144-8617(03)00180-2.

Iamareerat, B., Singh, M., Sadiq, M.B., Anal, A.K., 2018. Reinforced cassava starch based edible film incorporated with essential oil and sodium bentonite nanoclay as food packaging material. J. Food Sci. Technol. 55 (5), 1953–1959. https://doi.org/10.1007/s13197-018-3100-7.

Ismail, S., Mansor, N., Man, Z., 2017. A study on thermal behaviour of thermoplastic starch plasticized by [Emim] Ac and by [Emim] Cl. Procedia Eng. 184, 567–572. https://doi.org/10.1016/j.proeng.2017.04.138.

James, M.G., Denyer, K., Myers, A.M., 2003. Starch synthesis in the cereal endosperm. Curr. Opin. Plant Biol. 6 (3), 215–222. https://doi.org/10.1016/S1369-5266(03)00042-6.

Jiang, T., Duan, Q., Zhu, J., Liu, H., Yu, L., 2020. Starch-based biodegradable materials: challenges and opportunities. Adv. Ind. Eng. Polym. Res. 3 (1), 8–18. https://doi.org/10.1016/j.aiepr.2019.11.003.

Karnwal, A., Rauf, A., Jassim, A.Y., Selvaraj, M., Al-Tawaha, A.R.M.S., Kashyap, P., Kumar, D., Malik, T., 2025. Advanced starch-based films for food packaging: innovations in sustainability and functional properties. Food Chem. X 29, 102662. https://doi.org/10.1016/j.fochx.2025.102662.

Kaur, H., Dutt, D., Alam, T., Gaikwad, K.K., 2026. Agro-waste-based cushioning material for sustainable protective packaging: a comprehensive review. J. Packag. Technol. Res. 10 (1), 1–17. https://doi.org/10.1007/s41783-025-00198-1.

Keskin, B., Altay, B.N., Akyol, M., Meral, G., Uyar, O., & Fleming, P.D., 2018. Global packaging trends. Uluslararası Matbaa Teknolojileri Sempozyumu, 1–3.

Keskin, B., Altay, A., Kurt, A., Fleming, P.D., 2020. Sustainability in paper-based packaging. Gazi Mühendislik Bilimleri Dergisi (GMBD) 6 (2), 129–137.

Krogars, K., Antikainen, O., Heinämäki, J., Laitinen, N., Yliruusi, J., 2002. Tablet film-coating with amylose-rich maize starch. Eur. J. Pharmaceut. Sci. 17 (1–2), 23–30. https://doi.org/10.1016/S0928-0987(02)00134-3.

Kulkarni, A., Emrich, J., Narayan, R., 2024. Humidity resistant biodegradable starch foams reinforced with polyvinyl Butyral (PVB) and chitosan. Polymers 16 (23). https://doi.org/10.3390/polym16233402.

Kumar, R., Ghoshal, G., Goyal, M., 2021. Biodegradable composite films/coatings of modified corn starch/gelatin for shelf life improvement of cucumber. J. Food Sci. Technol. 58 (4), 1227–1237. https://doi.org/10.1007/s13197-020-04685-y.

Kusuma, I., Ma'arif, A., Rahmadhia, S.N., 2024. EDP Sciences Indonesia Starch based for intelligent food packaging: a review. BIO Web of Conferences 146. https://doi.org/10.1051/bioconf/202414601099.

Lauer, M.K., Smith, R.C., 2020. Recent advances in starch-based films toward food packaging applications: physicochemical, mechanical, and functional properties. Compr. Rev. Food Sci. Food Saf. 19 (6), 3031–3083. https://doi.org/10.1111/1541-4337.12627.

Lim, L.T., Auras, R., Rubino, M., 2008. Processing technologies for poly(lactic acid). Prog. Polym. Sci. 33 (8), 820–852. https://doi.org/10.1016/j.progpolymsci.2008.05.004.

Lin, D., Kuang, Y., Chen, G., Kuang, Q., Wang, C., Zhu, P., Peng, C., Fang, Z., 2017. Enhancing moisture resistance of starch-coated paper by improving the film forming capability of starch film. Ind. Crop. Prod. 100, 12–18. https://doi.org/10.1016/j.indcrop.2017.02.013.

Liu, H., Xie, F., Yu, L., Chen, L., Li, L., 2009. Thermal processing of starch-based polymers. Prog. Polym. Sci. 34 (12), 1348–1368. https://doi.org/10.1016/j.progpolymsci.2009.07.001.

Liu, D., Zhao, P., Chen, J., Yan, Y., Wu, Z., 2022. Recent advances and applications in starch for intelligent active food packaging: a review. Foods 11 (18). https://doi.org/10.3390/foods11182879.

Luchese, C.L., Spada, J.C., Tessaro, I.C., 2017. Starch content affects physicochemical properties of corn and cassava starch-based films. Ind. Crop. Prod. 109, 619–626. https://doi.org/10.1016/j.indcrop.2017.09.020.

Ma, X., Yu, J., Kennedy, J.F., 2005. Studies on the properties of natural fibers-reinforced thermoplastic starch composites. Carbohydr. Polym. 62 (1), 19–24. https://doi.org/10.1016/j.carbpol.2005.07.015.

MacNeill, G.J., Mehrpouyan, S., Minow, M.A.A., Patterson, J.A., Tetlow, I.J., Emes, M.J., 2017. Starch as a source, starch as a sink: the bifunctional role of starch in carbon allocation. J. Exp. Bot. 68 (16), 4433–4453. https://doi.org/10.1093/jxb/erx291.

Medina-Jaramillo, C., Ochoa-Yepes, O., Bernal, C., Famá, L., 2017. Active and smart biodegradable packaging based on starch and natural extracts. Carbohydr. Polym. 176, 187–194. https://doi.org/10.1016/j.carbpol.2017.08.079.

Mironescu, M., Lazea-Stoyanova, A., Barbinta-Patrascu, M.E., Virchea, L.-I., Rexhepi, D., Mathe, E., Georgescu, C., 2021. Green design of novel starch-based packaging materials sustaining human and environmental health. Polymers 13 (8), 1190. https://doi.org/10.3390/polym13081190.

Moorthy, S.N., 2002. Physicochemical and Functional Properties of Tropical Tuber Starches: A Review. Wiley-VCH Verlag. https://doi.org/10.1002/1521-379X(200212)54:12ESCAPED_CHARACTER_LESS_THAN559::AID-STAR2222559>3.0.CO;2-F.

Müller, C.M.O., Laurindo, J.B., Yamashita, F., 2011. Effect of nanoclay incorporation method on mechanical and water vapor barrier properties of starch-based films. Ind. Crop. Prod. 33 (3), 605–610. https://doi.org/10.1016/j.indcrop.2010.12.021.

Muñoz-Gimena, P.F., Oliver-Cuenca, V., Peponi, L., López, D., 2023. A review on reinforcements and additives in starch-based composites for food packaging. Polymers 15 (13). https://doi.org/10.3390/polym15132972.

Narancic, T., Verstichel, S., Reddy Chaganti, S., Morales-Gamez, L., Kenny, S.T., De Wilde, B., Babu Padamati, R., O'Connor, K.E., 2018. Biodegradable plastic blends create new possibilities for end-of-life management of plastics but they are not a panacea for plastic pollution. Environ. Sci. Technol. 52 (18), 10441–10452. https://doi.org/10.1021/acs.est.8b02963.

Nath, R., Dutta, K., 2024. Sources of starch and its morphology: a review. In: Interdisciplinary Research in Technology and Management. CRC Press, pp. 402–411. https://doi.org/10.1201/9781003430469-47.

"Navigating the Pros and Cons of Starch-Based Food Packaging - NguyenStarch." Accessed: Jul. 07, 2025. Online. Available: https://nguyenstarch.com/advantages-and-disadvantages-of-starch-as-food-packaging-materials/.

Nawaz, H., Waheed, R., Nawaz, M., Shahwar, D., 2020. Physical and Chemical Modifications in Starch Structure and Reactivity. IntechOpen. https://doi.org/10.5772/intechopen.88870.

Ning, J., Fu, B., Tang, X., Yuan, H., Zhang, Y., Wang, X., 2025. Starch retrogradation in starch-based foods: mechanisms, influencing factors, and mitigation strategies. Int. J. Biol. Macromol. 300, 140354. https://doi.org/10.1016/j.ijbiomac.2025.140354.

Niranjana Prabhu, T., Prashantha, K., 2018. A review on present status and future challenges of starch based polymer films and their composites in food packaging applications. Polym. Compos. 39 (7), 2499–2522. https://doi.org/10.1002/pc.24236.

Onyeaka, H., Obileke, K., Makaka, G., Nwokolo, N., 2022. Current research and applications of starch-based biodegradable films for food packaging. Polymers 14 (6). https://doi.org/10.3390/polym14061126.

"Paper & Paperboard Packaging market Size, Share, Trend Analysis by 2033," https://www.emergenresearch.com/industry-report/paper-and-paperboard-packaging-market?utm_source=chatgpt.com.

Pei, J., Palanisamy, C.P., Srinivasan, G.P., Panagal, M., Kumar, S.S.D., Mironescu, M., 2024. A comprehensive review on starch-based sustainable edible films loaded with bioactive components for food packaging. Int. J. Biol. Macromol. 274. https://doi.org/10.1016/j.ijbiomac.2024.133332.

Phukan, M., and Nongkhlaw, F.T. "STARCH: AN OVERVIEW OF ITS CONVENTIONAL AND UNCONVENTIONAL SOURCES, PHARMACEUTICAL AND INDUSTRIAL APPLICATIONS." Online. Available: www.dibru.ac.in/ctprISSN.

Polman, E.M.N., Gruter, G.J.M., Parsons, J.R., Tietema, A., 2021. Comparison of the aerobic biodegradation of biopolymers and the corresponding bioplastics: a review. Sci. Total Environ. 753. https://doi.org/10.1016/j.scitotenv.2020.141953.

Reyniers, S., Ooms, N., Gomand, S.V., Delcour, J.A., 2020. What makes starch from potato (Solanum tuberosum L.) tubers unique: a review. Compr. Rev. Food Sci. Food Saf. 19 (5), 2588–2612. https://doi.org/10.1111/1541-4337.12596.

Ribba, L., Garcia, N.L., D'Accorso, N., Goyanes, S., 2017. Disadvantages of Starch-Based Materials, Feasible Alternatives in Order to Overcome These Limitations Starch-Based Materials in Food Packaging: Processing, Characterization and Applications. Elsevier Inc., Argentina, pp. 37–76. https://doi.org/10.1016/B978-0-12-809439-6.00003-0.

Rodrigues Arruda, T., Campos Bernardes, P., Robledo Fialho e Moraes, A., de Fátima Ferreira Soares, N., 2022. Natural bioactives in perspective: the future of active packaging based on essential oils and plant extracts themselves and those complexed by cyclodextrins. Food Res. Int. 156. https://doi.org/10.1016/j.foodres.2022.111160.

Sadeghizadeh-Yazdi, J., Habibi, M., Kamali, A.A., Banaei, M., 2019. Application of edible and biodegradable starch-based films in food packaging: a systematic review and meta-analysis. Curr. Res. Nutr. Food Sci. 7 (3), 624–637. https://doi.org/10.12944/CRNFSJ.7.3.03.

Samsudin, H., Hani, N.M., 2017. Use of Starch in Food Packaging Starch-Based Materials in Food Packaging: Processing, Characterization and Applications. Elsevier Inc., Malaysia, pp. 229–256. https://doi.org/10.1016/B978-0-12-809439-6.00008-X.

Soykeabkaew, N., Thanomsilp, C., Suwantong, O., 2015. A review: starch-based composite foams. Compos. Appl. Sci. Manuf. 78, 246–263. https://doi.org/10.1016/j.compositesa.2015.08.014.

Talja, R.A., Helén, H., Roos, Y.H., Jouppila, K., 2007. Effect of various polyols and polyol contents on physical and mechanical properties of potato starch-based films. Carbohydr. Polym. 67 (3), 288–295. https://doi.org/10.1016/j.carbpol.2006.05.019.

Talja, R.A., Peura, M., Serimaa, R., Jouppila, K., 2008. Effect of amylose content on physical and mechanical properties of potato-starch-based edible films. Biomacromolecules 9 (2), 658–663. https://doi.org/10.1021/bm700654h.

Tapia-Blácido, D.R., Aguilar, G.J., de Andrade, M.T., Rodrigues-Júnior, M.F., Guareschi-Martins, F.C., 2022. Trends and challenges of starch-based foams for use as food packaging and food container. Trends Food Sci. Technol. 119, 257–271. https://doi.org/10.1016/j.tifs.2021.12.005.

Tayade, R., Kulkarni, K.P., Jo, H., Song, J.T., Lee, J.D., 2019. Insight into the prospects for the improvement of seed starch in legume—a review. Front. Plant Sci. 10. https://doi.org/10.3389/fpls.2019.01213.

Tharanathan, R.N., 2003. Biodegradable films and composite coatings: Past, present and future. Trends Food Sci. Technol. 14 (3), 71–78. https://doi.org/10.1016/S0924-2244(02)00280-7.

"(PDF) Trends in Packaging Sector," https://www.researchgate.net/publication/358988609_Trends_in_Packaging_Sector.

"Understanding Starch-Based Plastics in Sustainable Packaging," https://www.jacobwhite.com/news/starch-based-plastics-sustainable-packaging.

Warren, R.A.J., 1996. Microbial hydrolysis of polysaccharides. Annu. Rev. Microbiol. 50 (1), 183–212. https://doi.org/10.1146/annurev.micro.50.1.183.

Wittaya, T., 2012. Rice starch-based biodegradable films: properties enhancement. In: Structure and Function of Food Engineering. InTech. https://doi.org/10.5772/47751.

Yu, X., Chen, L., Jin, Z., Jiao, A., 2021. Research progress of starch-based biodegradable materials: a review. J. Mater. Sci. 56 (19), 11187–11208. https://doi.org/10.1007/s10853-021-06063-1.

CHAPTER THREE

Starch processing techniques for packaging applications

Afreen Sultana
Department of Food, Nutrition and Packaging Sciences, Clemson University, Clemson, SC, United States

3.1 Introduction

Starch has the chemical formula $C_6H_{10}O_5$, meaning it is a polymer made up of long chains of glucose molecules joined by glycosidic bonds. Granules, which can be round, oval, or irregular in shape and range in diameter from roughly 0.1–200 μm depending on the source and maturation stage, are the most common type of natural starch. The two types of microstructures seen in starch are branched and linear. Amylose is known for its linear structure, which consists of α-1,4 linked glucose units, while amylopectin is known for its highly branched structure, which consists of short α-1,4 chains connected by α-1,6 bonds. Tapioca starch and other starches have been found to contain long-branched structures. Amylose has a molecular weight of roughly ×106, which is 10 times that of common synthetic polymers like polyethylene, polypropylene, and polystyrene. In contrast, amylopectin has a molecular weight that is significantly higher than amylose. Amylopectin's molecular weights could reach millions based on light-scattering tests. Amylose has demonstrated behavior that is comparable to that of traditional synthetic polymers due to its linear shape. Since amylopectin's molecular weight is far larger than that of typical synthetic polymers, the polymer chains' mobility is decreased, leading to extremely high viscosity (Jiang et al., 2020).

Multiple steps are involved in the manufacturing of flexible starch-based packaging, each specifically designed to handle a particular difficulty. These stages include raw material collection, conditioning, characterization, formulation, extrusion, molding, printing, packing, and storage. One low-cost technique is solvent casting, which involves spreading a polymer solution on a substrate and then removing the solvent. The potential for adding active compounds to the final films is one aspect of this approach. Similar methods include tape-casting, which forms a homogeneous film

Starch-Based Sustainable Food Packaging
ISBN: 978-0-443-36346-7
https://doi.org/10.1016/B978-0-443-36346-7.00003-5

by drying the mixture on a moving belt with an adjustable blade to control thickness. For the film to form steadily in both situations, plasticizers are necessary. However, even though these low-cost methods might appear perfect, they frequently produce films with inadequate mechanical strength and water resistance, which makes them less appropriate for real-world uses in food packaging. Other methods, such as extrusion, use heat and pressure to transform starch into a more robust plastic material with reduced water vapor permeability. Because of the tension created during blowing, these methods necessitate considerable mechanical strength (Garavito et al., 2024).

Nevertheless, the absence of defined production techniques and inconsistent raw materials in poor nations continues to cause scalability problems. In the food business, where safety and sustainability are crucial, packaging is especially important. Starch is inexpensive, biodegradable, and simple to mix with both inorganic and organic reinforcements. Because starch-based solutions require less sophisticated infrastructure than cellulose or other bioplastics, they can be produced on a modest scale in developing nations. This chapter discusses the common, conventional, and novel techniques used for the production of starch-based packaging as well as the challenges faced during the commercialization of starch-based packaging.

3.2 Conventional methods

Traditional preparation methods usually involve film casting, coating, foaming, and extrusion techniques, which are further discussed in detail.

3.2.1 Film casting

Film casting is a popular technique for developing films based on starch. Water is utilized as the solvent in the starch film-forming solution in the solvent-casting process. Initially, a film-forming solution is prepared by heating and stirring a mixture of film-forming materials and solvent, which is a film suspension of gelatinized starch. When heated, the starch would dissolve in water, causing the grains to swell and explode. This will result in the loss of the semi-crystalline structure and the formation of a network when the small amylose particles start to seep out of the granule. The network will compress water, making the solution more viscous (Marichelvam et al., 2019). This solution is then poured onto a smooth surface, such as a glass plate or plastic plate. The spread solution is then dried in an oven or at room temperature to evaporate the solvent. Finally, the prepared films are removed from the plate by peeling. It is suggested to carry out degassing of the film suspensions of gelatinized starch before pouring onto the plate to avoid the presence of air bubbles in the final film. Because starch has a strong hydrophilic nature, pure starch-based films have a number of disadvantages, including low mechanical strength. In addition, the strong intermolecular hydrogen connection between the amylose and amylopectin macromolecular network chains makes the starch-based films often fragile. This disadvantage will restrict its use in food packaging. A plasticizer is typically added to the starch-based films in order to solve this issue. Plasticizers can increase the mobility of polymer chains by decreasing the intermolecular forces in the film matrix. Small molecules called polyols (sorbitol and glycerol, e.g.) that intercalate and intersperse among the polymer chains, breaking the hydrogen bond and causing the chains to spread apart, make plasticizers (Ben et al., 2022). Typically, water and a plasticizer are used prior to the gelatinization process. Polyols, mono, di-, and oligosaccharides are the most widely used plasticizers. For instance, most hydrocolloid films, including starch films, almost consistently include polyols like glycerol. By reducing intermolecular tensions and disrupting amylose packing, a glycerol plasticizer will enhance the mechanical characteristics of films. However, the hydrophilicity of the film is increased by the addition of glycerol (Othman et al., 2023).

3.2.1.1 Advantages of film casting

Solvent casting is the most used technique for film preparation since it is an easy and inexpensive manufacturing process. The solvent casting technology is adaptable and suitable for both laboratory and industrial settings. When multiple drying steps are carried out in solvent casting, films with multiple layers can be produced. This enables the loading of active ingredients in various concentrations and film layers, as well as the incorporation of active ingredients with physical-chemical incompatibilities (Musazzi et al., 2020).

3.2.1.2 Limitations of film casting

The film casting method is simple and inexpensive, but it has some limitations. In particular, it is difficult to use on a large commercial scale and to control film thickness and uniformity accurately. Although the film solution casting method is simple to operate, it is difficult to realize large-scale continuous production due to its high-water content, long drying time and high energy consumption, which increases the production cost.

The applications of the solution casting technology are, therefore, limited to laboratory-scale production and are impractical at the industrial level (Gupta et al., 2024).

3.2.2 Tape casting

The tape-casting process has similarities to solvent-casting; however, the support on which the film-forming fluid spreads enables drying control by the use of infrared, heat convection, and/or heat conductivity. The movement of a scraper blade (batch process) or conveyor tape (continuous process) drives this process. One benefit of this method is that it enables the thickness to be controlled using an adjustable blade located at the bottom of the apparatus where the solution is applied. Typically, the films range in thickness from 20 μm to 1 mm. By repeating the operations on the film, tape-casting also makes it possible to create multi-layer films. In order to maintain proper flow conditions and simultaneously avoid undesired flow and sedimentation, the tape-casting technique necessitates a film-forming solution with shear-thinning behavior and sufficient apparent viscosity values. Technology transfer is challenging because traditional plastics companies do not adopt this technique despite permitting a continuous process (do Val Siqueira et al., 2021).

3.2.3 Coating

These coatings can be applied to the product's surface using a variety of methods, including dipping, spraying, fluidized bed, and panning. The type of product to be coated, its surface characteristics, and the coating's main goal all influence the deposition techniques used for coating applications on food items. The food business often uses spraying and panning techniques, but the dipping method is widely employed in laboratories because of its ease of use, affordability, and ability to provide good coverage on irregular surfaces. Following application, the solution is dehydrated, creating a thin film of adhering material that covers the food surface and is either functional or protective. The physical and rheological features of the coating solutions, as well as the attributes of the coated product, directly impact how well each application technique works (Luciano et al., 2022).

The fluidized bed technique has been extensively researched for coating in the food, chemical, and pharmaceutical industries. It is utilized as functional components and additives in a variety of food products. It seeks to create consistent coatings that are impacted by the physical state and droplet formation. A coating solution is sprayed onto fluidized particles using nozzles to create a shell structure. Droplets are atomized with air by binary or pneumatic nozzles, which promote solvent evaporation and viscosity. Fluidization is achieved when velocity supports particles without carrying them away, causing low-density or small goods to move upward in a high-pressure airflow. Coating solutions are sprayed on top, bottom, or rotating for consistent coverage.

The dipping technique is frequently utilized since it is convenient and provides a thorough and uniform covering of complex or uneven surfaces. In this three-step process, the product to be coated is submerged in the coating solution for a predetermined amount of time at a steady pace. The dwelling guarantees that there is an adequate amount of solution and that the coating and the product will interact. After that, the extra solution is drained, and a thin coating layer is formed on the product's surface via the deposition process. The solvent is then removed from the product's surface by a drying process that can be done with heat or at room temperature. Alternating cycles of immersion and drying is also feasible when needed.

The spraying approach is frequently utilized for coating deposition because it may be applied to a wide range of macromolecules and irregular shapes and sizes. It permits several consecutive applications if necessary. This technique creates smaller droplets, boosting the liquid's surface area and guaranteeing uniform coating dispersion. Nozzles help with automation and uniform application in continuous manufacturing. It is essential to comprehend the coating solution's rheological behavior since droplet formation depends on its physical characteristics. Benefits of spray coatings include multilayer applications, controlled thickness, homogeneity, decreased cross-contamination, and industrial scalability. However, one important factor influencing coating quality is atomization pressure.

In the panning process, a coating solution is poured or sprayed over items inside a revolving bowl, or "pan." Forced air is used to dry the tumbling product after it has been evenly coated by the pan's rotation. To use cool air to remove heat produced by friction, temperature control is essential. Frequently utilized for round or oval-shaped extruded products, it is also used for small things like as nuts. It is widely used for coating tablets in the pharmaceutical business and for sugar-panned candies in the food industry. Research has demonstrated that it can create extremely flexible, glossy, and transparent coatings. Although this process coatings big quantities efficiently, water evaporation can make it time-consuming and expensive.

Fruits and vegetables are frequently preserved via surface coating, and the preservation principle has various facets that fall into two categories: intrinsic qualities and functional features. All edible coatings have inherent qualities, however functional qualities are added by adding appropriate functional ingredients based on particular needs. An edible coating can create a barrier to keep the outside environment from coming into direct contact with the surface of fruits and vegetables. Consequently, adverse elements in the external environment, like dust and microbes, are difficult to directly affect fruits and vegetables, minimizing damage and safeguarding their nutritional value and organizational integrity. Furthermore, edible coatings typically include a certain level of mechanical strength and toughness that can reduce the mechanical damage to fruits and vegetables during storage and transit by acting as a buffer and support (Dai et al., 2024). Fruits and vegetables will gradually lose water during storage, transportation, and circulation due to respiration and the effects of external environmental factors like temperature and humidity. This will result in undesirable outcomes like quality loss, wilting branches and leaves, and peel shrinkage (Hoffmann et al., 2021). By lowering the partial pressure differential between the inside and outside, the edible coating prevents the transmission of gases and subsequently acts as a partial gas barrier. The development of a gas environment with low O_2 and high CO_2 in the coating is facilitated by this partial barrier, which prevents respiration, lowers transpiration loss, and delays the ripening and aging of fruits and vegetables (Zhang et al., 2023).

3.2.4 Foaming

The thermodynamic process of foaming a polymer is made up of thermal stabilization, structural growth, and cell nucleation. Thermodynamic instability in a polymer saturated with gas at high temperatures and pressures triggers nucleation. Extrusion, solvent extraction, thermo-pressing, microwave heating, and supercritical fluid extrusion can cause a polymer to foam. Because of their low density, high resistance, and affordability, expanded polystyrene foamed packaging is the most widely used. But because they could take a while to break down, these are not particularly environmentally friendly (García-Guzmán et al., 2022).

Foams must, therefore, be created using biodegradable substances like starch. Starch is an unsuitable material for foaming due to its low mechanical resistance, high hydrophilicity, limited resistance to retrogradation, and low thermal stability. Starch blends with polymers and plasticizers have been employed to enhance the resultant foams' mechanical qualities. used CO_2-assisted supercritical extrusion ($scCO_2$) to create a biodegradable foamed material with PLA and TPS. They found that an 80:20 PLA and TPS mixture produced a foam with a consistent structure and that the mixture's expansion and porosity behaved like pure PLA with high porosity (96%) in their investigation (Chauvet et al., 2021).

Similarly, foams were created using $scCO_2$-assisted extrusion of TPS with glyceraldehyde. Further integrating crosslinking agents like sodium hexametaphosphate and nano-silica particles enhanced the foam's expansion ratio, resilience rate, complex dynamic moisture-proof viscosity, and crystallization rate. In contrast, the researchers employed glycerol and water as plasticizers to create foams using a blend of thermoplastic corn starch (TPS) and polystyrene (PS) by melt extrusion. They discovered that adding 20% PS to TPS resulted in 60% more foam expansion than TPS and six times more foam expansion than PS but 50% less moisture absorption than TPS (Zabihi et al., 2021).

3.2.5 Extrusion

Extrusion is a continuous process that deforms the solid raw material into the desired shape under pressure and high temperature (Fig. 3.1). Extrusion is a five-step process involving feed loading, mixing and deforming, shape formation, solidification and cutting. In the first step, one or multiple raw material/s are loaded into the hopper while maintaining the uniform flow speed. In the second step, raw material gets mixed into a homogenous mixture and, under high temperature, ranging between 150 and 315°C, gets converted into a molten stage. There are three kinds of processes: extrusion injection, extrusion blowing, and extrusion compression molding.

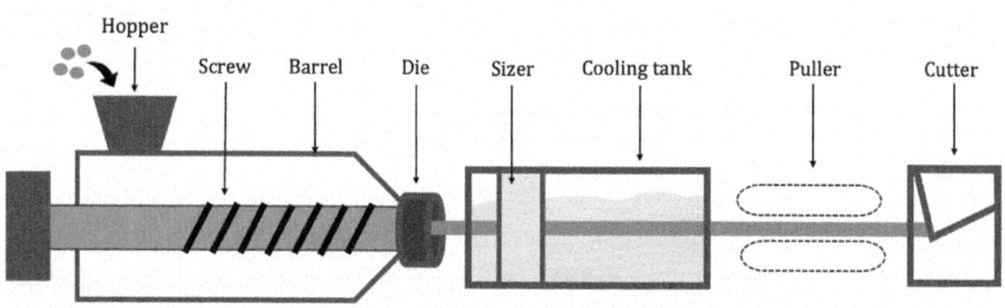

Figure 3.1 Diagrammatic representation of components and step-up for the extrusion process.

Extrusion blowing for the extruded film-forming material consists of three steps: blowing, pressing, and cooling. The films are produced by heating, pressing, and cooling the extruded film-forming material during the extrusion compression molding process. Extrusion injection molding and extrusion compression molding are similar since all methods use extrusion and molding during processing. On the other hand, extrusion injection molding creates the product by injecting molten raw materials into the mold and then cooling it. Extrusion injection molding is widely used in polymer processing because it can produce a large number of complex plastic goods at a moderate cost (do Val Siqueira et al., 2021).

Using the co-extrusion process, an extrudate made up of several thermoplastic melt streams is produced. The complex structures of coextruded films include many functional layers, such as tie layers that are intended to join adjacent layers with varying degrees of compatibility to create a multilayer film. Co-extrusion, a technology that uses extrusion processes to create multilayer films, involves extruding two or more polymer materials together. Unlike multilayer bio-based biodegradable films, this technique produces bilayers with intense interfacial adhesion. The composition utilized in the creation of bilayers determines their biodegradability. Bilayers have been employed to provide structures with more firmness, improve physicochemical characteristics, and increase antibacterial potency.

3.2.5.1 Advantages of extrusion

The films produced by the film extrusion process have morphologies similar to those produced by the casting method, and the process has a high production efficiency (Yudhistira et al., 2024). Extrusion has several benefits over solvent casting, including solvent-free processing and the ability to handle high viscosity polymers with ease under a variety of processing parameters (temperature and pressure). It also improves the film's transparency and mechanical qualities (elongation and strength) (Singaram et al., 2024).

3.2.5.2 Disadvantages of extrusion

Despite being extremely effective, shear forces at high temperatures during extrusion can cause polymers to deteriorate. This ultimately causes the macromolecular chains to break, altering the product's mechanical characteristics (Yamaguchi et al., 2023). In some cases, bonding between the starch and compatibilizer is able to maintain stability even after several rounds up to five, but the mechanical strength reduces significantly (Tavanaie and Ghahari, 2021).

3.3 Novel methods

The need for materials with better properties than reinforced and functionalized materials has prompted the creation of novel material processing technologies such as electrospinning, 3D printing, and reactive extrusion. These production methods produce materials with unique structural, chemical, physical, biological, and spatial distribution properties. These polymeric changes enable the production of third-generation materials by giving us materials with memory, the ability to produce and store energy, and the ability to regenerate tissues.

3.3.1 3D printing

Early advancements in topography and photo-sculptures in the late 18th century are where 3D printing got its start. Blanter later presented a method for making topographical maps in 1892, which involved layering cut pieces to create 3D structures. Charles Hull created the additive process of 3D printing in 1986. It involves layering materials to create items according to computer-guided plans. Rapid prototyping of packaging is made possible by 3D printers. Designers may quickly test forms and functionality by iterating through several iterations (Witek-Krowiak et al., 2024).

In order to create a complex, freeform 3D structure, 3D printing entails depositing the provided material layer by layer, as shown in Fig. 3.2. Phase transitions or chemical reactions aid in the fusing of deposited layers during the additive manufacturing process. 3D geometry in STL format is designed using computer-aided software such as SketchUp, Tinkercad, and AutoCAD. The design is then divided into separate layers by slicing software using G and M machine codes so that it may be printed (Nida et al., 2025). Seven standardized procedures have been defined by the American Society for Testing and Materials International as the foundation for additive manufacturing (AM) and 3D printing (3D). These methods include vat photopolymerization, sheet lamination, binder jetting, powder bed fusion, material extrusion, and directed energy deposition (Thakur et al., 2024).

The inkjet process uses a specialized, individually controlled multinozzle material dispenser to deposit photopolymer droplets onto a substrate. These droplets then develop into thin layers that solidify when various curing techniques, including UV light or chemical/ionic crosslinking, are used to produce a three-dimensional object (Li et al., 2019). Examples of typical inkjets are thermal and piezoelectric. A heat actuator in the chamber

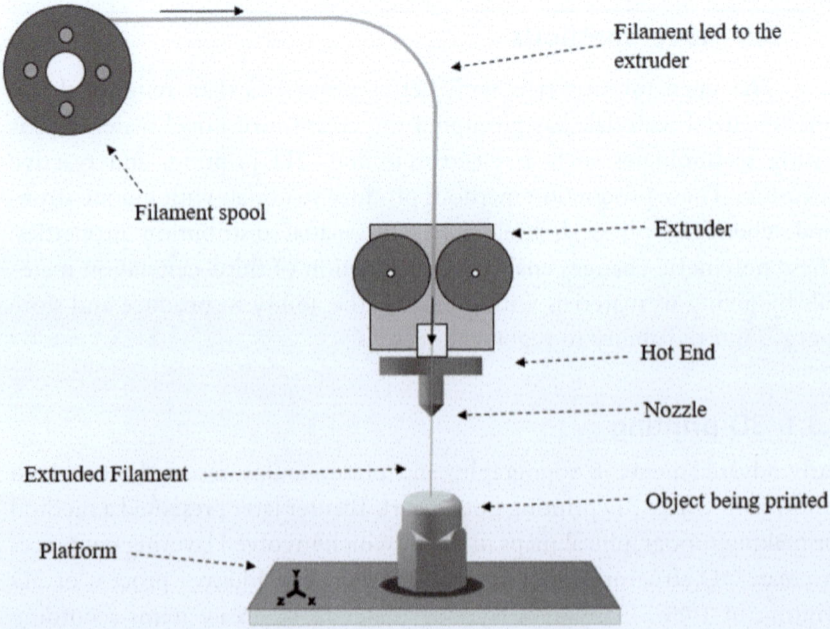

Figure 3.2 Pictorial representation of the essential components for 3D printing (Seok et al., 2023).

creates heat bubbles in an inkjet by compressing ink inside the nozzle and releasing a droplet, usually 20 μm in size. Heating the bubble to temperatures between 250 and 350°C is usually required to produce a substantial expansion. The liquid temperature in the chamber only rises 4–10°C above ambient temperature because of the short heating procedure. The liquid is compressed and deformed by the piezoelectric material in a piezoelectric printer, which causes droplets to be ejected. Its excellent control over droplet size reduces material waste, and its high-speed, completely automated process guarantees efficient output. Furthermore, inkjet printing makes it possible to create various deposition patterns, reducing the possibility of material contamination and supporting intricate functional material release profiles. Because of its versatility, the method can be used to process a wide range of materials, such as proteins, DNA, suspensions, nanoparticles, functional polymers, and sol-gel materials. Inkjet printing is a viable option for developing biopolymer 3DP in functional food packaging because of its accuracy and adaptability (Thakur et al., 2024).

The process of powder bed fusion (PBF) combines three printing methods: electron beam melting (EBM), which uses a high-energy laser to fuse or melt powdered material (usually metals or thermoplastics) layer by layer to create three-dimensional objects, and selective heat sintering (SHS) and selective laser sintering (SLS). While polymer materials are primarily processed using the SLS technique, metals are the major material for which the EBM and SHS processes are used. Each PBF process involves applying a thin layer of powder to a build platform, which is then selectively melted or sintered using heat or a laser source. Repeating the process until the entire item is produced, a new coating of powder is applied when each layer is completed (Bareen et al., 2022).

Vat polymerization, also known as lithography-based additive manufacturing, is a technique that produces three-dimensional structures by solidifying a liquid or paste substance with a laser beam or light pattern, respectively. A light source (UV, infrared, and visible light) selectively cures the material layer by layer in accordance with the desired shape while a construction platform is immersed in a vat of liquid or paste material. The object is progressively built up by the cured layers adhering to the construction platform. After printing is finished, the item is usually post-cured to guarantee its ultimate stability and strength and washed in a solvent to get rid of any leftover material. There is a challenge, with regard to the ink viscosity appropriate for vat polymerization, which ranges from 1 to 300 mPa·s.

Mechanical properties, printability, heat resistance, and gelling behavior are some of the variables that determine whether starch-based materials are suitable for 3D printing applications. Printability can often be improved by combining starch-based materials with other materials or by chemically altering them (Maniglia et al., 2021; Anwajler et al., 2024).

3.3.1.1 Advantages of 3D printing

With 3D printing technology, almost any shape may be employed. There are no restrictions, not even with regard to intricate shapes or geometry (Witek-Krowiak et al., 2024). 3D printer filaments are made to resemble many materials, including carbon fiber, glass, and wood. This enables designers to experiment with multiple prototypes and select the best materials. From concept to production, the full process can be completed internally. This minimizes the chance of a new design leaking, cuts costs, and shortens product lead times (Anwajler et al., 2024).

3.3.1.2 Limitations of 3D printing

3D printing processing technique is gaining attention but has several limitations, such as high prices, slow printing rates, small part sizes, and poor strength (Iftekar et al., 2023). There are still limitations with porosity, residual stress, and post-processing complications that prevent it from being widely used. Such as when the speed is too slow or the power is too high, the layers may deform. Warping causes sections to become warped and, in severe cases, stops the process and makes it impossible to apply new layers. The balling could be another outcome of an uncorrected combination of characteristics. An excessively high scan speed and power might result in balling, which is characterized by minute concentrations of fused particles (Dananjaya et al., 2024).

3.3.2 Electrospinning

Electrospinning is a versatile and highly efficient technique for fabricating nanofibers, producing scaffolds and membranes for diverse applications in different fields, including biomedicine, tissue engineering, environmental science, energy storage, sensors, electronics, textiles, optical applications, and advanced material sciences (Shabani et al., 2025). This method creates nano- or macrometric-scale fibers by stretching and elongating a viscoelastic jet that is produced from a polymer solution or by melting using electrostatic forces. It is possible to create electrospun nanofibers from natural or synthetic polymers or even a combination of the two (de Souza et al., 2021).

The primary electrospinning parameters are flow (feed rate), voltage, and the distance between the needle tip and the collector. As shown in Fig. 3.3, the electrospinning device is an easy assembly consisting of a

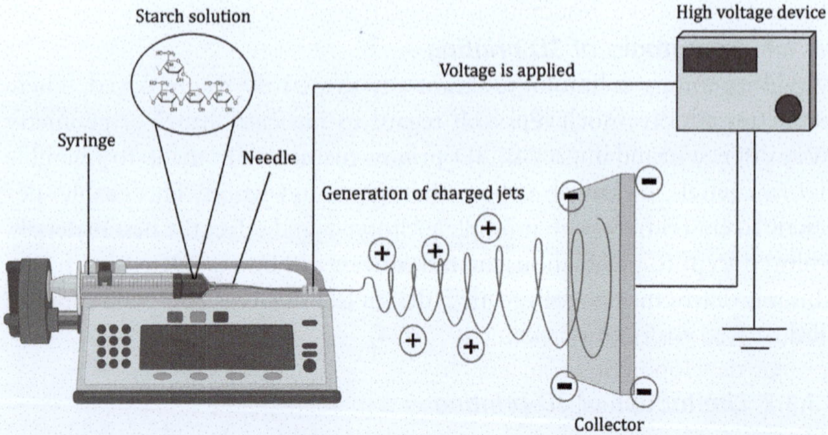

Figure 3.3 Diagrammatic representation of components and step-up for the electrospinning process.

grounded collector plate, a high-voltage power source, and an infusion pump with a plastic syringe attached to a needle (spinneret) (Fonseca et al., 2024). Three stages could be used to categorize the electrospinning process briefly: When the electric field force is sufficient to overcome the viscoelastic force of the solution, (1) the charged droplet of polymer solution at the needle forms a conical shape (also known as a "tailor cone") in the electric field; (2) a solid jet is deposited on the collector's surface and produces a large number of fibers (Zhu et al., 2022).

One study examined the optimization of electrospun yam starch fiber operating conditions using a 2^3 multivariate design. The fibers responded with an average diameter of 134.1 ± 16.7 nm, uniform shape, and no beads when the optimal conditions were voltage of 20 kV, distance of 20.5 cm, and flow of 1 mL/h (de Souza et al., 2021). Some of the studies on starch to develop fibers are mentioned below (Table 3.1):

The two parameters that affect the electrospinnability of starch-based solutions are viscosity and electrical conductivity. Fiber formation depends on the polymeric solution, which varies depending on the polymers utilized, their concentrations, and the properties of the polymeric solution. The most important factor is viscosity, which is influenced by the concentration of polymers. It is closely related to the properties of fibers, such as their diameter and shape. Electrical conductivity is another characteristic that is affected by the solvent, the kind and concentration of the polymer, and the presence of ionizable chemicals. As the electrical conductivity of the polymeric solutions carried during electrospinning decreases, so does their charge density (Fonseca et al., 2024).

3.3.2.1 Advantages of electrospinning

Higher productivity and large-scale manufacturing could be achieved using multi-needle or non-needle electrospinning devices in comparison to traditional film-forming procedures. Furthermore, electrospinning is a non-thermal process, so it helps preserve the structure's stability, particularly for additives with poor thermal durability at high temperatures (Zhu et al., 2022). Lightweight, high surface area-to-volume ratio, versatility in producing materials in a wide range of forms and sizes, and the capacity to manipulate material composition to obtain desired outcomes depending on characteristics and functions are some benefits of electrospun fibers. Because of their nanoscale size, starch-based films containing nanofibers have extremely high surface activity, making them ideal for active food packaging. Furthermore, the shape and structure of the electrospun starch

Table 3.1 Some of the recent studies on nanofiber development using the electrospinning technique.

Raw material	Solvent	Electrospinning parameter	Nanofiber diameter	References
Octenyl succinylated starch-pullulan	Deionized water	V = 18 kV; d = 11 cm; f = 0.2 mL/h	120–260 nm	Shi et al. (2025)
Taro	Formic acid	V = 17 kV; d = 15 cm; f = 0.8 mL/h	298–759 nm	Sawitri et al. (2025)
Taro, yam, white sweet potato, and potato	Formic acid (75%, w/v)	V = 23 kV; d = 20 cm; f = 0.60 mL/h	133–849 nm	Fonseca et al. (2025)
Corn starch with iron and tannic acid	Distilled water	V = 20 kV; d = 15 cm; f = 1 mL/h	368–406 nm	Yang et al. (2024)
Octenylsuccinylated starch, pullulan and cinnamaldehyde essential oil	Distilled water	V = 18 kV; d = 15 cm; f = 0.4 mL/h	211–244 nm	Huang et al. (2024)
Cassava starch	Formic acid	V = 23 kV; d = 20 cm; f = 0.6 mL/h	165–301 nm	Santos et al. (2024)
Tapioca starch with glycerol, silver nanoparticle and guar gum solution	Distilled water	V = 15 kV; d = 10 cm; f = 0.25 mL/h	103.9 nm	Alqahtani (2024)

Potatoes, tapioca, green beans, wheat, normal corn and waxy corn	Formic acid	V = 15–20 kV; d = 15 cm; f = 0.2 mL/h	126–843 nm	Cao et al. (2022)
Pea and cassava	Formic acid (75%, w/v)	V = 13–27 kV; d = 16–25 cm; f = 0.2–1.0 mL/h	101–394 nm	Tochihuitl-Vázquez et al. (2022)
Potato	Formic acid (75%, w/v)	V = 13–27 kV; d = 16–25 cm; f = 0.2–1.0 mL/h	134.1 nm	Fonseca et al. (2020)
Potato	Formic acid (75%, w/v)	V = 15 kV; d = 15 cm; f = 0.8 mL/h	128–143 nm	Fonseca et al. (2019)
Corn	Formic acid (75%, w/v)	V = 15–18 kV; d = 15 cm; f = 0.6 and 0.8 mL/h	70–264 nm	Fonseca et al. (2019)
Corn	Dimethyl sulfoxide	V = 6–10 kV; d = 5–8 cm f = 2–4 mL/h	3.35–22.35 μm	Kong and Ziegler (2013)

fibers might be readily altered to preserve different active components and increase their mechanical and barrier qualities (Fonseca et al., 2024). Furthermore, they enable the use of mathematical models to explain how variables behave, enabling the prediction of the ideal circumstances for achieving the intended result (de Souza et al., 2021).

3.3.2.2 Limitations of electrospinning

For the electrospinning process, strong chemicals like formic acid and dimethyl sulfoxide (DMSO) are used to pre-treat the starch. The technique involves adding plasticizers and synthetic polymers to the combination of starch-based biopolymers to enhance the nanofibers' functionality. The sensitivity of the process parameters, such as flow rate, voltage, and distance between the spinneret and collector, as well as solvent compatibility and bead formation, are some of the variables that affect electrospinning, making it challenging to scale to commercial levels (Thakur et al., 2024). Chemically modifying starch can be a useful strategy for creating functional fibers. The production of starch fibers with diameters smaller than 100 nm, which are referred to as nanofibers and have a high surface area-to-volume ratio and a larger capacity to capture and release phenolic chemicals, presents another difficulty (Fonseca et al., 2024).

3.3.3 Reactive extrusion

Reactive extrusion is a manufacturing method that changes polymers via grafting, polycondensation, cross-linking, and polymerization during extrusion by enzymatic or chemical processes. Because reactive extrusion combines the thermomechanical energy needed to break down the natural structure of the starch granule, it allows the reaction between starch and organic acids to occur in a single step without the need for additional reagents. Additionally, reactive extrusion provides a fast reaction time. It is also a commercially viable, ongoing process that is easy to scale up to an industrial level (González-Seligra et al., 2022). This method is a viable substitute for producing esterified and cross-linked starches. The reactive extrusion method was utilized to successfully cross-link starch chains using sodium tripolyphosphate as a cross-linker. The phosphating (cross-linking) method employed during reactive extrusion and the hierarchical structure of the starch resulted in the production of more hydrophilic materials. This is because more OH groups were accessible to absorb water as a result of a decrease in hydrogen (H)-bonding interactions. Type A and type V crystal structures both had their interplanar d-spacings increased by the adjustment, but the kind of crystal structure remained unchanged (Gutiérrez and Valencia, 2021). The type V structure, which is associated with H-bonding, is weakened by the strong type A crystalline domain exerted by the starch's hierarchical structure. Viscosity is significantly higher in materials derived from corn starch nanocrystals than in products derived from corn starch. Amylopectin levels in the former were higher than those in the latter, which helped to explain this. In general, those more crystalline films slowed down the films' biodegradation. This change in the starch's hierarchical structure led to the production of more crystalline solids. For instance, cellulose acetate/corn starch mixtures were utilized to produce biodegradable films using this technique (Herniou-Julien et al., 2019).

Co-rotating twin screw extruders are typically used for this process because of their great mixing capacity and continuous operation. Because of its great mixing capacity and continuous operation, this process is typically performed in co-rotating twin screw extruders. One major benefit of this extrusion process is that the polymerization processes can be conducted without solvents or with very little of them. Crosslinking, glycolysis, hydrolysis, compatibilization, and other quick reactions that take one to 15 minutes are typically handled with this method. Reactive epoxy compatibilizer (REC) was used to create films by reactively extruding thermoplastic starch and PBAT. According to the findings, REC successfully increases the TPS/PBAT mixes' compatibility and lengthens the PBAT polymeric chains by enhancing the films' mechanical qualities (Bai et al., 2021). Salicylated starch-based food packaging materials made by reactive extrusion/thermo-molding were processed in a single step (mechanochemistry) using tartaric acid to aid in the aforementioned reaction. They used tartaric acid, a food-grade green organocatalyst, and salicylic acid as the active ingredient to be esterified in order to investigate the impact of the salicylyl group's organocatalytic esterification on hermoplastic corn starch. This study showed that starch hydrolysis in the presence of either salicylic acid or tartaric acid was determined by low water concentrations, which are fundamentally related to the feedstocks (starch and glycerol) (Troncoso et al., 2025).

The reactive extrusion technique can also be used for the development of active and intelligent packaging systems. In order to create nonmigratory active and intelligent packaging using a solvent-free, effective, and continuous process, curcumin was grafted via radical grafting to polypropylene via reactive extrusion (Redfearn et al., 2023). Reactive extrusion used a twin-screw extruder with chromium octanoate as a catalyst and thermo-molded native and carboxymethylated plantain flour/polystyrene blends with glycerol as a plasticizer to develop the films (Troncoso et al., 2025).

3.3.3.1 *Advantages of reactive extrusion*

Reactive extrusion is a fast and low-cost method. The lack or low concentration of solvent as a reaction medium in extruders as reactors also has the significant benefit of removing the need for extra technological processes like solvent recovery or final product purification. These are the primary elements influencing the ongoing advancement of reactive extrusion in the synthesis and processing of different polymers, polymer blends, and composites (including biodegradable ones) in both the laboratory and the industrial setting (Formela et al., 2018).

3.3.3.2 *Drawbacks of reactive extrusion*

The primary factor is the residence time, which is only a few minutes in an extruder. Because of this, reactive extrusion can only produce quick reactions, even though reagent concentration and temperature are frequently more crucial in a molten state than in a solution. Reactive extrusion reactions with high exothermicity may be challenging to control due to an extruder's limited cooling capacity (Vergnes and Berzin, 2006).

3.4 Challenges in the development of starch-based packaging at an industrial scale

Food packaging manufactured from biobased materials has not been able to fully compete with the plastic packaging market despite 10 years of research due to a number of problems. Increased environmental consciousness, the depletion of petrochemical resources, corporate and governmental requirements, and suitable technology are the main factors driving the development of packaging materials based on biodegradable starch. Below is a discussion of some of the more important issues.

3.4.1 Lack of homogeneity

Achieving homogeneity in the starch-based film/coating production process is difficult. The variable appearance and color of non-homogeneous films may be an unfavorable feature for customers. The reinforcing agent may have an impact on the color of the starch films; in one study, adding guar gum increased the overall color difference of the film, which was thought to be caused by the natural color of the raw materials (Saberi et al., 2016). A non-homogeneous coating or film may alter the mechanical properties, changing the film's strength and pliability.

3.4.2 Production cost

The cost of packaging may rise as a result of the production of starch-based packaging. The goal of the lean supply chain methodology is to reduce expenses by getting rid of non-value-added tasks. By cutting down on time, labor, inventory, and equipment wasted on operations that don't improve the supply chain, this strategy aids companies in cutting waste (Asim et al., 2022). Effective packaging design might assist cut down on the quantity of raw materials and product transportation, which would lessen packaging costs (Zhu et al., 2022b). Packaging production costs may be lowered by using waste or byproducts from the agro-industry and agriculture as a raw material (Sadh et al., 2018).

3.4.3 Incompatible expectations and ineffective collaboration between industry and academia

Because the conditions and parameters used at the laboratory and industrial scales differ greatly, it is difficult to duplicate the commercial production of starch bio-composite. Because of the mechanical properties of the starch bio-composite, natural biopolymer films are typically not suited to industrial processing methods such as extrusion or film blowing, which restricts their industrial processability and applications (Nilsen-Nygaard et al., 2021). The production process of biopolymers could be forecasted using simulation models (Al-Raeei and El-Daher, 2021). Research on starch bio-composite films for food preservation is scarce. In order to solve the issue of food shelf life and guarantee product quality, research on packaging performance and compatibility with various food kinds can be conducted (Nilsen-Nygaard et al., 2021).

By enhancing material performance and refining processing conditions, starch bio-composites can improve their low heat resistance and stability (Amrutha et al., 2023). Therefore, proper collaboration between academia and industry is necessary to better understand the limitations and facilitate the upscaling process.

3.5 Conclusions

Since commercial starch-based materials lack the optical, barrier, and mechanical qualities of their synthetic counterparts, their widespread production for the food industry is still pending. Nevertheless, commercial starch-based materials could undoubtedly play a significant role in the food packaging industry as an environmentally friendly and sustainable alternative to synthetic packaging. To address these issues, some research has been done on ways to improve and modify standard thermoplastic processing methods, alter starch chemically and physically, and blend it with other biopolymers, plasticizers, and functional nano/micro fillers. Investigating how various processing methods affect the multi-scale structure of starch is crucial for the logical control of its structural characteristics and helps to comprehend the production of multi-scale packaging made from starch. This chapter presents the current status of processing techniques of starch-based packaging and its composites. Advancements in technology have led to the creation of new techniques and pathways to produce starch-based packaging in a cost-effective way. Different approaches are being used to overcome the inherent shortcomings of biopolymers.

References

Al-Raeei, M., El-Daher, M.S., 2021. Temperature dependence of the specific volume of Lennard-Jones potential and applying in case of polymers and other materials. Polym. Bull. 78 (3), 1453–1463. https://doi.org/10.1007/s00289-020-03166-8.

Alqahtani, T., 2024. Silver nanoparticles impregnated in tapioca starch biofilm made using the electrospinning technique: a cutting-edge material for food packaging. Biomass Convers. Biorefinery 14 (22), 28155–28162. https://doi.org/10.1007/s13399-022-03344-w.

Amrutha, S.R., Rejimon, P.K., Suja, N.R., Mart, A., 2023. Thermal properties of biopolymers. In: Thomas, S., AR, A., Jose Chirayil, C., Thomas, B. (Eds.), Handbook of Biopolymers. Springer, Singapore. https://doi.org/10.1007/978-981-16-6603-2_12-1.

Anwajler, B., Szołomicki, J., Noszczyk, P., Baryś, M., 2024. The potential of 3D printing in thermal insulating composite materials—experimental determination of the impact of the geometry on thermal resistance. Materials 17 (5), 1202. https://doi.org/10.3390/ma17051202.

Asim, Z., Shamsi, I.R.A., Wahaj, M., Raza, A., Abul Hasan, S., Siddiqui, S.A., Aladresi, A., Sorooshian, S., Seng Teck, T., 2022. Significance of sustainable packaging: a case-study from a supply chain perspective. Appl. Syst. Innov. 5 (6). https://doi.org/10.3390/asi5060117.

Bai, J., Pei, H., Zhou, X., Xie, X., 2021. Reactive compatibilization and properties of low-cost and high-performance PBAT/thermoplastic starch blends. Eur. Polym. J. 143, 110198. https://doi.org/10.1016/j.eurpolymj.2020.110198.

Bareen, M.A., Sahu, J.K., Prakash, S., Bhandari, B., 2022. 3D printing: technologies, fundamentals, and applications. In: Food Industries Smart and Sustainable Food Technologies. Springer Nature, India, pp. 197–234. https://doi.org/10.1007/978-981-19-1746-2_7.

Ben, Z.Y., Samsudin, H., Firdaus Yhaya, M., 2022. Glycerol: its properties, polymer synthesis, and applications in starch based films. Eur. Polym. J. 175, 111377. https://doi.org/10.1016/j.eurpolymj.2022.111377.

Cao, P., Wu, G., Yao, Z., Wang, Z., Li, E., Yu, S., Liu, Q., Gilbert, R.G., Li, S., 2022. Effects of amylose and amylopectin molecular structures on starch electrospinning. Carbohydr. Polym. 296, 119959. https://doi.org/10.1016/j.carbpol.2022.119959.

Chauvet, M., Sauceau, M., Baillon, F., Fages, J., 2021. Blending and foaming thermoplastic starch with poly (lactic acid) by CO_2-aided hot melt extrusion. J. Appl. Polym. Sci. 138 (14). https://doi.org/10.1002/app.50150.

Dai, L., Wang, X., Mao, X., He, L., Li, C., Zhang, J., Chen, Y., 2024. Recent advances in starch-based coatings for the postharvest preservation of fruits and vegetables. Carbohydr. Polym. 328, 121736. https://doi.org/10.1016/j.carbpol.2023.121736.

Dananjaya, S.A.V., Chevali, V.S., Dear, J.P., Potluri, P., Abeykoon, C., 2024. 3D printing of biodegradable polymers and their composites – current state-of-the-art, properties, applications, and machine learning for potential future applications. Prog. Mater. Sci. 146, 101336. https://doi.org/10.1016/j.pmatsci.2024.101336.

de Souza, E.J.D., Nardo dos Santos, F., Pires, J.B., Kringel, D.H., Silva, W.M.F.da, Meinhart, A.D., Dias, A.R.G., Zavareze, E.da R., 2021. Production and optimization of ultrafine fiber from yam starch by electrospinning method using multivariate analysis. Starch - Stärke 73 (3–4). https://doi.org/10.1002/star.202000174.

do Val Siqueira, L., Arias, C.I.L.F., Maniglia, B.C., Tadini, C.C., 2021. Starch-based biodegradable plastics: methods of production, challenges and future perspectives. Curr. Opin. Food Sci. 38, 122–130. https://doi.org/10.1016/j.cofs.2020.10.020.

Fonseca, L.M., da Silva, F.T., Antunes, M.D., Halal, S.L.M.El, Lim, L.-T., Dias, A.R.G., 2019. Aging time of soluble potato starch solutions for ultrafine fibers formation by electrospinning. Starch - Stärke 71 (1–2). https://doi.org/10.1002/star.201800089.

Fonseca, L.M., de Oliveira, J.P., de Oliveira, P.D., da Rosa Zavareze, E., Dias, A.R.G., Lim, L.T., 2019. Electrospinning of native and anionic corn starch fibers with different amylose contents. Food Res. Int. 116, 1318–1326. https://doi.org/10.1016/j.foodres.2018.10.021.

Fonseca, L.M., Radünz, M., dos Santos Hackbart, H.C., da Silva, F.T., Camargo, T.M., Bruni, G.P., Monks, J.L.F., da Rosa Zavareze, E., Dias, A.R.G., 2020. Electrospun potato starch nanofibers for thyme essential oil encapsulation: antioxidant activity and thermal resistance. J. Sci. Food Agric. 100 (11), 4263–4271. https://doi.org/10.1002/jsfa.10468.

Fonseca, L.M., Cruz, E.P.d., Crizel, R.L., Jansen-Alves, C., Dias, A.R.G., Zavareze, E.d. R., 2024. New advances of electrospun starch fibers, encapsulation, and food applications: a review. Trends Food Sci. Technol. 147. https://doi.org/10.1016/j.tifs.2024.104467.

Fonseca, L.M., Nardo dos Santos, F., Buchveitz Pires, J., Cruz, E.P.da, Pacheco, C.de O., Diaz, P.S., Guerra Dias, A.R., Zavareze, E.da R., 2025. Electrospun fibers from esterified starch of tubers: effect of concentration and source of starch on fibers performance. Starch - Stärke 77 (8). https://doi.org/10.1002/star.202400150.

Formela, K., Zedler, L., Hejna, A., Tercjak, A., 2018. Reactive extrusion of bio-based polymer blends and composites – current trends and future developments. Express Polym. Lett. 12 (1), 24–57. https://doi.org/10.3144/expresspolymlett.2018.4.

Garavito, J., Peña-Venegas, C.P., Castellanos, D.A., 2024. Production of starch-based flexible food packaging in developing countries: analysis of the processes, challenges, and requirements. Foods 13 (24), 4096. https://doi.org/10.3390/foods13244096.

García-Guzmán, L., Cabrera-Barjas, G., Soria-Hernández, C.G., Castaño, J., Guadarrama-Lezama, A.Y., Rodríguez Llamazares, S., 2022. Progress in starch-based materials for food packaging applications. Polysaccharides 3 (1), 136–177. https://doi.org/10.3390/polysaccharides3010007.

González-Seligra, P., Goyanes, S., Famá, L., 2022. Effect of the incorporation of rich-amylopectin starch nano/micro particles on the physicochemical properties of starch-based nanocomposites developed by flat-die extrusion. Starch/Staerke 74 (1–2). https://doi.org/10.1002/star.202100080.

Gupta, D., Lall, A., Kumar, S., Patil, T.D., Gaikwad, K.K., 2024. Plant-based edible films and coatings for food-packaging applications: recent advances, applications, and trends. Sustain. Food Technol. 2 (5), 1428–1455. https://doi.org/10.1039/d4fb00110a.

Gutiérrez, T.J., Valencia, G.A., 2021. Reactive extrusion-processed native and phosphated starch-based food packaging films governed by the hierarchical structure. Int. J. Biol. Macromol. 172, 439–451. https://doi.org/10.1016/j.ijbiomac.2021.01.048.

Herniou-Julien, C., Mendieta, J.R., Gutiérrez, T.J., 2019. Characterization of biodegradable/non-compostable films made from cellulose acetate/corn starch blends processed under reactive extrusion conditions. Food Hydrocoll. 89, 67–79. https://doi.org/10.1016/j.foodhyd.2018.10.024.

Hoffmann, T.G., Ronzoni, A.F., da Silva, D.L., Bertoli, S.L., de Souza, C.K., 2021. Cooling kinetics and mass transfer in postharvest preservation of fresh fruits and vegetables under refrigerated conditions. Chem. Eng. Trans. 87, 115–120. https://doi.org/10.3303/CET2187020.

Huang, X., Teng, Z., Xie, F., Wang, G., Li, Y., Liu, X., Li, S., 2024. Loading of cinnamon essential oil into electrospun octenylsuccinylated starch-pullulan nanofiber mats: electrospinnability evaluation, structural characterization, and antibacterial potential. Food Hydrocoll. 148, 109426. https://doi.org/10.1016/j.foodhyd.2023.109426.

Iftekar, S.F., Aabid, A., Amir, A., Baig, M., 2023. Advancements and limitations in 3D printing materials and technologies: a critical review. Polymers 15 (11), 2519. https://doi.org/10.3390/polym15112519.

Jiang, T., Duan, Q., Zhu, J., Liu, H., Yu, L., 2020. Starch-based biodegradable materials: challenges and opportunities. Adv. Ind. Eng. Polym. Res. 3 (1), 8–18. https://doi.org/10.1016/j.aiepr.2019.11.003.

Kong, L., Ziegler, G.R., 2013. Quantitative relationship between electrospinning parameters and starch fiber diameter. Carbohydr. Polym. 92 (2), 1416–1422. https://doi.org/10.1016/j.carbpol.2012.09.026.

Li, V.C.F., Kuang, X., Hamel, C.M., Roach, D., Deng, Y., Qi, H.J., 2019. Cellulose nanocrystals support material for 3D printing complexly shaped structures via multi-materials-multi-methods printing. Addit. Manuf. 28, 14–22. https://doi.org/10.1016/j.addma.2019.04.013.

Luciano, C.G., Caicedo Chacon, W.D., Valencia, G.A., 2022. Starch-based coatings for food preservation: a review. Starch/Staerke 74 (5–6). https://doi.org/10.1002/star.202100279.

Maniglia, B.C., Pataro, G., Ferrari, G., Augusto, P.E.D., Le-Bail, P., Le-Bail, A., 2021. Pulsed electric fields (PEF) treatment to enhance starch 3D printing application: effect on structure, properties, and functionality of wheat and cassava starches. Innov. Food Sci. Emerg. Technol. 68, 102602. https://doi.org/10.1016/j.ifset.2021.102602.

Marichelvam, M.K., Jawaid, M., Asim, M., 2019. Corn and rice starch-based bio-plastics as alternative packaging materials. Fibers 7 (4), 32. https://doi.org/10.3390/fib7040032.

Musazzi, U.M., Khalid, G.M., Selmin, F., Minghetti, P., Cilurzo, F., 2020. Trends in the production methods of orodispersible films. Int. J. Pharm. 576, 118963. https://doi.org/10.1016/j.ijpharm.2019.118963.

Nida, S., Moses, J.A., Anandharamakrishnan, C., 2025. 3D printed food package casings from sugarcane bagasse: a waste valorization study. Biomass Convers. Biorefinery 15 (2), 1835–1845. https://doi.org/10.1007/s13399-021-01982-0.

Nilsen-Nygaard, J., Fernández, E.N., Radusin, T., Rotabakk, B.T., Sarfraz, J., Sharmin, N., Sivertsvik, M., Sone, I., Pettersen, M.K., 2021. Current status of biobased and biodegradable food packaging materials: impact on food quality and effect of innovative processing technologies. Compr. Rev. Food Sci. Food Saf. 20 (2), 1333–1380. https://doi.org/10.1111/1541-4337.12715.

Othman, S.H., Syafiqah Kahar, N., Nordin, N., Ruzanna, A.Shapi'i, 2023. Properties and food packaging applications of solvent casting-made starch-based films incorporated with essential oils: a review. Starch - Stärke 75 (5–6). https://doi.org/10.1002/star.202200152.

Redfearn, H.N., Warren, M.K., Goddard, J.M., 2023. Reactive extrusion of nonmigratory active and intelligent packaging. ACS Appl. Mater. Interfaces 15 (24), 29511–29524. https://doi.org/10.1021/acsami.3c06589.

Saberi, B., Thakur, R., Vuong, Q.V., Chockchaisawasdee, S., Golding, J.B., Scarlett, C.J., Stathopoulos, C.E., 2016. Optimization of physical and optical properties of biodegradable edible films based on pea starch and guar gum. Ind. Crop. Prod. 86, 342–352. https://doi.org/10.1016/j.indcrop.2016.04.015.

Sadh, P.K., Duhan, S., Duhan, J.S., 2018. Agro-industrial wastes and their utilization using solid state fermentation: a review. Bioresour. Bioprocess. 5 (1). https://doi.org/10.1186/s40643-017-0187-z.

Santos, F.N.d., Cruz, E.P.d., Fonseca, L.M., Pires, J.B., Diaz, P.S., Dias, A.R.G., Zavareze, E.d.R., 2024. Cassava starch esterification with formic acid for fabrication of electrospun fibers. Int. J. Biol. Macromol. 266. https://doi.org/10.1016/j.ijbiomac.2024.131182.

Sawitri, A., Luthfianti, H.R., Hapidin, D.A., Edikresnha, D., Palar, P.S., Surtiyeni, N., Wibowo, A., Khairurrijal, K., 2025. Synthesis and characterization of novel electrospun nanofibers based on taro starch: influence of solvent and isolation agent on morphology and diameter. Polym. Int. 74 (3), 217–230. https://doi.org/10.1002/pi.6709.

Seok, I., Kilula, D., Guo, Z., 2023. Micro/nanoscale 3-dimensional fabrication using multi-photons polymerization: review. ES Mater. Manuf. 21. https://doi.org/10.30919/esmm5f849.

Shabani, A., Al, G.A., Berri, N., Castro-Dominguez, B., Leese, H.S., Martinez-Hernandez, U., 2025. Electrospinning technology, machine learning, and control approaches: a review. Adv. Eng. Mater. 27 (7). https://doi.org/10.1002/adem.202401353.

Shi, L., Li, Z., Qing, S., Ren, Z., Li, P., Li, S., Weng, W., 2025. Underlying mechanism of electrospun starch-based nanofiber mats to adsorb the key off-odor compounds of oyster peptides. Food Chem. X 25, 102061. https://doi.org/10.1016/j.fochx.2024.102061.

Singaram, A.J.V., Guruchandran, S., Ganesan, N.D., 2024. Review on functionalized pectin films for active food packaging. Packag. Technol. Sci. 37 (4), 237–262. https://doi.org/10.1002/pts.2793.

Tavanaie, M.A., Ghahari, A.H., 2021. A study on melt recycling of bio-based polypropylene/thermoplastic starch compound. J. Appl. Polym. Sci. 138 (43). https://doi.org/10.1002/app.51282.

Thakur, D., Bareen, M.A., Gupta, A., Saha, S., Sahu, J.K., 2024. Frontiers in 3D printing for biobased food packaging. Food Sci. Biotechnol. https://doi.org/10.1007/s10068-024-01770-2.

Tochihuitl-Vázquez, D., Ramírez-Bon, R., Yáñez-Limón, J.M., Martínez-Bustos, F., 2022. Reactive extrusion as a pretreatment in cassava (Manihot esculenta Crantz) and pea (Pisum sativum L.) starches to improve spinnability properties for obtaining fibers. Molecules 27 (18), 5944. https://doi.org/10.3390/molecules27185944.

Troncoso, V.T., Hernandez-Hernandez, O., Victoria Alvarez, M., Ponce, A.G., Mendieta, J.R., Gutiérrez, T.J., 2025. Organocatalytically salicylated starch-based food packaging obtained via reactive extrusion/thermo-molding. Int. J. Biol. Macromol. 320 (Pt 4), 146116. https://doi.org/10.1016/j.ijbiomac.2025.146116.

Vergnes, B., Berzin, F., 2006. Modeling of reactive systems in twin-screw extrusion: challenges and applications. C. R. Chim. 9 (11–12), 1409–1418. https://doi.org/10.1016/j.crci.2006.07.006.

Witek-Krowiak, A., Szopa, D., Anwajler, B., 2024. Advanced packaging techniques—a mini-review of 3D printing potential. Materials 17 (12), 2997. https://doi.org/10.3390/ma17122997.

Yamaguchi, A., Arai, S., Arai, N., 2023. Molecular insight into toughening induced by core-shell structure formation in starch-blended bioplastic composites. Carbohydr. Polym. 315, 120974. https://doi.org/10.1016/j.carbpol.2023.120974.

Yang, Z., Zhang, D., Cai, J., 2024. Robust design of starch composite nanofibrous films for active food packaging: towards improved mechanical, antioxidant, and antibacterial properties. Int. J. Biol. Macromol. 260 (Pt 1), 129329. https://doi.org/10.1016/j.ijbiomac.2024.129329.

Yudhistira, B., Husnayain, N., Punthi, F., Gavahian, M., Chang, C.K., Hsieh, C.W., 2024. Progress in the application of emerging technology for the improvement of starch-based active packaging properties: a review. ACS Food Sci. Technol. 4 (9), 1997–2012. https://doi.org/10.1021/acsfoodscitech.4c00260.

Zabihi, A., Fasihi, M., Rasouli, S., 2021. Microstructural and physical properties of thermoplastic corn starch/polystyrene blend foams affected by different contents and combinations of plasticizers. Res. Sq. https://doi.org/10.21203/rs.3.rs-599880/v1.

Zhang, Q., Yang, W., Zhang, S., et al., 2023. Enhancing the applicability of gelatin-carboxymethyl cellulose films by cold plasma modification for the preservation of fruits. LWT 178, 114612. https://doi.org/10.1016/J.LW.

Zhu, W., Zhang, D., Liu, X., Ma, T., He, J., Qi, D., Din, Z.-ud, Zhou, J., Chen, L., Hu, Z., Cai, J., 2022a. Improving the hydrophobicity and mechanical properties of starch nanofibrous films by electrospinning and cross-linking for food packaging applications. LWT 169, 114005. https://doi.org/10.1016/j.lwt.2022.114005.

Zhu, Z., Liu, W., Ye, S., Batista, L., 2022b. Packaging design for the circular economy: a systematic review. Sustain. Prod. Consum. 32, 817–832. https://doi.org/10.1016/j.spc.2022.06.005.

CHAPTER FOUR

Thermoplastic starch and composites

Jerzy Szałapak, Tomasz Tadeusz Murawski and Zuzanna Żołek-Tryznowska
Faculty of Mechanical and Industrial Engineering, Warsaw University of Technology, Warsaw, Poland

4.1 Introduction

Thermoplastic starch (TPS) has emerged as a promising biodegradable alternative to conventional plastics, driven by the growing demand for sustainable materials. Starch, a naturally abundant polysaccharide, can be transformed into a thermoplastic form through a process known as plasticization, which involves disrupting its crystalline structure under heat and shear in the presence of plasticizers. These additives, such as polyols and other low-molecular-weight compounds, play a crucial role in reducing intermolecular forces and enhancing the material's flexibility. Beyond pure TPS, the development of TPS-based composites incorporating fibers, fillers, or other polymers has expanded their application potential, offering improved mechanical performance and tailored properties for diverse industrial uses. In addition to improving mechanical properties, the development of composites enables enhanced moisture resistance and dimensional stability, as well as modification of barrier properties, which are so important in food packaging. To obtain the required parameters, it is necessary to select suitable fillers, a second polymer matrix component, or modifiers. As a result, TPS composites cover a large group of materials that can be classified according to the type of additives, as well as to the mechanisms of their interaction.

4.2 Starch

Starch is a polysaccharide composed of carbon, hydrogen, and oxygen elements, forming glucose units, and thus has the molecular formula $(C_6H_{10}O_5)_n$. Agricultural plants, such as potatoes, rice, and corn, are the

Starch-Based Sustainable Food Packaging
ISBN: 978-0-443-36346-7
https://doi.org/10.1016/B978-0-443-36346-7.00009-6

primary sources of starch. Starch is a biopolymer consisting of two anhydroglucose polymers: linear amylose and branched amylopectin. Amylose is linked by α-D-(1,4) glycosidic bonds, resulting in a moderate molecular weight (~10^5) and a linear polymer, while amylopectin is linked with α-D-(1,4) and α-D-(1,6) glycosidic bonds, resulting in a highly branched structure with short-chain branches and higher molecular weights (~10^8) (Vilaplana et al., 2012). The chemical and schematic structures are shown in Fig. 4.1. The amylose and amylopectin ratio in starches depends on their origin and source, and it has a significant impact on the physicochemical properties of starch, such as dispersibility and rheological properties (Wang et al., 2024), which can affect starch processing in various industrial processes. The typical range of amylose is between 20% and 30% (Rashwan et al., 2024). Genetic modification can lead to the production of high-amylose starches or high-amylopectin, also known as waxy starches (Zhong et al., 2022).

Starch, as a raw material, has numerous advantages. Starch is odorless, cheap, biodegradable, and abundant. Starch is a renewable raw material

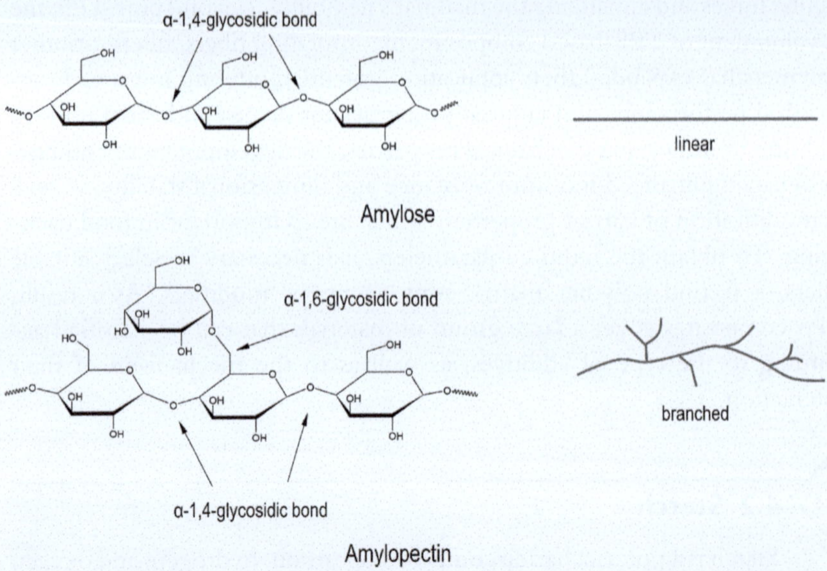

Figure 4.1 Chemical structure and schematic representation of amylose and amylopectin. *Based on Wang et al. (2024).*

because it can be extracted from plant roots (such as potatoes or cassava), tubers, or seeds (such as corn, wheat, or rice) without harming the plants. Native starch is semicrystalline and occurs in the form of granules. Their shape and size are related to their source, but their dimensions are generally between 2 and 100 μm. Starch crystallinity is a complex issue, and amylose and amylopectin exhibit semicrystalline and amorphous regions, respectively. At the same time, the starch granule shows alternating semicrystalline layers and amorphous background layers (Blennow et al., 2000).

Native starch poses certain challenges for industrial applications due to its granules' tendency to readily absorb water, swell, and break down rapidly, leading to a loss of viscosity and the formation of a cohesive paste (Olawoye et al., 2025). Native starch is insoluble in cold water, but it forms a suspension in water exhibiting a shear-thickening effect and does not show any thermoplastic properties. However, when heated, a starch solution in water tends to gelatinize. Gelatinization is a process, whereas the molecular order in the starch granules tends to be irreversibly disrupted by the melting of starch crystallites in a water solution. During this process, water molecules first enter amorphous regions and then move to disruptive forces in crystalline regions. At the same time, new bonds between starch molecules are formed. Gelatinization of starch can be observed at elevated temperatures – above 70 °C – and at water contents greater than 60% w/w (Donmez et al., 2021). At lower water contents (lower than 60% w/w) the melting of starch can be observed instead of gelatinization (Donmez et al., 2021). The gelatinization temperature is strongly influenced by the amylose content (Li, 2022), distribution of amylopectin chain lengths (Matsuki et al., 2003), as well as the type of starch source (Żołek-Tryznowska and Kałuża, 2021). The amylose content in starch has a parabolic effect on the gelatinization temperature. Rice starches with very high (>25%) or very low (<15%) amylose levels generally exhibit higher gelatinization temperatures than those containing an intermediate amount of amylose (15%–25%) (Li et al., 2022). The starch source has a strong influence on gelatinization temperatures. For example, corn starch shows gelatinization at 64.4 °C, whereas potato starch shows gelatinization at 58.2 °C (Żołek-Tryznowska and Kałuża, 2021). Starch gelatinization temperatures, along with amylose contents, are summarized in Table 4.1.

Table 4.1 Gelatinization temperature and amylose content of selected starches.

Starch	Gelatinization temperature in °C	Amylose content in %
Corn (maize)	70 (Liu et al., 2006) 64.4 (Jane et al., 1999)	26 (Liu et al., 2006)
Potato	60.6 to 66.2 (Jaramillo et al., 2023)	6.5%–32.2% (Singh et al., 2018)
Rice	56.4 to 71.1 (Lai, 2000)	32.6 (Varavinit et al., 2003)
Tapioca	60 to 95 (Fei Liu et al., 2016)	37 (Haryanti et al., 2014)
Wheat	58 to 84 (Li et al., 2020)	25 (De Arcangelis et al., 2021)

4.3 Thermoplastic starch

Thermoplastic starch (TPS), unlike native starch, exhibits thermoplastic behavior. To develop TPS, starch must be processed with polar plasticizers at an elevated temperature, accompanied by additional shear forces (Si and Zhang, 2024). Starch, a complex material with a heterogeneous structure, can be transformed into a thermoplastic form through the combined action of heat and mechanical forces (Foret et al., 2023). This action, when combined with water or other plasticizing agents, disrupts its granular architecture, rendering it more amorphous and more processable. Therefore, the hydrogen intra- and intermolecular in starch granules are disrupted, resulting in the formation of hydrogen bonds between starch and plasticizers. Plasticizers can penetrate the starch matrix and form with –OH groups on starch molecules of new hydrogen bonds. The formation of hydrogen bonds between the plasticizer and starch molecules can be confirmed by infrared spectroscopy, as evidenced by the shift of the broad band at approx. 3300 cm^{-1} related to the formation of inter- and intramolecular hydrogen bonds in and between starch molecules, into lower wavenumbers (Murawski et al., 2025).

The production of TPS is closely associated with the incorporation of various plasticizers, such as glycerol, sorbitol, urea, and other low-molecular-weight polar compounds. The appropriate selection of plasticizers enables achieving targeted properties, such as plasticity, flexibility, and reduced friction during production (Majeed et al., 2023), while also contributing to increased material flexibility (Abhitha et al., 2025). The development of TPS involves dissolving native starch in water, adding the plasticizer, and then heating above the gelatinization temperature. In this case, water is treated rather as a solution carrier that is subsequently evaporated from the material than as a plasticizer (Gao et al., 2021). It is essential to note that water molecules can also penetrate the starch matrix and form stable hydrogen bonds. It is also known that starch films can absorb water, and their properties are affected by the humidity in which they are stored (Basiak et al., 2018). Therefore, the properties of TPS are primarily due to the synergistic effects of water and the plasticizer.

4.3.1 Thermodynamics of plasticizing

According to the thermodynamic relation:

$$\Delta G = \Delta H - T\Delta S \tag{4.1}$$

The process is spontaneous when the free energy is zero or negative ($\Delta G \leq 0$).

For the mixing process thermodynamic relation (Eq. 4.1) is expressed as follows:

$$\Delta G^M = \Delta H^M - T\Delta S^M \tag{4.2}$$

where ΔG^{M} – Gibbs free energy of mixing, indicating whether the mixing is spontaneous; ΔH^{M} – enthalpy of mixing – the amount of heat energy exchanged during the mixing of components, and ΔS^M – entropy of mixing, which is a measure of the increase of disorder after mixing.

From a thermodynamic point of view, plasticizers lower the material's glass transition temperature ($\mathrm{T_g}$). In starch, this effect is achieved by disruption of the double-helix structure, the introduction of weaker hydrogen bonds, both inter- and intramolecular, and the formation of novel hydrogen bonds between the plasticizer and starch molecules, which leads to an increase in the system's entropy. The plasticizer penetrates between starch chains, loosening their ordered structure. Higher values of ΔS are related to "greater disorder" and greater segment mobility of chains.

Melting disrupts the crystalline structure of starch, resulting in gelatinization and an increase in entropy. The incorporation of polar plasticizers, such as water or glycerol, reduces the hydrogen bond strength, making the process spontaneous ($\Delta H < 0$) and enabling the formation of a homogeneous TPS system. External factors such as heat, shear, or microwaves accelerate gelatinization through three main steps: (i) swelling and segmentation of starch granules by plasticizers, (ii) breakdown of amylopectin double helices and promotes formation of starch–plasticizer bonds, and (iii) reorganization of chains into entangled topological structures, which upon cooling form an amorphous TPS matrix (Si and Zhang, 2024). Although the formation of TPS is thermodynamically favorable ($\Delta G < 0$ due to increased entropy), it is not spontaneous in practice. The process requires heat and often mechanical energy to overcome kinetic barriers associated with breaking the crystalline structure and enabling plasticizer diffusion.

According to Hilderbrand theory, the enthalpy of mixing can be expressed by following equation (Hansen and Poulsen, 2007):

$$\Delta H_M = \varphi_1 \varphi_2 V^M (\delta_1 - \delta_2)^2 \tag{4.3}$$

Where φ_1 and φ_2 are the molar fractions of the solvent and the solute; δ_1 and δ_2 are the solubility parameters of the solvent and the solute and V^M is the volume of the mixture.

The solubility parameter, δ, can be expressed as (Hansen and Poulsen, 2007):

$$\delta = \left(\frac{E_{coh}}{V} \right)^{1/2} \tag{4.4}$$

Where V^M is the molar volume of the solvent, and E_{coh} is the energy of vapourization, that is, the energy of evaporation of a saturated liquid to the ideal gaseous state, at $T =$ const. The cohesive energy of substances can be predicted using group additivity methods. There are several methods for calculating E_{coh} and solubility parameters for solvents and polymers. They are widely discussed in *Properties of Polymers* (*Van Krevelen and Te Nijenhuis, 2009*). The solubility parameter unit is $MPa^{½}$ or $(cal \cdot cm^{-3})^{1/2}$. The solubility parameter allows predicting the solubility of a polymer because knowing the polymer and solvent parameters determines whether the polymer will dissolve: if $\delta_1 \approx \delta_2$, the polymer should dissolve; however, if $\delta_1 >> \delta_2$ or $\delta_1 << \delta_2$, the polymer will not dissolve in the solvent. For native starch, the solubility parameter values range from 12 (Dogenski et al., 2020) to 14.5 (Si and Zhang, 2024) $(cal \cdot cm^{-3})^{1/2}$. The enthalpies of the starch gelatinization process are positive ($\Delta H > 0$), to ensure a spontaneous process of mixing, $\Delta G < 0$, high entropy, and high

Table 4.2 Solubility parameters of native starch, water, and glycerol.

Compound	Solubility parameter in $(cal \cdot cm^{-3})^{1/2}$
Native starch	12 (Dogenski et al., 2020)
	14.5 (Si and Zhang, 2024)
Water	23.4 (Hansen and Poulsen, 2007)
Glycerol	16.5 (Si and Zhang, 2024)
	17.6 (Hansen and Poulsen, 2007)

temperature are required. Water is not a good plasticizer (see Table 4.2), but it causes swelling of the starch granules. Due to its low molecular weight, it can also penetrate between the starch chains, breaking the hydrogen bonds. On the other hand, glycerol has a solubility parameter similar to that of native starch and is recognized as a good plasticizer. Consequently, the enthalpy of mixing (ΔH^M) for glycerol-plasticized TPS is lower than that for water-plasticized TPS, increasing the thermoplastic transition temperature (Si and Zhang, 2024).

4.3.2 Plasticizers

Starch-based films represent an eco-friendly packaging option for products with limited shelf life. In contrast to fossil-based packaging, they should be capable of degrading under home composting conditions after disposal. Nevertheless, to ensure their suitability for food packaging applications, some modifications are required to enhance mechanical strength and barrier properties. As mentioned earlier, plasticizers can increase the material's flexibility and elongation.

A plasticizer is a type of polymer additive that is non-volatile and characterized by a low molecular mass. By selecting the appropriate set of plasticizers, desired properties such as plasticity, flexibility, and reduced friction during manufacturing can be achieved (Majeed et al., 2023). These plasticizers, used in biodegradable materials, must also be biodegradable and environmentally friendly. Due to the molecular structure of starch and its tendency to form hydrogen bonds between starch molecules, compounds with polar groups, such as hydroxyl, carboxyl, and amide groups, can be successfully used as plasticizers. Water is found to be the primary plasticizer for TPS; however, due to its low boiling point, it can volatilize during processing (Montilla-Buitrago et al., 2021).

Therefore, less volatile compounds are introduced. Most commonly, glycerol is applied as a plasticizer for starch-based films. Glycerol (propane-1,2,3-triol) contains hydroxyl groups in the molecule, which are responsible for interaction with starch. Apart from glycerol, other polyols are used, for example sugars (glucose, fructose, maltose, sucrose) (Ploypetchara and Gohtani, 2018; Bayer et al., 2022)); sugar alcohols such as sorbitol (Rahmawati et al., 2019; Lim et al., 2020; Hazrati et al., 2021; Gurunathan et al., 2025), mannitol (Aydin and Ilberg, 2016; Saberi et al., 2017) and glycol derivatives (Żołek-Tryznowska and Cichy, 2018; Zhou and Wang, 2021; Biran Ay and Ertürk, 2022). Various mono-, di-, and polyalcohols were investigated as plasticizers for potato starch-based materials by Alee et al. (2023). According to Alee et al., the number of hydroxyl groups and the molecular architecture of alcohol affect the plasticizing performance. Greater compatibility and superior plasticizing efficiency are related to polyols containing multiple —OH groups. On the other hand, compounds with fewer —OH groups or extended alkyl chains tend to exhibit poor compatibility with starch. Zhang and Han investigated the impact of monosaccharides (mannose, glucose, fructose) and polyols (glycerol, sorbitol) on the mechanical, moisture-related, and thermomechanical properties of starch-based edible films (Zhang and Han, 2006). Films containing incorporated monosaccharides were stronger and more extensible than those containing polyols and exhibited lower water vapor permeability at higher plasticizer concentrations.

Furthermore, derivatives containing $-NH_2$ groups: amines (thriethanolamine, ethanolamine), amides (formamide), urea (Ma and Yu, 2004; Ma et al., 2004; Ma et al., 2006; Wang et al., 2014; Edhirej et al., 2017).

Last but not least, carboxylic acids such as citric, tartaric, and oxalic acids are effective crosslinking agents for developing TPS (Gebresas et al., 2023). Finally, for plasticizing purposes, ionic liquids or deep eutectic solvents can also be applied (Romano et al., 2025). Zdanowicz reported using deep eutectic solvents composed of urea, polyols such as glycerol and sorbitol, and monosaccharides like glucose and fructose to dissolve and plasticize potato starch-based materials (Zdanowicz, 2021). It was found that mixtures of urea and glycerol produced the most flexible films, achieving elongation at break values above 200%.

The synergic effect of plasticizers was also reported. Ma et al. reported that composite plasticizers exhibit greater plasticizing capacity than glycerol or erythritol alone (Ma et al., 2025). By selecting the appropriate set of plasticizers, desired properties such as plasticity, flexibility, and reduced friction during manufacturing can be achieved (Majeed et al., 2023). These plasticizers, used in biodegradable materials, must also be biodegradable and environmentally friendly. Due to the molecular structure of starch and its affinity to form hydrogen bonds between starch molecules and polyols, various polyols, particularly those with hydroxyl and other polar groups, can be successfully used as plasticizers.

The plasticization effect influences many material properties. However, in the literature, three main parameters are widely discussed: thermal properties (T_g), mechanical properties (tensile strength, elongation at break, etc.), and barrier properties (gas permeability and water vapor permeability). The plasticizer content influences the properties of starch films. Increasing plasticizer content generally enhanced the films' physical, thermal, and mechanical properties (Edhirej et al., 2017). Some plasticizers and their influence on TPS properties are shown in Table 4.3.

4.3.2.1 Antiplasticization effect

The addition of plasticizers can transform a glassy polymer into a thermoplastic elastomer by increasing elongation at break, while simultaneously decreasing the maximum tensile stress. The opposite phenomenon to plasticization is known as "antiplasticization", which occurs at low plasticizer concentrations and is characterized by an increase in tensile strength and a reduction in elongation at break (Mascia et al., 2020). The user can observe this effect as an increase in material stiffness rather than a decrease, which is the opposite of what plasticizers usually do. As mentioned earlier (see Chapter H4), the plasticizer reduces intermolecular forces in the polymer by increasing chain mobility and lowering the glass transition temperature (T_g). In the antiplasticization effect, low-molecular-weight plasticizers can fill the free volume between polymer chains. This effect reduces the chain's mobility, increases its density, and results in increased stiffness, brittleness, and reduced flexibility. The antiplasticization effect occurs at a specific concentration and temperature. At elevated temperatures and/or increased concentrations, these compounds exhibit plasticizing behavior (Job, 2018). Generally, for the fossil-based polymers such as polyvinylchloride, or polycarbonate, the antiplasticization effect is observed at low plasticizer concentrations, lower than $<15\%$ (Lourdin et al., 1997). The mechanism of the antiplasticization phenomenon is not fully understood; however, it can be explained by free-volume and interaction theories.

Table 4.3 Overview of plasticizer types and their influence on TPS materials.

Starch source	Plasticizer	Thermal properties	Mechanical properties	Barrier properties	Reference
Corn starch	Glycerol & sorbitol	Sorbitol-plasticized films show enhanced thermal stability compared to glycerol	Glycerol increases flexibility (elongation); sorbitol provides higher tensile strength and mechanical support	Sorbitol results in lower water vapor transmission (better barrier) compared to glycerol	Sirbu et al. (2024)
Potato starch	Malic acid, Isoleucine, Asparagine	Malic acid acts as a strong anti-retrogradation agent, stabilizing the amorphous phase	Malic acid inhibits recrystallization, preventing brittleness during storage.	Higher moisture uptake for malic acid compared to urea/glycerol blends.	Niazi et al. (2015)
Sugar palm starch	Glycerol & sorbitol	T_g slightly decreases as plasticizer concentration increases from 15% to 45%	Tensile strength decreases while elongation at break increases with increasing plasticizer content	Water vapor permeability (WVP) increases as plasticizer concentration is raised	Sanyang et al. (2015)

Wheat starch	Fructose, urea, sorbitol	Fructose showed higher thermal performance with lower mass loss during heating.	Fructose yields the highest tensile strength; urea leads to high elongation but high moisture content.	Films with fructose showed the lowest water absorption and improved water resistance.	Mohammed et al. (2023)
Cassava starch	Glycerol & Honey	Honey addition improves the flow of suspensions; T_g typically decreases with glycerol.	Honey increases tensile strength and resistance to stress but reduces extensibility.	Honey reduces water vapor permeability and moisture content compared to pure glycerol.	Mali et al. (2006)
Anchote starch	Ionic liquids (1-ethyl-3-methylimidazolium acetate)	Ionic liquids decrease molecular order, leading to lower T_g than traditional polyols	Lower tensile strength compared to sorbitol; ionic liquids significantly increase free volume and flexibility	Enhanced water absorption due to the highly hydrophilic nature of ionic liquids.	Abera et al. (2020)

The effect of the plasticizer on the free volume is shown schematically in Fig. 4.2.

Free volume is the space or pores between the amorphous structure of chains or molecules of the polymer. Those spaces allow for the movement of molecules within the bulk structure of the polymer material. Above the T_g temperature, the molecules can move, rotate, or bend; under the T_g temperature, the molecules are tightly packed, and movement is not possible, and they are "frozen" in the glassy state. The small concentration of a low-molecular-weight compound – a plasticizer – instead of pushing the molecules apart, sits in the spaces between the chains, reducing free volume and thereby decreasing chain movement.

At the same time, the plasticizer forms stable, novel hydrogen bonds with the polymer macromolecules, thereby restricting chain mobility. Plasticizer molecules bind to polymer chains at low concentrations, limiting short-range rotational motions.

The effect of antiplasticization is also observed in biobased and natural-based polymers such as TPS. The absence of an antiplasticization phenomenon reported in the literature for various plasticizers of TPS may be attributed to the limited range of diluent concentrations investigated, as low-molecular-weight compounds are typically employed at levels sufficient to observe film flexibility. Chang et al. demonstrated that glycerol exhibits antiplasticizing behavior on film modules when incorporated at a low concentration (2.5%) in dry starch films (water activity ≤ 0.22). Antiplasticization induced by water was evident exclusively in mechanical properties measured under high deformation conditions (i.e., tensile

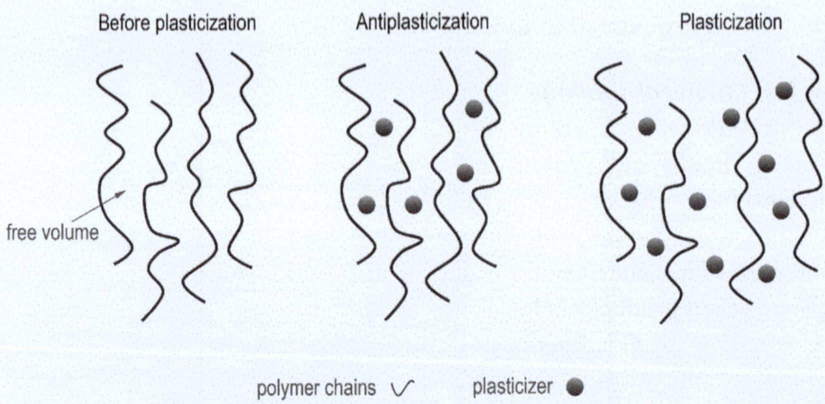

Figure 4.2 The effect of plasticizer on the free volume in polymer bulk material.

strength) with films containing 10–15% glycerol (Chang et al., 2006). Conversely, for TPS with a larger content of plasticizers (i.e., >50%), the amorphous regions attract more moisture and swell, leading to the formation of a more flexible materia*l* (Ben et al., 2022).

4.4 TPS-based composites – classification framework

Composite materials are developed by combining two or more materials to obtain a novel material with unique structural features. Generally, composite materials are developed from a matrix and reinforcements in which the matrix acts as an adhesive binding the filler materials (Mishra et al., 2022). Thermoplastic starch (TPS)-based composites are an important class of biodegradable materials, developed primarily to overcome the limitations of pure TPS: low mechanical strength, high moisture sensitivity, and limited dimensional stability. At the same time, it is essential to maintain high bio-based content and compatibility with melt processing technologies. In food packaging applications, TPS composites are not just a simple substitute for pure biodegradable polymers, but multiphase materials in which morphology, interphase interactions, and processing history are important (Avérous, 2004; Avérous and Pollet, 2012).

Early scientific publications on starch plasticization and TPS processing indicated that TPS can serve as a continuous matrix or a dispersed phase, depending on plasticizer content, composition, and shear conditions during extrusion. Research on composites made of natural fibers and biopolymers has clearly shown that interfacial quality often outweighs the intrinsic properties of the individual components (George et al., 2001). Subsequent studies have confirmed this observation, demonstrating that the properties of TPS-based materials must be analyzed using the principles of composite materials, as mechanical, barrier, and durability properties are primarily determined by phase continuity and interfacial adhesion (Avérous, 2004). For this reason, TPS-based composites for food packaging applications can be classified according to the nature and function of the filler introduced into the matrix. Such classification reflects not only the strict limits of the composite composition but also the dominant material design strategies, providing a framework for organizing the literature on TPS-based composites (Avérous and Pollet, 2012).

The first large group of materials includes TPS composites reinforced with natural fillers. The fillers in these composites are solid phases of biological origin, such as lignocellulosic fibers, agricultural waste, or polysaccharide-based nanofillers. The use of these materials is primarily aimed at improving mechanical properties: stiffness and strength, and modifying stress transfer mechanisms. The effectiveness of the reinforcement depends mainly on the filler's geometry and the length scale. It should be noted that fibrous, particulate, and nanostructured fillers have a clear impact on the morphology and other properties of the composite. In TPS composites intended for food packaging, cellulose- and starch-based fillers are predominant (George et al., 2001; Avérous and Pollet, 2012).

Another important group of materials is composites based on TPS blends with biodegradable polymers – these include primarily polylactic acid, poly(butylene adipate-co-terephthalate), poly(ε-caprolactone), or poly(butylene succinate). In composites, the blend itself serves as a matrix with modified properties. TPS ensures biodegradability and lower costs, while the polyester phase improves mechanical properties and reduces moisture sensitivity. Research shows that the properties of the blend depend primarily on interphase compatibility and phase morphology, and not solely on composition. Therefore, an essential element in the design and preparation of such blends is the use of reactive compatibilization and control of the physical state of TPS (Avérous, 2004; Avérous and Pollet, 2012).

The last group of materials includes TPS composites containing functional fillers. Such fillers are used primarily to modify performance properties other than mechanical parameters. These include mineral fillers, clays, and nanoclays, as well as hybrid organic-inorganic materials (e.g., nanocellulose), commonly used in biopolymer-based packaging (Avérous and Pollet, 2012; Rhim et al., 2013). Their action is primarily based on the control of morphology and interfaces, localizing at phase boundaries, thereby modifying the crystallization, barrier properties, thermal stability, and rheological behavior of TPS composites, even if they constitute a small percentage of the material (Jayarathna et al., 2022; Muñoz-Gimena et al., 2023). Their action primarily involves modifying diffusion paths and limiting chain mobility rather than a load-transfer mechanism (Surendren et al., 2022).

The above materials overlap, and in many cases, the designed composites are combinations of TPS/polyester mixtures with natural reinforcement and a functional filler. However, it should be emphasized that this categorization reflects the main design principles of melt-processed TPS-based composites used in food packaging, and their properties depend primarily on morphology and interfacial chemistry rather than on the chemical identity of the individual components. A more detailed analysis is presented below, including composites with natural fillers (section), TPS/polyester matrix blends (section), and TPS composites with functional fillers (section).

4.4.1 TPS-composites enriched with plant-based fillers

TPS, considered an environmentally friendly, renewable material, has been studied for several decades, especially in the context of food packaging industry applications. One of its main applications is food packaging. Due to material limitations, particularly in mechanical properties and moisture resistance, large-scale research has begun on the development of TPS-based composites. Biologically derived fillers are considered reinforcing phases due to their availability, low cost, and compatibility with sustainable management principles. In this chapter, the analysis of TPS-based composites is limited to materials for use in food packaging technologies. Such an approach ensures that both academic and industrial perspectives are considered.

Research on TPS-based composites for food packaging applications has primarily concentrated on a limited number of natural reinforcement types. These materials can be broadly classified into three main categories.

- Particulate fillers, for example, spherical
- Fibrous fillers
- Nanostructured fillers.

This classification allows us to show not only the differences arising from the materials' geometry, but also the various mechanisms of structural reinforcement in composites. Although alternative biological fillers, such as protein- or animal-derived materials, are sometimes discussed in broader review contexts, reports on their use as separate reinforcing phases in melt-processed.

As a result, the ways of improving such composites for food packaging are mostly limited to the main types of fillers, namely plant-based fillers such as lignocellulosic fibers, solid particles derived from agricultural waste, and nanomaterials, including nano- and microcellulose in the form of fibers and crystals (CNF, CNC, MFC) as well as starch nanocrystals. When incorporated into a TPS matrix, each reinforcement class enhances composite performance through distinct mechanisms, largely governed by filler morphology and interfacial interactions.

Fibrous fillers, which aim to improve stress transfer, are typically derived from lignocellulosic biomass. Particulate fillers are usually obtained from agro-industrial waste (e.g., wood dust) and can modify barrier properties and mechanical stiffness, as well as affect biodegradation. They also typically reduce the composite's final cost. Fibrous fillers improve mechanical properties mainly through effective load transfer, while particulate fillers act as rigid inclusions affecting stiffness, barrier performance, and degradation behavior. The last type – nanostructured fillers, including nanocellulose, owing to their high specific surface area and extensive interfacial contact, offer the highest reinforcement efficiency even at low filler contents. Even in small amounts, they can significantly affect molecular mobility, transport properties, and phase morphology.

Overall, such classification enables a consistent analysis of TPS-based composites from both scientific and industrial perspectives. Since additives derived from animal or protein waste are scarcely represented in this field, current developments in natural TPS-based composites are largely focused on polysaccharide- and lignocellulose-based materials. This scope provides a logical transition to the discussion of TPS-based composites containing functional fillers, addressed in the following sections of this chapter.

4.4.1.1 *Fiber fillers in TPS-based composites*

Fiber fillers are among the most frequently investigated reinforcement materials for TPS composites for use in food packaging technologies. This type of filler, usually lignocellulosic in nature, is typically derived from plant fibers, agricultural waste, or trees. In TPS composites, their main task is to improve mechanical parameters. The addition of fibers increases the modulus of elasticity and improves load transfer by leveraging the material's significant length-to-diameter ratio, which promotes better stress transfer from the TPS-based matrix to the filler.

Early research on TPS composites showed that the use of natural fibers, such as hemp or sisal, significantly increases the composite's tensile strength, together with stiffness, compared to the pure matrix material. The enhancement in mechanical performance was attributed to adequate filler dispersion in the matrix and adequate interfacial adhesion (Gironès et al., 2012). Similar effects were observed when using fibers derived from agricultural waste, such as sugarcane bagasse. These studies showed that fiber size and morphology have the primary influence on the mechanical properties of the composite (Dos Santos et al., 2018). It has also been shown that finer fibers allow for a more even distribution of stress. In comparison, thicker fibers can act as stress concentrators – composites of this type tended to crack prematurely.

Due to their chemical characteristics, wood fibers are considered a separate, functional subgroup within lignocellulosic fillers for use in TPS composites. Wood fibers usually provide higher stiffness and dimensional stability, but at the same time, reduce tensile strength (Grylewicz et al., 2019). The observed behavior can be attributed to the restricted mobility of the matrix polymer chains and of interactions among the fibers, starch, and the plasticizing system. The efficiency of wood fiber reinforcement in the composite is also strongly dependent on processing conditions, which can affect the wettability and interfacial adhesion in the TPS matrix.

A separate, important issue is the hydrophilic nature of lignocellulosic fibers, which increases water sensitivity and dimensional instability in humid conditions. At the same time, a high fiber content can negatively affect composite processability, leading to poorer surface quality and even fiber breakage during extrusion. As a result, despite the undoubted advantages of fibers as fillers in TPS composites, their content in composites is typically kept moderate. They are used in quantities that improve mechanical parameters while minimizing negative impacts on processability or environmental stability. Natural fiber fillers are therefore considered effective in enhancing mechanical performance, despite their sensitivity to moisture and processing conditions. This is possible with proper control of fiber morphology, their content in the composite, and analysis of interfacial interactions. However, due to limitations, alternative solutions, such as molecular or nanostructural systems, have attracted research. These are described in the following sections.

4.4.1.2 Particulate fillers in TPS-based composites

Particulate fillers are the second class of additives for TPS. They differ from fibrous fillers in terms of both their reinforcement mechanism and morphology. Particulate additives are typically characterized by a low length-to-diameter ratio, which means that they act mainly as rigid inclusions, evenly dispersed in the polymer matrix, and therefore do not transfer loads in the manner known from fibrous fillers. The properties of the composite can be influenced by particle size, chemical composition, and dispersion quality in the matrix.

In terms of TPS composites, agricultural waste such as nut shells, plant residues, and ground biomass is particularly often studied. Studies of walnut shell particles, which are rich in lignin, as additives to TPS polymers have shown that these fillers provide higher material stiffness and lower oil permeability. This makes such composites particularly interesting for the food packaging industry associated with fatty products. However, as was to be expected, an increase in the content of the particle filler negatively affects the material's resistance to stretching or tearing – this is the result of the presence of poorly deformable, rigid particles acting as stress concentrators in the material (Quilez-Molina et al., 2022).

In addition to filler content, particle size also significantly affects the composite's mechanical properties. As in the case of fibrous fillers, finer particles allow for better dispersion and, consequently, a more homogeneous material. This results in a more uniform stress distribution and a reduced density of structural defects, rather than in effective load transfer. Larger, coarse particles often promote the formation of pores, voids, and interfacial delamination, which impairs the mechanical properties of the composite. This is particularly evident in incompatible composites, where interfacial adhesion is mainly based on physical interactions (Surendren et al., 2022).

From the perspective of processors, particulate fillers are generally very attractive materials. In addition to their low cost, their agricultural waste origin makes them easier to extrude than fibrous materials. They are much less susceptible to cracking and orientation effects. However, if their share in the composite is too high, they can increase the melt viscosity, thereby impairing flow stability, especially when dealing with very irregular shapes or significant surface roughness. In addition, as with fiber additives, most particles are hydrophilic, so again, increased moisture absorption was observed, which negatively affects barrier properties and dimensional stability in humid conditions. It should be noted that the impact of particulate fillers on gas and vapor barrier properties strongly depends on dispersion quality and interfacial adhesion and may be either beneficial or detrimental. However, in general, natural particulate fillers are inexpensive – and therefore cost-effective – and sustainable material for modifying the mechanical properties (especially stiffness) and selected barrier properties of TPS composites. Their main limitation compared to fibrous fillers is their lower reinforcement efficiency, rather than processing constraints. For this reason, due to greater limitations than in the case of fibrous fillers, such as the need for careful optimization of particle size and roughness, content, and interaction with the matrix, as well as similar hydrophilicity, work has been carried out on other solutions. The next section presents composites based on nanostructured fillers, which aim to combine the processing advantages of solid particles with the mechanical properties of fibers.

4.4.1.3 Nanostructured fillers in TPS-based composites

Nanostructured fillers, the last of the types of materials described in this section, are among the most efficient materials in terms of the ratio of reinforcement efficiency to percentage content in TPS composites. They combine high mechanical property improvement efficiency with a low content ratio relative to the matrix material. Modifications of TPS with nanomaterials such as cellulose nanofibers (CNF), cellulose nanocrystals (CNC), microfibrillated cellulose (MFC), or starch nanocrystals – unlike the fibrous or particulate fillers described in previous sections – are characterized by a significant specific surface area and strong interfacial interactions with the TPS matrix (Zdanowicz and Sałasińska, 2023). As a result, they not only directly influence the mechanical properties of the composite but also modify its morphology, barrier properties, and moisture sensitivity.

Among the fillers, cellulose-based nanomaterials attract particular attention. Composites based on cellulose nanofibers or nanocrystals constitute the vast majority of the tested materials. Researchers consistently demonstrate an apparent increase in elastic modulus or strength with a filler content not exceeding 5% by weight. This behavior is generally attributed to the transfer of stresses via hydrogen bonds between cellulose surfaces and the TPS matrix. TPS/CNF and TPS/CNC consistently show a significant increase in elastic modulus and strength with a filler content typically below 5% by weight, which reflects the effective transfer of stresses through hydrogen bonds between hydroxyl groups on cellulose surfaces and the TPS matrix (Ghanbari et al., 2018). Interestingly, unlike micro-scale fibers, nanocellulose reinforces the composite without significantly reducing its plasticity parameters, as it can form continuous or partially permeable structures in the TPS phase rather than acting as stress concentrators. In addition, nanocellulose improves the barrier properties of TPS composites by creating tortuous diffusion paths in the matrix and reducing gas and vapor permeability, which is particularly important in food packaging applications. González et al. assume that this is due to both a physical barrier to the transport of permeating substances and the limited mobility of polymer chains resulting from strong interactions between the filler and the matrix (González et al., 2020).

Starch nanocrystals are the second type of filler in TPS composites. They are characterized by excellent compatibility with the matrix due to their chemical similarity, allowing for uniform dispersion under appropriate processing conditions (García et al., 2011). The presence of starch nanocrystals in the composite increases stiffness, improves barrier properties, and allows a high content of bio-based components to be maintained while avoiding the interfacial incompatibility problems common with inorganic fillers. However, the effectiveness remains strongly dependent on the level of dispersion and plasticization, which can affect the ability to form reinforcing structures, similarly to nanocellulose-based compositions (García et al., 2011).

In addition to the advantages of using nanostructured natural fillers in TPS composites, attention should also be paid to their limitations and the technological challenges associated with processing. First, it is worth noting the problem of nanoparticle agglomeration during extrusion, which impedes achieving uniform dispersion. The use of increased shear during the process improves dispersion, but at the same time, degrades both the matrix and the filler. Dispersion state and filler surface chemistry also significantly affect the effectiveness of moisture protection. Poorly dispersed nanofillers can increase moisture absorption by creating interfacial defects that facilitate water penetration. The mere presence of the filler also increases the viscosity of the molten composite, which complicates the extrusion of films or sheets on an industrial scale.

Overall, nanostructured natural fillers improve the parameters of TPS composites to the greatest extent among natural fillers. They enable simultaneous improvement of mechanical and barrier properties at low filler content. From a durability perspective, well-dispersed nanostructured fillers can reduce the moisture sensitivity of TPS composites by limiting water diffusion and stabilizing the polymer network. Thanks to these features, nanostructured additives, which sit on the boundary between conventional natural reinforcements and functional fillers, serve as a conceptual bridge to TPS-based composites containing multifunctional or hybrid fillers. They are discussed in the following sections of this chapter.

4.4.2 TPS composites enriched in polymer blends

In the context of TPS-based composites, mixtures of polymers with biodegradable polyesters are primarily considered as matrices whose morphology, interfacial properties, and mechanical properties significantly influence composite performance. TPS blends prepared by melt blending with biodegradable polyesters represent an important class of materials for the development of sustainable packaging systems. The creation of such polymer compositions appears to be an essential strategy for modifying the matrix's properties, even before the addition of reinforcing or functional fillers. In this context, TPS is most often combined with aliphatic and aliphatic-aromatic polyesters, with particular emphasis on.

- polylactic acid (PLA),
- poly(butylene adipate-co-terephthalate) (PBAT),
- poly(ε-caprolactone) (PCL)
- PCL (PBS).

Each of the above materials has different mechanical, thermal, and interfacial properties. The chemical structures of polyesters are shown in Fig. 4.3. These blends are not considered end materials but rather matrices for TPS-based composites for applications in the food packaging industry, due to their specific phase morphology, interfacial chemistry, and moisture sensitivity. These features significantly impact the functional performance and durability of the composites (Yamaguchi et al., 2024).

Rigid systems such as TPS/PLA provide a useful reference for examining interfacial incompatibility and highlight the need for reactive compatibilization strategies. In contrast, elastomeric blends, exemplified by TPS/PBAT, offer enhanced plasticity, albeit with greater moisture sensitivity. TPS/PCL blends demonstrate that high mechanical strength can be achieved without extensive chemical modification, relying instead on hydrogen bonding and effective plasticization, although their long-term stability remains closely linked to starch retrogradation and environmental conditions. TPS/PBS systems, meanwhile, typically require either reactive compatibilization or targeted modification of the TPS phase to achieve balanced mechanical and barrier properties. Within this context, TPS/polyester blends are best regarded as tunable matrix materials rather than final packaging products, providing a platform for incorporating and reinforcing with functional fillers, which are addressed in the following section.

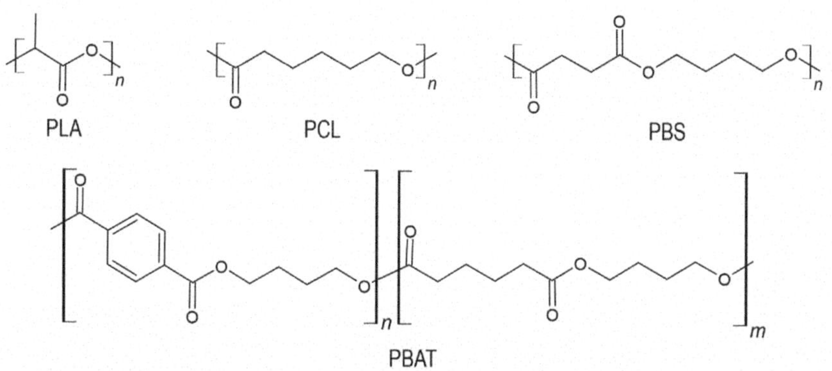

Figure 4.3 Chemical structures of biodegradable polyesters used in TPS composites.

4.4.2.1 *TPS/PLA blends – interfacial incompatibility and compatibilization strategies*

Early studies on TPS and polylactic acid (PLA) blends indicated a lack of intercompatibility between the two polymers. All attempts to process using commonly known melt-blending methods, such as twin-screw extrusion, resulted in a coarse morphology with phase separation and low interfacial adhesion, due to the thermodynamic immiscibility of TPS with PLA (Huneault and Li, 2007). Thermal analyses and scanning electron microscopy (SEM) revealed large, often micrometer-sized TPS domains dispersed throughout the PLA matrix, resulting in reduced stress transfer, lower mechanical parameters, and reduced plastic deformation capability (Shin et al., 2011). In addition, TPS's hydrophilicity led to increased moisture absorption, further deteriorating interfacial adhesion and accelerating property loss during long-term storage. The presence of TPS also significantly affected PLA crystallization, with changes in crystallinity and thermal transformations primarily regulated by phase morphology rather than by the mixture's composition (Li et al., 2023). These early studies indicated that obtaining materials for packaging applications – not only for food – was not possible by simply controlling the TPS-to-PLA weight ratio. Therefore, targeted control of interfacial interactions began to be considered.

To address the poor interfacial adhesion characteristic of TPS/PLA blends, work began on compatibilization to stabilize the morphology and improve mechanical properties. The most effective method proved to be compatibilization, acting directly at the TPS-PLA phase boundary during melt processing – thereby reducing interfacial tension and inhibiting the coalescence of TPS domains under shear (Huneault and Li, 2007).

Reactive compatibilization, based on the functionalization of maleic anhydride (MA), is the most widely studied and best documented for TPS/PLA blends. By grafting MA onto PLA chains or by introducing MA-containing compatibilizers during melt blending, esterification reactions with TPS hydroxyl groups are promoted, thereby forming interfacial copolymers. These reactions reduce the TPS domains and increase interfacial adhesion, leading directly to a marked improvement in mechanical properties such as plasticity and tensile strength (Huneault and Li, 2007). It is worth noting that the improvement in parameters relative to incompatible blends is mainly due to interfacial chemical interactions rather than changes in mass composition, indicating the dominant role of interfacial engineering.

The second most popular compatibilization method is the use of chain extenders with an epoxy function. Multifunctional epoxy compounds can react simultaneously with both TPS hydroxyl groups and PLA terminal carboxyl groups during melt processing. They then induce chain extension and partial branching. This mechanism, in addition to improving interfacial adhesion, increases the melt viscosity and strength, thereby enhancing morphological stability during extrusion and film processing (Li and Huneault, 2011). In addition, compared to maleic anhydride-based compatibilization, epoxy compatibilizers often offer broader processing options. However, excessive reactivity can negatively affect processability if not carefully controlled.

Milder compatibilization methods can be applied with active or weakly interacting additives such as poly(ethylene glycol) (PEG). PEG can improve phase dispersion and impact resistance by increasing chain mobility and promoting better interfacial contact without forming covalent bonds, as described by (Ferrarezi et al., 2013). However, the interfacial stabilization achieved by this approach is generally less durable than that from reactive compatibilization, especially with respect to long-term morphological stability and moisture resistance.

The development of an effective method for obtaining a TPS/PLA composition led to further intensive work on expanding the range of properties through research on two-component blends and, later, three-component or hybrid preparations. In two-component blends without fillers, mechanical properties are largely determined by the phase morphology. A higher TPS content usually reduces the tensile strength and stiffness of the material, while increasing its plasticity, depending, of course, on the effectiveness of compatibilization (Yokesahachart et al., 2024). However, a TPS content that is too high in the mixture negatively affects the blend's strength and film-forming ability. Therefore, very careful optimization of the processing and plasticization conditions of the plasticizer content is required (Shirai et al., 2013).

However, compatibilization did not solve the problem of moisture sensitivity – even in compatible compositions, this is a limitation, as water absorption can plasticize the starch phase and thus change the interfacial adhesion, leading to a gradual deterioration of mechanical parameters during aging (Shin et al., 2011). These limitations became one of the main reasons for the development of ternary blends containing a secondary, more flexible and biodegradable polymer. Early work on TPS/PLA blends modified with poly(eadipate-co-terephthalate) showed that the addition of a third component can reduce brittleness and increase elongation at break by changing phase continuity and energy dissipation mechanisms (Ren et al., 2009).

In summary, research on TPS/PLA blends has made them the benchmark for starch-based blends. Despite significant improvements in interfacial adhesion and mechanical properties through reactive compatibilization, limitations in moisture sensitivity and processing resistance remain significant problems. It is these limitations that are driving the search for other polyester matrices and hybrid composite structures. These are discussed in the following sections.

4.4.2.2 Fundamentals of TPS/PBAT blends

Initial tests of TPS and poly(butylene adipate-co-terephthalate) (PBAT) blends showed that these compositions are inherently immiscible. However, thanks to the elastomeric properties of PBAT, these compositions achieved significantly better mechanical resistance parameters than TPS compositions with more rigid polyesters. As in the case of TPS/PLA, two-component TPS/PBAT blends processed by injection molding typically exhibit a phase-separated morphology. In a continuous PBAT matrix, TPS forms dispersed domains, which also improves plasticity compared to compositions based on more rigid polyesters (Brandelero et al., 2010). Further studies of TPS/PBAT blends have shown that higher TPS content generally reduces tensile strength and stiffness due to weak interfacial adhesion between the hydrophilic TPS and the hydrophobic PBAT, resulting in a coarse phase morphology, especially at higher TPS content. At the same time, it is possible to maintain relatively high elongation at break, indicating the dominant role of PBAT in energy dissipation during deformation (Aversa and Barletta, 2022). Morphological analyses, primarily based on SEM images, confirm that mechanical properties are correlated with phase continuity and dispersion domain size, rather than composition alone. In incompatible compositions, TPS domains tend to increase in volume during melt processing, making it impossible to achieve a balance between stiffness and strength as TPS content increases in the blend.

Furthermore, as described by Ha et al. (2024), two-component TPS/PBAT films prepared by blow molding or cast extrusion indicate that these blends are sensitive to moisture due to the hydrophilic TPS component. The presence of TPS increases water adsorption and barrier permeability in the absence of effective interfacial control, resulting in lower dimensional stability. Even the beneficial effect of TPS on biodegradation and compostability does not compensate for these limitations.

Despite the composition of the two-component mixture, the chemistry of the starch phase is of great importance for interfacial interaction. Comparative studies of natural and chemically modified starch have shown that hydrophobic substitutions on the starch glucose chain significantly improve its compatibility with PBAT, to the extent that the phase morphology changes, mechanical properties—including stress transfer—and barrier properties improve. These results indicate that altering the surface chemistry of TPS enables regulation of interfacial exposure and reduces phase separation without altering the PBAT phase (Wongphan et al., 2022).

As in the case of TPS/PLA, significant improvements in the properties of TPS/PBAT were achieved through reactive compatibilization during melt processing. Reactive extrusion methods, combined with the simultaneous use of multifunctional compatibilizing agents, promote the formation of covalent or semi-covalent bonds between TPS hydroxyl groups and functional groups in the PBAT phase. As a result, a much finer and more stable morphology is obtained, leading to a significant increase in mechanical parameters, such as tensile strength, elastic modulus, and elongation at break. At the same time, the processability of the material has been maintained, making TPS/PBAT blends suitable for use in the packaging industry (Bai et al., 2021). According to Ren et al., PBAT can partially compensate for the brittleness of TPS/PLA systems induced by TPS addition (Ren et al., 2009). Compatibilization improves phase dispersion and increases moisture resistance.

In summary, the available literature presents TPS/PBAT blends as flexible, adjustable materials for starch-based food packaging, where morphology and performance are primarily determined by interfacial engineering rather than the blend composition.

4.4.2.3 TPS/PCL blends: Interfacial interactions, morphology, and durability issues

Poly(ε-caprolactone) (PCL) is a biodegradable polyester characterized by a low glass transition temperature, high chain flexibility, and strong affinity for polar functional groups. Early attempts to use it in compositions with TPS have already shown that such blends exhibit increased strength, dimensional stability, and moisture resistance compared to pure TPS, and that the material can be prepared by extrusion and injection molding across a wide range of compositions. Even a small addition of PCL ($\approx$10% by weight) significantly reduces the natural brittleness, shrinkage, and moisture sensitivity of TPS, while increasing flexibility (Averous et al., 2000). At the same time, studies have also shown limited thermodynamic miscibility between these polymers, with phase morphologies typical of incompatible systems – as in the case of TPS/PLA and TPS/PBAT blends. This means that the mechanical properties of TPS/PCL blends depend on the TPS: PCL and plasticizer ratios, as well as on their processing history. Analysis of aging tests showed progressive structural change after shaping – time-dependent rearrangements in the TPS phase may affect the long-term mechanical stability and dimensional integrity of the material.

Interestingly, subsequent research has shown that TPS/PCL blends can exhibit significantly higher interfacial interactions than other TPS systems with polyesters. Detailed morphological analyses have shown that, in the case of a sufficiently plasticized TPS phase, it is possible to achieve a high level of interaction in morphology as a result of strong hydrogen bonds between the ester groups in PCL and the hydroxyl groups in TPS. Optimal plasticization conditions ensured exceptional plasticity without the need for compatibilizers, even with a significant TPS content. This means that TPS/PCL behaves very differently from other TPS blends based on stiffer polyesters, and the mobility of the TPS chains enables better stress transfer (Li and Favis, 2010).

Environmental conditions and storage history strongly influence the stability of these beneficial interactions – studies of retrogradation have shown that, despite the initially superior mechanical properties of blends containing a significant amount of TPS, recrystallization in the starch phase will occur over time. A significant, and even dominant, influence of humidity on the blend's properties was also observed, indicating a strong effect of starch on TPS/PCL behavior. The polyester content is less important in this case (Mina Hernandez, 2021).

Furthermore, a small proportion of PCL in the blend showed threshold behavior – a small addition of polyester (PCL $\approx$ 5% by weight) significantly improved the blend's elasticity and dimensional stability without causing clear phase separation. A higher content of this material led to more pronounced phase segregation and reduced mechanical benefits. These results further emphasize the delicate balance between composition, plasticization, and morphology in TPS/PCL blends, which significantly increases elasticity and dimensional stability without causing significant phase separation. At the same time, a higher PCL content leads to more pronounced phase segregation and reduced mechanical benefits (Ortega-Toro et al., 2016b).

Given the beneficial effects of reactive compatibilization in other TPS-containing polyester blends, similar studies were conducted for PCL. The use of PCL containing maleic anhydride (MA) or glycidyl methacrylate functions promotes the formation of covalent bonds at the interface between starch and polyester, which also results in better mechanical and barrier properties in TPS/PCL blends (Ortega-Toro et al., 2016a). This shows that this material can be modified by appropriate plasticization and hydrogen bonding or by deliberate chemical compatibilization.

4.4.2.4 TPS/PBS blends: interfacial compatibilization and structure-property relationships

The last of the polyesters described in this chapter that are used in TPS-based blends is polybutylene succinate (PBS), which is interesting primarily because of its biodegradability. It is characterized by relatively high stiffness, thermal stability, and compostability, but, compared to the polyesters described in the previous sections, it exhibits even lower compatibility with TPS, leading to clear phase separation in the unmodified blend (Olonisakin et al., 2025). Early studies of TPS/PBS blends processed in the molten state showed coarse morphology, low interfacial adhesion, and a significant decrease in mechanical properties with increasing TPS content, especially under humid conditions. For this reason, interfacial engineering became a prerequisite for developing TPS/PBS materials with mechanical properties suitable for use in the food packaging industry.

As in the case of TPS/PLA, TPS/PBAT, and TPS/PCL, reactive compatibilization was crucial. For this purpose, maleic anhydride (MA) can be applied. As a result, melt-extruded TPS/PBS blends improved interfacial miscibility, reduced the dispersed-phase size, and yielded significantly better mechanical properties: tensile strength and elongation at break. As in the cases described above, an improvement in moisture resistance was also observed, reflecting reduced interphase voids and increased stress transfer at the phase boundary (Yin et al., 2015). Subsequent research focused, among other things, on alternative compatibilization methods, such as reactive processing and starch activation. Comparative studies of various starch-based compatibilisers have shown that pre-plasticized or de-structured TPS allows for a larger reactive surface area and, consequently, enables more effective interfacial coupling during the mixing of molten material than native starch (Fahrngruber et al., 2020). Such materials were characterized by finer TPS dispersion and, consequently, improved mechanical properties – tear resistance, tensile strength, and better film integrity. It is also worth mentioning attempts at compatibilization using low-molecular-weight organic acids, such as citric acid. Reactive extrusion with citric acid improved phase adhesion, modified crystallization behavior, and enhanced mechanical properties across the entire range of blend ratios (Patil et al., 2025).

The role of chemical grafting at the phase boundary has been described in studies on esterification reactions during reactive extrusion. Esterified TPS phases were successfully combined with PBS using maleic anhydride or tartaric acid as reactive agents. In the presence of appropriate catalysts, this enabled copolymers to be grafted at the phase boundaries. The materials obtained in this way exhibited a finer, more homogeneous morphology than those of incompatible blends. It is worth noting that the use of tartaric acid enabled better mechanical properties despite a similar macroscopic phase structure (Duc and Giang, 2021). Hence, in addition to morphology, interfacial chemistry also determines load transfer and mechanical reinforcement in TPS/PBS blends.

Another method tested to improve the properties of TPS/PBS involved modifying the TPS phase to function as an active elastomeric component. It has been shown that reactive extrusion of TPS with organic acids reduces the molecular weight of starch, lowers the melt viscosity, and improves the dispersion of TPS in PBS (Zhang et al., 2019). The resulting TPS-based bioelastomer enables a transition from the classic "sea-island" morphology to more homogeneous phase structures, thereby improving the material's impact strength.

TPS/PBS blends are susceptible to various processing methods. Unlike TPS/PCL blends, where hydrogen bonds can provide a certain degree of compatibility, TPS/PBS materials typically require deliberate reactive compatibilization or modification of the TPS phase to achieve enhanced mechanical or barrier properties. TPS/PBS properties can be tailored through interfacial engineering, providing a solid foundation for TPS-blend-based composites.

4.4.3 Functional fillers in TPS composites

Functional fillers in TPS-based composites intended for food packaging are used not only to reinforce the packaging structure but also to modify its physical and transport properties. Functional fillers – unlike the plant fillers discussed earlier, which improve mechanical properties by transferring loads – are designed to alter barrier properties, elastic modulus, thermal stability, and even the behavior of the material during processing in the molten state. In practice, this means that fillers are added not to a pure TPS matrix, but to a TPS/polyester blend matrix. As discussed in the previous section, such blends enable the production of a base material with sufficient mechanical properties, ready for further processing (Bangar et al., 2021).

In TPS-based composites, functional fillers are typically inorganic or hybrid materials. Typically, their effect is to influence interfacial interactions or polymer chain mobility (Olonisakin et al., 2025). Layered fillers (e.g., clays, nanoclays) can create tortuous diffusion paths, thereby reducing gas and vapor permeability. Mineral fillers (e.g., calcium carbonate) increase the stiffness and heat resistance of the composite by limiting matrix deformation and influencing crystallization behavior. Biological nanostructured fillers are treated as intermediate materials that simultaneously modify the mechanical properties and barrier properties of the composite. The effectiveness of these fillers is primarily a result of dispersion quality in the matrix, interphase compatibility, and extrusion processing conditions. It is worth remembering that improving one property often requires a compromise in another, e.g., strength, moisture sensitivity, or processability, which makes proper formulation adjustment a key issue.

The following sections present functional fillers in TPS-based composites, focusing on layered and mineral fillers and their impact on barrier properties, elastic modulus, and thermal stability, as well as cellulose fillers. These effects will be compared with those achieved by using the natural fillers discussed earlier.

4.4.3.1 Clays and nanoclays

Layered silicate fillers (e.g., clays, nanoclays) are among the most frequently studied functional fillers for use in TPS composites processed in the molten state. They are primarily used to modify transport properties, thermal stability, and phase morphology by restricting interfacial contact and creating tortuous diffusion paths. Since barrier properties and dimensional stability are key in the food packaging industry, often even more critical than absolute mechanical strength, clays are particularly attractive and are frequently studied materials.

Early studies on melt-extruded clay-based TPS composites demonstrated that the microstructure of the starch matrix can be influenced by adding unprocessed or organically modified clays, provided adequate dispersion is achieved during processing (Dean et al., 2007). Depending on the type of clay used and the processing conditions, intercalated or partially exfoliated morphologies were observed, leading to greater composite stiffness and improved thermal resistance. At this stage of the research, the agglomeration of clay and its incomplete dispersion in the matrix proved problematic, limiting the composite reinforcement's effectiveness and potentially affecting its strength.

When a TPS/polyester mixture, and especially TPS/PBAT or TPS/PLA, is used as a matrix, an improvement in elastic modulus and dimensional stability is often achieved, even with a relatively low filler content due to nanoclay's tendency to localize in the TPS-rich phase or at the polymer-polymer interface, owing to its hydrophilic surface chemistry. Selective filler localization improves phase morphology and limits TPS domain coalescence during melt processing, thereby indirectly stabilizing the blend's morphology under shear (López et al., 2015).

A significant advantage of composites with clay additives is the substantial increase in thermal stability. Layered silicates act as a physical barrier to volatile decomposition products and reduce the mobility of polymer chains at elevated temperatures. This effect is significant in TPS-based composites, as the starch phase is highly susceptible to thermal degradation. In TPS/clay and TPS/polyester/clay composites, moderate increases in the degradation onset temperature and improvements in heat resistance have been observed. From an application perspective, especially in the food packaging industry, these properties represent a key advantage when higher processing or service temperatures are required (Dean et al., 2007).

When it comes to limitations in the use of clays and nanoclays in TPS-based composites, attention should primarily be paid to problems arising from high filler content or poor dispersion. These can lead to brittleness, increased melt viscosity, or reduced processability, especially regarding film extrusion. In addition, some clays are hydrophilic, which can further improve the composite's moisture sensitivity, mainly when interfacial adhesion is poorly controlled. For this reason, relatively low clay filler contents are used, and their functional advantages are maximized within a narrow application window.

In general, clays are used primarily as morphology and transport modifiers in TPS-based composites, rather than as fillers to improve mechanical properties. Compared to the natural fillers described earlier, such as fibrous or particulate fillers, they offer better barrier properties and higher thermal stability. For this reason, they are an attractive functional filler option for food packaging applications. At the same time, it is necessary to continuously monitor the dispersion state and processing conditions, which directly leads to comparisons with mineral fillers, as described in the next section.

4.4.3.2 Mineral fillers

Another type of functional filler in TPS composites is mineral fillers, including talc, calcium carbonate ($CaCO_3$), and silica. They differ from clays and nanoclays in terms of both geometry and reinforcement mechanism – unlike clays and nanoclays, mineral fillers are typically isotropic or weakly anisotropic in morphology. They mainly increase stiffness, modify rheology, and improve thermal stability. They do not affect interfacial restrictions or tortuosity of the diffusion path. Compared to the natural fillers discussed earlier, mineral fillers provide more stable processing, predictable property modification, and lower moisture sensitivity. However, they do not increase the content of renewable raw materials – they only offer a slight improvement in mechanical properties or biodegradability. In TPS-based composites, they should be treated as functional modifiers whose primary purpose is to increase stiffness, thermal resistance, or processing reliability.

As functional fillers, mineral fillers have a lower impact on barrier properties than layered fillers due to their geometry, which is less conducive to extending the effective diffusion path. Improved barrier properties to gases or water vapor are usually the result of increased crystallinity or reduced free volume, rather than of transport mechanisms. These mechanisms are used in ternary TPS biodegradable polymer blends with mineral fillers, such as talc. Paquette et al. reported improvements in water vapor barrier properties resulting from increased crystallinity and matrix stiffness (Paquette et al., 2026).

Talc, along with calcium carbonate, is one of the most frequently studied mineral fillers in TPS and TPS/polyester composites. Talc, thanks to its plate-like structure and relatively good dispersion, has a positive effect on the modulus of elasticity and heat resistance while maintaining sufficiently good processability. In composites based on a TPS/PBAT matrix, a significant increase in Young's modulus and heat deflection temperature was achieved due to the limited mobility of the polymer chains and the increased crystallinity of the polyester phase. When comparing talc with nanoclays, a similar increase in stiffness requires a higher proportion in the composite. However, it provides better processing stability and lower sensitivity to dispersion quality (Scharnowski et al., 2022).

Calcium carbonate, on the other hand, is a filler that works much better in composites used for the production of flexible packaging. Its particles act mainly as rigid inclusions, increasing the material's stiffness and modifying its behavior upon impact. It is worth noting that their effect is strongly correlated with particle size and surface treatment. It is precisely the improvement in impact resistance, while maintaining flow properties, that is the main reason for using $CaCO_3$ in TPS/PBAT-based composites. However, it does not significantly affect the composite's barrier properties, especially compared to platelet fillers. In contrast to talc, calcium carbonate does not increase thermal resistance or crystallinity. However, its low cost is noteworthy, which, combined with the aforementioned processing stability, makes it attractive for larger-scale applications (Scharnowski et al., 2022).

Other fillers tested for biodegradable polyester blends containing TPS were silica and calcium nanocarbonate, primarily for morphology modification. Silica particles can increase stiffness and affect the rheological properties of the composite. Still, due to problems related to interfacial adhesion and dispersion control (especially in the absence of adequate surface modification), their reinforcing effectiveness in composites containing a significant proportion of TPS is low. On the other hand, calcium carbonate nanoparticles, in combination with reactive compatibilizers, enable a finer phase morphology and improved stress transfer. However, as with silica, these effects are limited when TPS constitutes a significant proportion of the matrix (Negaresh et al., 2024).

4.4.3.3 Cellulose-based functional fillers

Cellulose-based fillers, with particular emphasis on cellulose nanomaterials such as cellulose nanofibers (CNF) and cellulose nanocrystals (CNC), are categorized between natural and functional fillers in TPS composites. Due to their nanometric dimensions, high aspect ratio, and strong interfacial activity, they are treated as functional fillers, despite their biological origin. However, the effects of their use go beyond their impact on mechanical parameters. In TPS-based matrices and TPS/polyester blends, cellulose nanofillers act mainly at the interfacial level. Unlike conventional fibrous fillers, they affect the mobility of polymer chains, phase morphology, and transport phenomena. CNF and CNC form percolating networks with a relatively low share in the composite, limit the segmental mobility of the matrix, and modify the material's rheological properties. In TPS/PBAT blends, the addition of CNF or CNC results in higher composite flexibility, a higher elastic modulus, and increased deformation resistance, reflecting the formation of a load-bearing network supported by nanomaterials (Fourati et al., 2021).

Compared to clays, nanoclays, and mineral fillers, nanocellulose-based fillers combine the characteristics of renewable raw materials with functional efficiency. Nanoclays are often used to improve the thermal stability of composites, while nanocellulose offers better compatibility with TPS, making it more effective in composites with high bio-based content. However, as with clays and nanoclays, attention should be paid to potential issues arising from dispersion, agglomeration, or increased material viscosity, which may affect the extrusion or film blowing process (Madhumitha et al., 2018).

One of the main functional applications of CNC and CNF is to improve the barrier properties of composites. Thanks to their high aspect ratio and strong hydrogen-bonding ability, these nanomaterials effectively create tortuous diffusion paths in materials with high TPS content, thereby reducing water vapor and gas permeability. In TPS- and TPS/PBAT-based composites, a significant improvement in water vapor and oxygen barrier properties was achieved with a much lower proportion of nanocellulose in the composite than when using talc or mineral fillers (Aouay et al., 2024).

In addition to permeability, nanocellulose fillers also reduce the moisture sensitivity of TPS-based composites. By immobilizing starch chains and partially shielding hydrophilic domains, they limit water absorption and moisture-induced plasticization. The effect is particularly pronounced when using lignin-containing cellulose nanofibers – the lignin residues impart a hydrophobic effect, further reducing water sorption by the material (Aouay et al., 2024). Cellulose-based nanofillers are a hybrid class of functional fillers in TPS-based composites. They enable simultaneous improvements in barrier properties, rheological stability, and moisture resistance while maintaining a high bio-based content in the composite. For these reasons, they are regarded as among the most promising functional fillers for TPS-based composites used in the food packaging industry.

4.5 Summary and implications for the design of TPS-based composites

Thermoplastic starch can be a green alternative for single-use packaging and a solution for reducing plastic packaging waste. TPS and its composite can be successfully applied to certain food packaging as an alternative to fossil-based non-degradable materials. However, thermoplastic starch has several limitations, among which the greatest is its low mechanical properties, water sensitivity, and gas pearmability. To improve TPS materials, various plastcizers can be used. Among them, the most common is glycerol, whose hydroxyl groups can form hydrogen bonds with hydrophilic starch molecules. Development of TPS composite with plant and natural fillers, or development of blends with other biodegradable polymers, such as PLA, PCL, or PBAT, can allow the design of materials with given or desired properties.

The available peer-reviewed literature indicates that no universal filler material for TPS-based composites exists that meets the wide range of requirements of the food packaging industry. Instead, the optimal composite composition must result from a careful balance among morphology, interfacial interactions, and processing constraints. So, fillers should be viewed not as independent supplements but as elements that adjust the matrix's behavior, either before or in parallel with a structural reinforcement strategy. Analysis of composites containing functional fillers shows that the choice of filler affects not only mechanical properties, but also barrier properties, rheological behavior, thermal stability, and moisture sensitivity. Unlike the natural fillers discussed earlier, which mainly affect mechanical properties, functional fillers primarily act at the interfacial and transport levels, influencing molecular mobility, phase continuity, and the long-term stability of TPS-based composites.

Layered silicates and nanoclays, due to their plate-like morphology and preferential localization in the starch-rich phase or at polymer-polymer phase boundaries, are used to improve the barrier properties of TPS-based composites. They create highly tortuous diffusion paths and promote morphological stabilization, while simultaneously increasing the material's viscosity and its sensitivity to processing. Therefore, their content in the material must be carefully selected, and dispersion must be optimized to maintain adequate industrial processability.

Conventional mineral fillers (including talc, calcium carbonate, and silica) belong to another group of functional fillers. Their presence in TPS-based composites increases stiffness and thermal resistance. They also improve the modulus of elasticity and heat deflection temperature, while having only a limited effect on the material's barrier properties and moisture resistance. Importantly, their reinforcing effects are primarily due to particle geometry and dispersion quality, rather than interfacial chemistry. For these reasons, their main applications are in composites requiring dimensional stability and mechanical stiffness.

Due to their dimensions, high aspect ratios, and strong hydrogen-bonding ability, cellulose-based nanofillers can simultaneously improve barrier properties, rheological stability, and resistance to moisture-induced plasticization. Compared to the inorganic fillers described above, they are more compatible with TPS-based matrices. They also retain a high content of bio-based components, although due to their impact on viscosity and dispersion requirements in the material, they are significantly more challenging to process.

The above perspective combines analyses of natural additives, polyester-based TPS blends, and functional fillers. It provides a coherent framework for selecting appropriate fillers in TPS-based composites for food packaging applications.

References

Abera, G., Woldeyes, B., Dessalegn Demash, H., Miyake, G., 2020. The effect of plasticizers on thermoplastic starch films developed from the indigenous Ethiopian tuber crop Anchote (Coccinia abyssinica) starch. Int. J. Biol. Macromol. 155, 581–587. https://doi.org/10.1016/j.ijbiomac.2020.03.218.

Abhitha, S., Amulya, P., Dubey, A., Praveen, B.V.S., Madhuri, P., 2025. Comparative analysis of bioplastic sheets derived from Biowaste using various plasticizers. Arabian J. Sci. Eng. 50 (13), 9831–9843. https://doi.org/10.1007/s13369-024-09891-w.

Alee, M., Fu, J., Duan, Q., Yang, M., Liu, H., Zhu, J., Xianyang, B., Chen, L., Yu, L., 2023. Plasticizing effectiveness and characteristics of different mono-alcohols, di-alcohols, and polyols for starch-based materials. Starch/Staerke 75 (11–12). https://doi.org/10.1002/star.202300059.

Aouay, M., Magnin, A., Lancelon-Pin, C., Putaux, J.-L., Boufi, S., 2024. Mitigating the water sensitivity of PBAT/TPS blends through the incorporation of lignin-containing cellulose nanofibrils for application in biodegradable films. ACS Sustain. Chem. Eng. 12 (29), 10805–10819. https://doi.org/10.1021/acssuschemeng.4c02245.

Avérous, L., 2004. Biodegradable multiphase systems based on plasticized starch: a review. J. Macromol. Sci. Polym. Rev. 44 (3), 231–274. https://doi.org/10.1081/mc-200029326.

Avérous, L., Pollet, E., 2012. Biodegradable polymers. In: Avérous, L., Pollet, E. (Eds.), Environmental Silicate Nano-Biocomposites. Springer London, London, pp. 13–39. https://doi.org/10.1007/978-1-4471-4108-2_2.

Averous, L., Moro, L., Dole, P., Fringant, C., 2000. Properties of thermoplastic blends: starch-polycaprolactone. Polymer 41 (11), 4157–4167. https://doi.org/10.1016/S0032-3861(99)00636-9.

Aversa, C., Barletta, M., 2022. Addition of thermoplastic starch (TPS) to binary blends of poly(lactic acid) (PLA) with poly(butylene adipate-co-terephthalate) (PBAT): extrusion compounding, cast extrusion and Thermoforming of home compostable materials. J. Polym. Sci. 40 (10), 1269–1286. https://doi.org/10.1007/s10118-022-2734-0.

Aydin, A.A., Ilberg, V., 2016. Effect of different polyol-based plasticizers on thermal properties of polyvinyl alcohol:starch blends. Carbohydr. Polym. 136, 441–448. https://doi.org/10.1016/j.carbpol.2015.08.093.

Bai, J., Pei, H., Zhou, X., Xie, X., 2021. Reactive compatibilization and properties of low-cost and high-performance PBAT/thermoplastic starch blends. Eur. Polym. J. 143, 110198. https://doi.org/10.1016/j.eurpolymj.2020.110198.

Bangar, S.P., Scott Whiteside, W., Omodunbi Ashogbon, A., Kumar, M., 2021. Recent advances in thermoplastic starches for food packaging: a review. Food Packag. Shelf Life 30, 100743. https://doi.org/10.1016/j.fpsl.2021.100743.

Basiak, E., Lenart, A., Debeaufort, F., 2018. How glycerol and water contents affect the structural and functional properties of starch-based edible films. Polymers 10 (4), 412. https://doi.org/10.3390/polym10040412.

Bayer, G., Shayganpour, A., Zia, J., Bayer, I.S., 2022. Polyvinyl alcohol-based films plasticized with an edible sweetened gel enriched with antioxidant carminic acid. J. Food Eng. 323, 111000. https://doi.org/10.1016/j.jfoodeng.2022.111000.

Ben, Z.Y., Samsudin, H., Yhaya, M.F., 2022. Glycerol: its properties, polymer synthesis, and applications in starch based films. Eur. Polym. J. 175, 111377. https://doi.org/10.1016/j.eurpolymj.2022.111377.

Biran Ay, S., Ertürk, N., 2022. Effect of ethylene glycol and glycerol concentrations on properties of Rye-based films TT - Etilen Glikol ve Gliserol Konsantrasyonlarının Çavdar Esaslı Filmlerin Özelliklerine Etkisi. Avrupa Bilim ve Teknoloji Dergisi (34), 705–710. https://doi.org/10.31590/ejosat.1083197.

Blennow, A., Bay-Smidt, A.M., Olsen, C.E., Møller, B.L., 2000. The distribution of covalently bound phosphate in the starch granule in relation to starch crystallinity. Int. J. Biol. Macromol. 27 (3), 211–218. https://doi.org/10.1016/s0141-8130(00)00121-5.

Brandelero, R.P.H., Yamashita, F., Grossmann, M.V.E., 2010. The effect of surfactant Tween 80 on the hydrophilicity, water vapor permeation, and the mechanical properties of cassava starch and poly(butylene adipate-co-terephthalate) (PBAT) blend films. Carbohydr. Polym. 82 (4), 1102–1109. https://doi.org/10.1016/j.carbpol.2010.06.034.

Chang, Y.P., Abd Karim, A., Seow, C.C., 2006. Interactive plasticizing–antiplasticizing effects of water and glycerol on the tensile properties of tapioca starch films. Food Hydrocoll. 20 (1), 1–8. https://doi.org/10.1016/j.foodhyd.2005.02.004.

De Arcangelis, E., Carmela Trivisonno, M., Angelicola, M., Quiquero, M., Di Nardo, V., Falasca, L., Sestili, F., Cristina Messia, M., Marconi, E., 2021. Milling and rheological properties of high amylose wheat. J. Cereal. Sci. 102, 103335. https://doi.org/10.1016/j.jcs.2021.103335.

Dean, K., Yu, L., Wu, D.Y., 2007. Preparation and characterization of melt-extruded thermoplastic starch/clay nanocomposites. Compos. Sci. Technol. 67 (3–4), 413–421. https://doi.org/10.1016/j.compscitech.2006.09.003.

Dogenski, M., Gurikov, P., Baudron, V., de Oliveira, J.V., Smirnova, I., Ferreira, S.R.S., 2020. Starch-based aerogels obtained via solvent-induced gelation. Gels 6 (3), 1–12. https://doi.org/10.3390/gels6030032.

Donmez, D., Pinho, L., Patel, B., Desam, P., Campanella, O.H., 2021. Characterization of starch–water interactions and their effects on two key functional properties: starch gelatinization and retrogradation. Curr. Opin. Food Sci. 39, 103–109. https://doi.org/10.1016/j.cofs.2020.12.018.

Dos Santos, B.H., De Souza Do Prado, K., Jacinto, A.A., Da Silva Spinacé, M.A., 2018. Influence of sugarcane bagasse fiber size on biodegradable composites of thermoplastic starch. J. Renew. Mater. 6 (2), 176–182. https://doi.org/10.7569/JRM.2018.634101.

Duc, V.M., Giang, N.C., 2021. Morphology and properties of polybutylene succinic and cassava starch blend: effect of esterification catalysts on graft copolymer formation. Vietnam J. Chem. 59 (1), 90–97. https://doi.org/10.1002/vjch.202000127.

Edhirej, A., Sapuan, S.M., Jawaid, M., Zahari, N.I., 2017. Effect of various plasticizers and concentration on the physical, thermal, mechanical, and structural properties of cassava-starch-based films. Starch - Stärke 69 (1–2). https://doi.org/10.1002/star.201500366.

Fahrngruber, B., Fortea-Verdejo, M., Wimmer, R., Mundigler, N., 2020. Starch/Poly (butylene succinate) compatibilizers: effect of different Reaction-Approaches on the properties of thermoplastic starch-based compostable films. J. Polym. Environ. 28 (1), 257–270. https://doi.org/10.1007/s10924-019-01601-0.

Fei Liu, Y., Laohasongkram, K., Chaiwanichsiri, S., 2016. Effects of heat-moisture treatment on molecular interactions and physicochemical properties of tapioca starch.

MOJ Food Processing & Technology 3 (Issue 3). https://doi.org/10.15406/MOJFPT.2016.03.00072.

Ferrarezi, M.M.F., de Oliveira Taipina, M., da Silva, L.C.E., Gonçalves, M.d.C., 2013. Poly (Ethylene glycol) as a compatibilizer for poly(lactic acid)/thermoplastic starch blends. J. Polym. Environ. 21 (1), 151–159. https://doi.org/10.1007/s10924-012-0480-z.

Foret, S., Mazian, B., Bekas, V., Martins, F.C.B., Campanella, O.H., Perré, P., Augusto, P. E.D., 2023. Thermoplastic starch biocomposites reinforced with hemp shives obtained via extrusion. Ind. Crop. Prod. 206, 117707. https://doi.org/10.1016/j.indcrop.2023.117707.

Fourati, Y., Tarrés, Q., Delgado-Aguilar, M., Mutjé, P., Boufi, S., 2021. Cellulose nanofibrils reinforced PBAT/TPS blends: mechanical and rheological properties. Int. J. Biol. Macromol. 183, 267–275. https://doi.org/10.1016/j.ijbiomac.2021.04.102.

Gao, W., Zhu, J., Kang, X., Wang, B., Liu, P., Cui, B., Abd El-Aty, A.M., 2021. Development and characterization of starch films prepared by extrusion blowing: the synergistic plasticizing effect of water and glycerol. LWT 148, 111820. https://doi.org/10.1016/j.lwt.2021.111820.

García, N.L., Ribba, L., Dufresne, A., Aranguren, M., Goyanes, S., 2011. Effect of glycerol on the morphology of nanocomposites made from thermoplastic starch and starch nanocrystals. Carbohydr. Polym. 84 (1), 203–210. https://doi.org/10.1016/j.carbpol.2010.11.024.

Gebresas, G.A., Szabó, T., Marossy, K., 2023. A comparative study of carboxylic acids on the cross-linking potential of corn starch films. J. Mol. Struct. 1277, 134886. https://doi.org/10.1016/j.molstruc.2022.134886.

George, J., Sreekala, M.S., Thomas, S., 2001. A review on interface modification and characterization of natural fiber reinforced plastic composites. Polym. Eng. Sci. 41 (9), 1471–1485. https://doi.org/10.1002/pen.10846.

Ghanbari, A., Tabarsa, T., Ashori, A., Shakeri, A., Mashkour, M., 2018. Preparation and characterization of thermoplastic starch and cellulose nanofibers as green nanocomposites: extrusion processing. Int. J. Biol. Macromol. 112, 442–447. https://doi.org/10.1016/j.ijbiomac.2018.02.007.

Gironès, J., López, J.P., Mutjé, P., Carvalho, A.J.F., Curvelo, A.A.S., Vilaseca, F., 2012. Natural fiber-reinforced thermoplastic starch composites obtained by melt processing. Compos. Sci. Technol. 72 (7), 858–863. https://doi.org/10.1016/j.compscitech.2012.02.019.

González, K., Iturriaga, L., González, A., Eceiza, A., Gabilondo, N., 2020. Improving mechanical and barrier properties of thermoplastic starch and polysaccharide nanocrystals nanocomposites. Eur. Polym. J. 123, 109415. https://doi.org/10.1016/j.eurpolymj.2019.109415.

Grylewicz, A., Spychaj, T., Zdanowicz, M., 2019. Thermoplastic starch/wood biocomposites processed with deep eutectic solvents. Compos. Appl. Sci. Manuf. 121, 517–524. https://doi.org/10.1016/j.compositesa.2019.04.001.

Gurunathan, M.K., Navasingh, R.J.H., Selvam, J.D.R., Čep, R., 2025. Development and characterization of starch bioplastics as a sustainable alternative for packaging. Sci. Rep. 15 (1). https://doi.org/10.1038/s41598-025-00221-0.

Ha, P.T.T., Son, N.T., Khoi, N.V., Trang, P.T., Duc, N.T., Anh, P.N., Linh, N.N., Tung, N.T., 2024. Poly(butylene adipate-co-terephthalate)/thermoplastic canna starch (Canna edulis ker.) (PBAT/TPS) blend film – a novel biodegradable material. J. Indian Chem. Soc. 101 (9). https://doi.org/10.1016/j.jics.2024.101245.

Hansen, C.M., Poulsen, T.S., 2007. Hansen Solubility Parameters - Biological Materials Hansen Solubility Parameters: A Users Handbook, second ed. CRC Press, Denmark, pp. 269–292. https://doi.org/10.1201/9781420006834.

Haryanti, P., Setyawati, R., Wicaksono, R., 2014. Pengaruh suhu dan lama pemanasan suspensi pati serta konsentrasi butanol terhadap karakteristik fisikokimia pati tinggi amilosa Dari tapioka. Jurnal Agritech 34 (03), 308. https://doi.org/10.22146/agritech.9459.

Hazrati, K.Z., Sapuan, S.M., Zuhri, M.Y.M., Jumaidin, R., 2021. Effect of plasticizers on physical, thermal, and tensile properties of thermoplastic films based on Dioscorea hispida starch. Int. J. Biol. Macromol. 185, 219–228. https://doi.org/10.1016/j.ijbiomac.2021.06.099.

Huneault, M.A., Li, H., 2007. Morphology and properties of compatibilized polylactide/thermoplastic starch blends. Polymer 48 (1), 270–280. https://doi.org/10.1016/j.polymer.2006.11.023.

Jane, J., Chen, Y.Y., Lee, L.F., McPherson, A.E., Wong, K.S., Radosavljevic, M., Kasemsuwan, T., 1999. Effects of amylopectin Branch chain length and amylose content on the gelatinization and pasting properties of starch. Cereal Chem. 76 (5), 629–637. https://doi.org/10.1094/cchem.1999.76.5.629.

Jaramillo, D.A., Peña-González, M.A., Paredes-Escobar, M., Pérez-González, A., Lazo-Vélez, M.A., 2023. In Vitro Digestibility and physicochemical properties of Isolated starch from potatoes (Solanum tuberosum and Solanum subsp. andigena) Varieties Commercialized in the Andean region: Ecuador. Starch/Staerke 75 (1–2). https://doi.org/10.1002/star.202200171.

Jayarathna, S., Andersson, M., Andersson, R., 2022. Recent advances in starch-based blends and composites for bioplastics applications. Polymers 14 (21). https://doi.org/10.3390/polym14214557.

Job, U., 2018. Plasticization and antiplasticization in amorphous food systems. Curr. Opin. Food Sci. 21, 72–78. https://doi.org/10.1016/j.cofs.2018.05.007.

Lai, 賴鳴鳳 (Vivian M F and Lii), 呂政義 (Cheng-Yi, 2000) 'Gelatinization Characterization of Rice Starches', 中央研究院化學研究所集刊, (47), pp. 55–64. https://doi.org/10.6522/BICAS.2000.47.6.

Li, C., 2022. Recent progress in understanding starch gelatinization - an important property determining food quality. Carbohydr. Polym. 293, 119735. https://doi.org/10.1016/j.carbpol.2022.119735.

Li, G., Favis, B.D., 2010. Morphology development and interfacial interactions in polycaprolactone/thermoplastic-starch blends. Macromol. Chem. Phys. 211 (3), 321–333. https://doi.org/10.1002/macp.200900348Canada.

Li, H., Huneault, M.A., 2011. Effect of chain extension on the properties of PLA/TPS blends. J. Appl. Polym. Sci. 122 (1), 134–141. https://doi.org/10.1002/app.33981.

Li, H., Dhital, S., Flanagan, B.M., Mata, J., Gilbert, E.P., Gidley, M.J., 2020. High-amylose wheat and maize starches have distinctly different granule organization and annealing behaviour: a key role for chain mobility. Food Hydrocoll. 105, 105820. https://doi.org/10.1016/j.foodhyd.2020.105820.

Li, E., Cao, P., Cao, W., Cheng, L., 2022. Relations between starch fine molecular structures with gelatinization property under different moisture content. Carbohydr. Polym. 278, 118955. https://doi.org/10.1016/j.carbpol.2021.118955.

Li, D., Luo, C., Zhou, J., Dong, L., Chen, Y., Liu, G., Qiao, S., 2023. The role of the interface of PLA with thermoplastic starch in the Nonisothermal crystallization behavior of PLA in PLA/thermoplastic starch/SiO2 composites. Polymers 15 (6), 1579. https://doi.org/10.3390/polym15061579.

Lim, W.S., Ock, S.Y., Park, G.D., Lee, I.W., Lee, M.H., Park, H.J., 2020. Heat-sealing property of cassava starch film plasticized with glycerol and sorbitol. Food Packag. Shelf Life 26, 100556. https://doi.org/10.1016/J.FPSL.2020.100556.

Liu, H., Yu, L., Xie, F., Chen, L., 2006. Gelatinization of cornstarch with different amylose/amylopectin content. Carbohydr. Polym. 65 (3), 357–363. https://doi.org/10.1016/j.carbpol.2006.01.026.

López, O., Castillo, L., Zaritzky, N., Barbosa, S., Villar, M., Alejandra García, M., 2015. Talc nanoparticles influence on thermoplastic corn starch film properties. Procedia Mater. Sci. 8, 338–345. https://doi.org/10.1016/j.mspro.2015.04.082.

Lourdin, D., Bizot, H., Colonna, P., 1997. Antiplasticization'" in starch-glycerol films? J. Appl. Polym. Sci. 63, 1047–1053. https://doi.org/10.1002/(SICI)1097-4628(19970222)63:8.

Ma, X., Yu, J., 2004. Formamide as the plasticizer for thermoplastic starch. J. Appl. Polym. Sci. 93 (4), 1769–1773. https://doi.org/10.1002/app.20628.

Ma, X., Yu, J., Feng, J., 2004. Urea and formamide as a mixed plasticizer for thermoplastic starch. Polym. Int. 53 (11), 1780–1785. https://doi.org/10.1002/pi.1580.

Ma, X.F., Yu, J.G., Wan, J.J., 2006. Urea and ethanolamine as a mixed plasticizer for thermoplastic starch. Carbohydr. Polym. 64 (2), 267–273. https://doi.org/10.1016/j.carbpol.2005.11.042.

Ma, C., Wang, G., Xia, C., Guo, L., Cui, B., Du, X., Wang, J., Sun, C., 2025. Effects of different ratios of glycerol to erythritol on the structure and properties of starch straws during long term storage. Food Chem. 464 (Pt 3), 141862. https://doi.org/10.1016/j.foodchem.2024.141862.

Madhumitha, G., Fowsiya, J., Mohana Roopan, S., Thakur, V.K., 2018. Recent advances in starch–clay nanocomposites. Int. J. Polym. Anal. Char. 23 (4), 331–345. https://doi.org/10.1080/1023666X.2018.1447260.

Majeed, T., Dar, A.H., Pandey, V.K., Dash, K.K., Srivastava, S., Shams, R., Jeevarathinam, G., Singh, P., Echegaray, N., Pandiselvam, R., 2023. Role of additives in starch-based edible films and coating: a review with current knowledge. Prog. Org. Coating 181, 107597. https://doi.org/10.1016/j.porgcoat.2023.107597.

Mali, S., Grossmann, M.V.E., García, M.A., Martino, M.N., Zaritzky, N.E., 2006. Effects of controlled storage on thermal, mechanical and barrier properties of plasticized films from different starch sources. J. Food Eng. 75 (4), 453–460. https://doi.org/10.1016/j.jfoodeng.2005.04.031.

Mascia, L., Kouparitsas, Y., Nocita, D., Bao, X., 2020. Antiplasticization of polymer materials: structural aspects and effects on mechanical and diffusion-controlled properties. Polymers 12 (4), 769. https://doi.org/10.3390/polym12040769.

Matsuki, J., Yasui, T., Kohyama, K., Sasaki, T., 2003. Effects of environmental temperature on structure and gelatinization properties of wheat starch. Cereal Chem. 80 (4), 476–480. https://doi.org/10.1094/cchem.2003.80.4.476.

Mina Hernandez, J.H., 2021. Effect of the incorporation of polycaprolactone (Pcl) on the retrogradation of binary blends with cassava thermoplastic starch (tps). Polymers 13 (1), 1–19. https://doi.org/10.3390/polym13010038.

Mishra, T., Mandal, P., Kumar Rout, A., Sahoo, D., 2022. A state-of-the-art review on potential applications of natural fiber-reinforced polymer composite filled with inorganic nanoparticle. Composites Part C: Open Access 9, 100298. https://doi.org/10.1016/j.jcomc.2022.100298.

Mohammed, A.A.B.A., Hasan, Z., Omran, A.A.B., Elfaghi, A.M., Khattak, M.A., Ilyas, R. A., Sapuan, S.M., 2023. Effect of various plasticizers in different concentrations on physical, thermal, mechanical, and structural properties of wheat starch-based films. Polymers 15 (1). https://doi.org/10.3390/polym15010063.

Montilla-Buitrago, C.E., Gómez-López, R.A., Solanilla-Duque, J.F., Serna-Cock, L., Villada-Castillo, H.S., 2021. Effect of plasticizers on properties, retrogradation, and processing of extrusion-obtained thermoplastic starch: a review. Starch/Staerke 73 (9–10). https://doi.org/10.1002/star.202100060.

Muñoz-Gimena, P.F., Oliver-Cuenca, V., Peponi, L., López, D., 2023. A review on reinforcements and additives in starch-based composites for food packaging. Polymers 15 (13). https://doi.org/10.3390/polym15132972.

Murawski, T.T., Szałapak, J., Tryznowski, M., Świtała, M., Gręda, D., Bednarczyk, E., Ryszkowska, J., Żołek-Tryznowska, Z., 2025. Long-term stability and properties of potato starch films plasticized with sugar–glycerol mixtures. Int. J. Biol. Macromol. 331 (Pt 2), 148444. https://doi.org/10.1016/j.ijbiomac.2025.148444.

Negaresh, M., Javadi, A., Garmabi, H., 2024. Poly(lactic acid)/poly(ε-caprolactone) blends: the effect of nanocalcium carbonate and glycidyl methacrylate on interfacial characteristics. Front. Mater. 11. https://doi.org/10.3389/fmats.2024.1377340.

Niazi, M.B.K., Zijlstra, M., Broekhuis, A.A., 2015. Influence of plasticizer with different functional groups on thermoplastic starch. J. Appl. Polym. Sci. 132 (22). https://doi.org/10.1002/app.42012.

Olawoye, B., Popoola-Akinola, O., Fagbohun, O.F., Adetola, D.B., Adeloye, A.E., Abdulwahab, A.A., Ogunjemilusi, M.A., Olaoye, I.O., 2025. Starch modification using sustainable processing Technology: impacts and applications. Starch/Staerke 77 (3). https://doi.org/10.1002/star.202400115.

Olonisakin, K., Mohanty, A.K., Thimmanagari, M., Misra, M., 2025. Recent advances in biodegradable polymer blends and their biocomposites: a comprehensive review. Green Chem. 27 (38), 11656–11704. https://doi.org/10.1039/d5gc01294e.

Ortega-Toro, R., Santagata, G., Gomez d'Ayala, G., Cerruti, P., Talens Oliag, P., Amparo Chiralt Boix, M., Malinconico, M., 2016a. Enhancement of interfacial adhesion between starch and grafted poly(ε-caprolactone). Carbohydr. Polym. 147, 16–27. https://doi.org/10.1016/j.carbpol.2016.03.070.

Ortega-Toro, R., Muñoz, A., Talens, P., Chiralt, A., 2016b. Improvement of properties of glycerol plasticized starch films by blending with a low ratio of polycaprolactone and/or polyethylene glycol. Food Hydrocoll. 56, 9–19. https://doi.org/10.1016/j.foodhyd.2015.11.029.

Paquette, S., Menossi, M., Pesaranhajiabbas, E., Khalil, H., Misra, M., Mohanty, A.K., 2026. Impact of organic and inorganic fillers on the performance of ternary biodegradable poly(butylene succinate-Co-butylene adipate), poly(butylene adipate-Co-terephthalate), and poly(3-Hydroxybutyrate-Co-3-Hydroxyvalerate) blends. Polym. Compos. 47 (5), 4509–4526. https://doi.org/10.1002/pc.70442.

Patil, H.B., Kumar, V., Kumavat, D.J., Mishra, D., Rao, A.R., 2025. Effect of citric acid on the blends of poly butylene succinate and thermoplastic corn starch. Int. J. Biol. Macromol. 311. https://doi.org/10.1016/j.ijbiomac.2025.143537.

Ploypetchara, T., Gohtani, S., 2018. Effect of sugar on starch edible film properties: plasticized effect. J. Food Sci. Technol. 55 (9), 3757–3766. https://doi.org/10.1007/s13197-018-3307-7.

Quilez-Molina, A.I., Paul, U.C., Merino, D., Athanassiou, A., 2022. Composites of thermoplastic starch and lignin-rich agricultural waste for the packaging of fatty foods. ACS Sustain. Chem. Eng. 10 (47), 15402–15413. https://doi.org/10.1021/acssuschemeng.2c04326.

Rahmawati, M., Arief, M., Satyantini, W.H., 2019. The effect of sorbitol addition on the characteristic of Carrageenan edible film. IOP Conf. Ser. Earth Environ. Sci. 236 (1), 012129. https://doi.org/10.1088/1755-1315/236/1/012129.

Rashwan, A.K., Younis, H.A., Abdelshafy, A.M., Osman, A.I., Eletmany, M.R., Hafouda, M.A., Chen, W., 2024. Plant starch extraction, modification, and green applications: a review. Environ. Chem. Lett. 22 (5), 2483–2530. https://doi.org/10.1007/s10311-024-01753-z.

Ren, J., Fu, H., Ren, T., Yuan, W., 2009. Preparation, characterization and properties of binary and ternary blends with thermoplastic starch, poly(lactic acid) and poly(butylene adipate-co-terephthalate). Carbohydr. Polym. 77 (3), 576–582. https://doi.org/10.1016/j.carbpol.2009.01.024.

Rhim, J.W., Park, H.M., Ha, C.S., 2013. Bio-nanocomposites for food packaging applications. Prog. Polym. Sci. 38 (10–11), 1629–1652. https://doi.org/10.1016/j.progpolymsci.2013.05.008.

Romano, S., De Santis, S., Frezza, C., Orsini, M., Sotgiu, G., Feroci, M., Rocco, D., 2025. Ionic liquids as starch plasticizers: the state of the art. Molecules 30 (5), 1035. https://doi.org/10.3390/molecules30051035.

Saberi, B., Chockchaisawasdee, S., Golding, J.B., Scarlett, C.J., Stathopoulos, C.E., 2017. Physical and mechanical properties of a new edible film made of pea starch and guar gum as affected by glycols, sugars and polyols. Int. J. Biol. Macromol. 104 (Pt A), 345–359. https://doi.org/10.1016/j.ijbiomac.2017.06.051.

Sanyang, M.L., Sapuan, S.M., Jawaid, M., Ishak, M.R., Sahari, J., 2015. Effect of plasticizer type and concentration on tensile, thermal and barrier properties of biodegradable films based on sugar palm (Arenga pinnata) starch. Polymers 7 (6), 1106–1124. https://doi.org/10.3390/polym7061106.

Scharnowski, J.B., Rodriguez-Uribe, A., Pal, A.K., Wang, T., Snowdon, M.R., Misra, M., Mohanty, A.K., 2022. Biocomposites from thermoplastic Postindustrial waste starches filled with mineral fillers for single-use flexible packaging. Macromol. Mater. Eng. 307 (6). https://doi.org/10.1002/mame.202100960.

Shin, B.Y., Jang, S.H., Kim, B.S., 2011. Thermal, morphological, and mechanical properties of biobased and biodegradable blends of poly(lactic acid) and chemically modified thermoplastic starch. Polym. Eng. Sci. 51 (5), 826–834. https://doi.org/10.1002/pen.21896.

Shirai, M.A., Grossmann, M.V.E., Mali, S., Yamashita, F., Garcia, P.S., Müller, C.M.O., 2013. Development of biodegradable flexible films of starch and poly(lactic acid) plasticized with adipate or citrate esters. Carbohydr. Polym. 92 (1), 19–22. https://doi.org/10.1016/j.carbpol.2012.09.038.

Si, W., Zhang, S., 2024. The green manufacturing of thermoplastic starch for low-carbon and sustainable energy applications: a review on its progress. Green Chem. 26 (3), 1194–1222. https://doi.org/10.1039/d3gc03671e.

Singh, N., Kaur, A., Shevkani, K., Ezekiel, R., Kaur, P., Isono, N., Noda, T., 2018. Structural, morphological, thermal, and pasting properties of starches from diverse Indian potato Cultivars. Starch/Staerke 70 (3–4). https://doi.org/10.1002/star.201700130.

Sirbu, E.E., Dinita, A., Tănase, M., Portoacă, A.I., Bondarev, A., Enascuta, C.E., Calin, C., 2024. Influence of plasticizers concentration on thermal, mechanical, and physicochemical properties on starch films. Processes 12 (9). https://doi.org/10.3390/pr12092021.

Surendren, A., Mohanty, A.K., Liu, Q., Misra, M., 2022. A review of biodegradable thermoplastic starches, their blends and composites: recent developments and opportunities for single-use plastic packaging alternatives. Green Chem. 19 (11). https://doi.org/10.1039/d2gc02169b.

Van Krevelen, D.W., Te Nijenhuis, K., 2009. Cohesive properties and solubility. In: Properties of Polymers:Their Correlation with Chemical Structure; Their Numerical Estimation and Prediction from Additive Group Contributions. Elsevier, Amsterdam, pp. 189–227. https://doi.org/10.1016/B978-0-08-054819-7.00007-8.

Varavinit, S., Shobsngob, S., Varanyanond, W., Chinachoti, P., Naivikul, O., 2003. Effect of amylose content on gelatinization, retrogradation and pasting properties of Flours from different Cultivars of Thai rice. Starch - Stärke 55 (9), 410–415. https://doi.org/10.1002/star.200300185.

Vilaplana, F., Hasjim, J., Gilbert, R.G., 2012. Amylose content in starches: toward optimal definition and validating experimental methods. Carbohydr. Polym. 88 (1), 103–111. https://doi.org/10.1016/j.carbpol.2011.11.072.

Wang, J.L., Cheng, F., Zhu, P.X., 2014. Structure and properties of urea-plasticized starch films with different urea contents. Carbohydr. Polym. 101 (1), 1109–1115. https://doi.org/10.1016/j.carbpol.2013.10.050.

Wang, Y., Ou, X., Ali Al-Maqtari, Q., He, H.-J., Othman, N., 2024. Evaluation of amylose content: structural and functional properties, analytical techniques, and future prospects. Food Chem. X 24, 101830. https://doi.org/10.1016/j.fochx.2024.101830.

Wongphan, P., Panrong, T., Harnkarnsujarit, N., 2022. Effect of different modified starches on physical, morphological, thermomechanical, barrier and biodegradation properties of cassava starch and polybutylene adipate terephthalate blend film. Food Packag. Shelf Life 32. https://doi.org/10.1016/j.fpsl.2022.100844.

Yamaguchi, A., Arai, S., Arai, N., 2024. Design strategy for blends of biodegradable polyester and thermoplastic starch based on a molecular dynamics study of the phase-separated interface. Carbohydr. Polym. 333, 122005. https://doi.org/10.1016/j.carbpol.2024.122005.

Yin, Q., Chen, F., Zhang, H., Liu, C., 2015. Fabrication and characterisation of thermoplastic starch/poly(butylene succinate) blends with maleated poly(butylene succinate) as compatibiliser. Plast, Rubber Compos. 44 (9), 362–367. https://doi.org/10.1179/1743289815Y.0000000031.

Yokesahachart, C., Khanoonkon, N., Yoksan, R., 2024. Effect of thermoplastic starch/poly (lactic acid) weight fraction on phase morphology and performance of biodegradable blends and their jute fiber composites. Int. J. Biol. Macromol. 283 (Pt 2), 137705. https://doi.org/10.1016/j.ijbiomac.2024.137705.

Zdanowicz, M., 2021. Deep eutectic solvents based on urea, polyols and sugars for starch treatment. Int. J. Biol. Macromol. 176, 387–393. https://doi.org/10.1016/j.ijbiomac.2021.02.039.

Zdanowicz, M., Sałasińska, K., 2023. Characterization of thermoplastic starch plasticized with ternary urea-polyols deep eutectic solvent with two selected fillers: Microcrystalline cellulose and Montmorillonite. Polymers 15 (4), 972. https://doi.org/10.3390/polym15040972.

Zhang, Y., Han, J.H., 2006. Plasticization of pea starch films with monosaccharides and polyols. J. Food Sci. 71 (6), E253. https://doi.org/10.1111/j.1750-3841.2006.00075.x.

Zhang, S., He, Y., Yin, Y., Jiang, G., 2019. Fabrication of innovative thermoplastic starch bio-elastomer to achieve high toughness poly(butylene succinate) composites. Carbohydr. Polym. 206, 827–836. https://doi.org/10.1016/j.carbpol.2018.11.036.

Zhong, Y., Qu, J.Z., Liu, X., Ding, L., Liu, Y., Bertoft, E., Petersen, B.L., Hamaker, B.R., Hebelstrup, K.H., Blennow, A., 2022. Different genetic strategies to generate high amylose starch mutants by engineering the starch biosynthetic pathways. Carbohydr. Polym. 287, 119327. https://doi.org/10.1016/j.carbpol.2022.119327.

Zhou, L., Wang, Y., 2021. Physical and antimicrobial properties of zein and methyl cellulose composite films with plasticizers of oleic acid and polyethylene glycol. LWT 140, 110811. https://doi.org/10.1016/j.lwt.2020.110811.

Żołek-Tryznowska, Z. and Cichy, Ł. (2018) Glycerol derivatives as a modern plasticizers for starch films. doi:10.24867/GRID-2018-p27.

Żołek-Tryznowska, Z., Kałuża, A., 2021. The influence of starch origin on the properties of starch films: packaging performance. Materials 14 (5), 1–11. https://doi.org/10.3390/ma14051146.

CHAPTER FIVE

Enhancing shelf-life

Arif Ozcan[1, 3], Zuzanna Żołek-Tryznowska[2]
[1]Marmara University, Faculty of Applied Sciences, Department of Printing Technologies,Istanbul, Turkey
[2]Faculty of Mechanical and Industrial Engineering, Warsaw University of Technology, Warsaw, Poland
[3]Rochester Institute of Technology, College of Engineering Technology, Department of Packaging and Graphic Media Science, Rochester, NY, United States

5.1 Introduction – the impact of packaging

Packaging has been known since around 7000 BCE, whereas the first primitive glass or pottery-based were used (Teixeira-Costa and Andrade, 2022). Nowadays, packaging plays an important role in the food industry. Food packaging must meet specific criteria and fulfill several functions (Rodríguez-Rojas et al., 2019). The primary functions of food packaging are to contain, protect, and inform. The food packaging must ensure food preservation and safe delivery to the customer. The packed product must be protected from external environmental factors such as oxygen, light, temperature, and microorganisms (Fan et al., 2023). At the same time, the packaging must provide the customer with the necessary product information and have a psychological impact on a potential customer through appropriate product presentation.

Over the years, various materials were introduced for food packaging, starting from ceramic or pottery, metals, paper and cardboard, and various plastics. The food packaging industry is intensely interested in developing packaging with different materials that are less toxic and harmful to humans and the environment. Therefore, attention is paid to biodegradable materials produced from natural resources for large-scale applications, such as starch-based films.

Apart from food protection from external factors and improvement in storage and transportation, it also maintains food freshness and extends the shelf life of the food product (Wang et al., 2024b). Packaging also plays a crucial role in preventing food waste during supply chains or households. Non-processed food: meat, fish, dairy, fresh fruits, and vegetables are exposed to various microorganisms due to high water content causing food waste (Teixeira-Costa and Andrade, 2022). Nowadays, food waste and food spoilage are some of the world's leading issues that need to be

Starch-Based Sustainable Food Packaging
ISBN: 978-0-443-36346-7
https://doi.org/10.1016/B978-0-443-36346-7.00010-2

solved. Annually, approx. 1.3 billion tons – one-third of the world's food production is wasted (Programme, 2024). This problem is common all over the world in developed and developing economies. According to Eurostat, the total food waste in the European Union in 2022 was nearly 60 million tons of fresh mass measured in 2022, of which households generated 54% (Eurostat, 2022). Worldwide, approx. 631 million tons were wasted in 2022, according to (Programme, 2024). On the other hand, minimizing food waste is critical for environmental, social, and economic reasons (Urugo et al., 2024). There are several ways to prevent food waste, deterioration, and spoilage; one of the methods is to extend the shelf life and quality of food products.

5.1.1 Understanding shelf-life

The term "shelf-life" can be defined in several ways depending on the consumer or producer's perspective. However, all these definitions show that the shelf life is the finite period during which a given product can be safely consumed (Soro et al., 2021). During this period, a food product retains its desired sensory features, such as taste and texture, chemical, microbiological, and functional properties. The shelf-life is specific to a given food product and is influenced by several factors such as ingredient composition, processing methods, type of packaging, and storage conditions factors such as humidity, temperature, light exposure, etc (Moschopoulou et al., 2019). In the EU, according to the European Council regulations, the shelf life of the food product must be communicated to the consumer through "use by" or "best before" date marking (Man, 2015). This date indicates that the food is safe for consumption when stored according to the recommended conditions.

Various reactions occur during food storage and affect food quality and safety. Food deterioration is related to physical, chemical, and microbiological processes. Food spoilage may be caused by any or all of these mechanisms, or the onset of one may trigger the onset of another mechanism. During food spoilage, the quality and safety are changed due to flavor, texture, odor, appearance, and various microbes, including dangerous ones, changes, making it unfit for consumption.

5.1.1.1 Food spoilage mechanism

Foods are organic materials that are eaten for their nutritional value. They are made up of fluids, proteins, lipids, carbohydrates, minerals, and other organic substances and come from either plants or animals (Mafe et al., 2024). A variety of nutrients, such as vitamins, lipids, minerals, proteins, carbs, water, and fiber, are found in food (Vigar et al., 2020). Foods can lose their nutritional content, color, texture, and edibility due to microbiological, chemical, or physical processes that cause spoiling (Karanth et al., 2023). Food preservation is therefore essential to retaining its quality over time (Nwosu et al., 2022).

Spoilage is the destruction of compounds such as proteins, carbohydrates, fats, and various organic acids, alcohols, aldehydes, cellulose and pectin in their structures, resulting in foods gaining sensory properties that are undesirable to the consumer. It occurs through changes that will limit the use of a foodstuff, reduce or eliminate its consumability.

Food spoilage is a negative food quality change that is manifested by undesirable changes in appearance, texture and especially odor, which make food risky and unhealthy for consumption by humans or animals (Lianou et al., 2016). It is a complex situation that generally occurs due to microbiological, physical and chemical reasons (Petruzzi et al., 2017). Although new systems and technologies continue to emerge every day, food spoilage is a serious problem that stands out worldwide (Odeyemi et al., 2020). Producers (farmers), retailers, and consumers suffer major economic losses from food spoilage (Snyder and Worobo, 2018). The rate of spoilage depends on factors such as storage conditions (Odeyemi et al., 2018; Ndraha et al., 2018; Mageswari et al., 2015), presence of water or moisture, pH, temperature (Bradford et al., 2018), some microflora such as fungi or bacteria (Ioannidis et al., 2018), food processing method (Wang et al., 2018), transportation method and conditions (Hammond et al., 2015). Spoiled food carries a high risk for food insecurity, especially waste and economic losses (Illikoud et al., 2019).

The most frequent cause of food spoiling is microbes, which are present everywhere. Bacteria and yeasts, with the exception of molds, are too small to be detected by the human eye and may go undetected when they colonize exposed foods (Hammond et al., 2015). Depending on the food type, spoilage bacteria can colonize food in a variety of ways. In contrast to foods with low water content, foods with a high water content, including meat, milk, and seafood, are more susceptible to bacterial deterioration, whereas molds or yeasts typically cause the deterioration of staple foods. However, the problem is challenging to resolve because of the intricacy of food spoiling and the interdependence of the elements that contribute to it.

In food spoilage, noticeable or unnoticeable changes in consistency, smell, and taste may occur. Not every change indicates that the food has spoiled, the degree of change is important. Spoiled food may not be suitable for consumption, but it may not be harmful to health either.

When food is harvested, sliced, or produced, it deteriorates. Some foods spoil quickly, but some can be stored for longer periods of time but may also have a very short shelf life before being consumed, undergoing physiological, chemical and biological changes (Karanth et al., 2023). Some people prefer aged steak and cheeses or extremely ripe fruits, so not all of these dietary changes are bad (Edo et al., 2024). Additionally, sugars must be converted into alcohol in order to produce wine and beer, and milk must be soured in order to produce cheese (Siddiqui et al., 2023). Food spoilage occurs when food loses its original flavor, texture and/or nutritional value and is no longer safe for the consumer (Amit et al., 2017). Bacteria, yeasts, and molds are the organisms that cause food to spoil. Food deterioration is typically inferred from the food's appearance, flavor, or aroma.

5.1.1.2 Microbial spoilage

Foods that undergo microbial food deterioration undergo organoleptic alterations that make them unfit for human consumption. The development of microbes and the enzymatic activity they produce are typically the source of microbial food spoiling, but it can also result from microbial enzymes that are present in food even after the relevant microorganisms have been rendered inactive. Foods can undergo a wide range of microbially driven organoleptic alterations, including as color defects, unfavorable tastes, odors, and physical and textural changes (such swelling and curdling). Because a single spoilage organism can cause several alterations or, less frequently, because multiple spoilage organisms grow in a food item, microbiologically damaged food products can have multiple faults (Snyder et al., 2024). Foods carrying microbial pathogens rarely exhibit simultaneous organoleptic spoilage signs, and foods that exhibit spoilage signs rarely cause foodborne infections or poisoning (despite being able to trigger a "gag reflex"). Microbial food safety and microbial food spoilage issues can be confused with each other in the public opinion and sometimes in the literature. Therefore, the only topic covered in this review is microbiological food spoiling.

Food safety and environmental sustainability are significantly impacted negatively by microbial food spoilage on a global scale (Gram et al., 2002). Food spoiling can also have serious effects on the food sector, including unfavorable customer reactions that may result in lower profitability or market withdrawals (Snyder et al., 2019). Furthermore, even if only a tiny percentage of the product is impacted, food firms may have to conduct expensive recalls or remove it from the market as a result of individual consumers sharing food rotting instances on social media.

Numerous microorganisms, such as bacteria and fungi, can cause microbial food spoilage. Bacteriophages, bacteria, and fungi are some of the possible sources of microbial food loss. Aerobic and anaerobic spore-forming molds and yeasts from the classes *Bacillales* and *Clostridiales*, non-spore-forming lactic acid bacteria, and all kinds of Gram-positive bacteria and bacteria of the genus *Pseudomonas* are the bacterial groups commonly encountered in food spoilage situations.

5.1.1.3 Chemical spoilage

The breakdown or reactivity of the meal's chemical components, such as proteins, lipids, and carbohydrates, can result in food deterioration in food products. The rate at which chemical reactions occur depends on temperature, water activity, pH, and exposure to light or oxygen. The ideal circumstances for each reaction vary. Every chemical interaction has the ability to alter a food product's color, flavor, aroma, and texture. When chemicals are stored for extended periods of time or are transported in inappropriate conditions, the process of chemical spoiling starts. Common chemical spoiling processes include oxidation or degradation of proteins, enzymatic activity in fruits and vegetables, non-enzymatic browning, chemical breakdown of carbohydrates, gelatinization, oligosaccharides in foods, and lipid degradation (Pothakos et al., 2015; Huis in't Veld, 1996).

Enzymatic activity or reactions with other proteins can be involved in protein breakdown. Complex proteins called enzymes function as catalysts to significantly speed up chemical reactions. Numerous enzyme types, many of which are created by microorganisms, react with various chemical components in food. Protein oxidation is another possibility. When myoglobin and oxymyoglobin in meats are exposed to too much oxygen they can oxidize to metmyoglobin, turning the meat brown. This color change is not what the consumer wants.

Tissues in fruits and vegetables may become softer and darker due to enzymatic activity. Phenol oxidase enzymes usually catalyze these reactions, combining with phenol molecules and oxygen to produce unwanted brown colors. When the cells are easily peeled, sliced, or crushed, these reactions happen quickly (Navina et al., 2023; von Neubeck et al., 2015).

Proteins (amino groups) and reducing sugars react to produce non-enzymatic browning, also known as Maillard browning. To create volatiles and dark colors, the processes go via polymerization and Strecker degradation (Murata, 2021). This causes the food to become brown in color. Additionally, a loss of nutritional content is typically linked to Maillard browning. During non-enzymatic browning, lysine, an important amino acid that combines easily with reducing sugars, is quickly lost. Additionally, Maillard browning reactions are highly pH dependent and infrequently take place at low pH values. Additionally, metal ions like copper and iron stimulate the process (Kathuria et al., 2023).

Gelatinization/recycling reactions and browning reactions are two examples of the chemical breakdown of carbohydrates. Carbohydrates can also undergo other browning processes, like caramelization, but these processes ask for temperatures greater than those that products will normally encounter during storage and distribution (Kurzyna-Szklarek et al., 2022). Gelatinization is the swelling of starch granules with water and the resulting loss of crystallinity. The extent of gelatinization depends on the ambient temperature and the amount of water, as well as the number of other substances such as lipids or sugars added to it. Re-association or re-crystallization of starches is known as retrogradation.

Hydrolytic processes can also break down food's polysaccharides and oligosaccharides. Using acid and enzymatic hydrolysis to turn starch into corn syrup is one of the most frequent hydrolytic processes in food. Enzymes have an impact on the amount of starch in food processes like fermentation and brewing. Similar hydrolytic processes can be carried out by pectin lyase and polygalacturonase to soften the texture of fruits and vegetables.

The majority of lipid degradation is caused by oxidation events, lipolytic enzyme activity, and other hydrolytic processes. The most significant process for breaking down fats and oils is lipid oxidation, also known as oxidative degradation, which happens in a variety of food items, including nuts, fried foods, meats, margarine, dried fruits, coffee and powdered milk. When lipids undergo oxidation, oxygen targets unsaturated fats. Color shifts, odd smells, and occasionally even hazardous substances are the results of this. The reagent may dissolve in the oil or food product, stay in the headspace, or flow through the packing while being stored. To reduce the oxidation of lipids, appropriate packaging should be used to prevent oxygen transmission in food products. The rate of oxidation is affected by the amount and location of double bonds on fatty acids or triglycerides. Other crucial factors include heat and light, since they have the ability to catalyze the oxidation process. The product's other trace components may have an impact on the rate and degree of oxidation. Iron and copper are examples of trace metals that can significantly catalyze the oxidation process. However, by reacting with the product's radical oxygen, tocopherols and other antioxidants can slow down or stop oxidation. In these situations, natural antioxidants are occasionally added in an effort to lessen oxidation (Singh and Anderson, 2004).

Another method of lipid degradation is hydrolytic degradation, which can occur by chemical processes or the activity of lipolytic enzymes. In the presence of water, free fatty acids are separated from the triglyceride molecules. The released free fatty acids have shorter chain lengths, lower taste thresholds, and sometimes off-flavors and odors. Reactions can be decreased by reducing moisture, and heating beyond 60°C can inactivate the majority of lipolytic enzymes (Kristott, 2000).

5.1.1.4 Physical spoilage

When food deteriorates physically, there are bodily changes or imbalances. This can be described as physical harm, such fresh fruits and vegetables getting bruised or dry, brittle things breaking. When fruits and vegetables are dropped or transported and distributed, they may sustain bruises. The product is rendered useless when there is significant physical damage. Cell disintegration and typically a drop in water content at the site of bruise or damage are the causes of enzymatic browning. Decomposition also harms cells and promotes the proliferation of microorganisms. The value and reputation of brands like ready-to-eat cereals, crackers, pie crusts, potato chips, and frozen foods are harmed when dry, brittle products break. Using well-designed packing techniques that shield goods from mechanical damage and vibration during distribution and transit can help minimize decay and breakage (Raposo et al., 2017).

The most frequent physical alteration or imbalance is caused by moisture or mass transfer of food ingredients. Food products frequently spoil due to changes in their water content, such as water migration, gain, or loss. A change in moisture by itself can make a product unfit for consumption, but it can also lead to other issues like chemical or microbiological spoiling (Skibsted et al., 2010). Meals made up of materials having different water activity. For instance, moisture frequently moves from fruit pieces to cereals that are ready to eat and from moist fillings to drier food product shells. Many fruits and vegetables can be significantly impacted by changes in humidity. Unwrapped leafy veggies wilt, grow, and lose moisture.

Temperature has a significant impact on the freshness of fresh fruits and vegetables. For maximum postharvest life, each crop has a certain respiration rate and optimal temperature range. Furthermore, as they ripen, climacteric fruits produce more ethylene. Ethylene is a potent regulator of plant growth, and most exposures cause produce to age more quickly. Therefore, limiting or removing ethylene exposure will prolong the post-harvest life of most products by delaying ripening and aging (Jaya et al., 2002).

Since most product can be damaged by freezing when the temperature is gradually lowered and partially frozen, temperature is crucial. This destroys the product by rupturing the cells. Most tropical fruits and vegetables, as well as other produce, are vulnerable to freezing damage, which often happens at temperatures between 5 and 15°C prior to freezing. Pitting, water soaking, discoloration, the formation of bad tastes, faster aging, and ripening or overripening are all consequences of freezing. These consequences, nevertheless, can occasionally go undetected for weeks following the harm (Ozmen and Langrish, 2002).

Among other physical alterations that can take place in food is crystal formation. Ice crystal formation in frozen foods like ice cream is one example of this. Foods that have seen several freeze/thaw cycles or delayed freezing processes are more prone to develop large extracellular ice crystals.

During storage, unpackaged materials with high water content lose water by releasing water vapor into the atmosphere. While the drying that occurs during this time plays a role in preventing the development of microorganisms, certain negative characteristics occur in food. These negative changes are frequently observed in foodstuffs that are stored in the cold for a very long time. An example of this is freezer burns. Freezer burns are seen in frozen meat, offal, poultry and fish. Freezer burns are frequently encountered in unpackaged foodstuffs with low humidity and in direct contact with the outside air. Freezer burns are hygienically harmless (Subramaniam and Wareing, 2016).

5.2 Enhancing shelf-life of food

Nowadays, the increasing consumer demand for minimally processed foods that are prone to spoilage jeopardizes food safety and quality (Ju et al., 2019; Sofi et al., 2018; Suppakul et al., 2003). Food spoilage caused by microbial growth or activity is the most common cause of food loss, making food unsafe for consumption. There are several ways to extend the shelf life of food while minimizing food waste in the supply chains and increasing food safety. One of them includes food processing such as refrigeration, freezing, drying, fermentation, canning, pasteurization, sterilization, etc (Prokopov and Tanchev, 2007; Reyes et al., 2024). For some products, such as fresh fruits, vegetables, meat, fish, or seafood, food processing is not a solution to extend the shelf's life. Therefore, in this case, appropriate packaging is adapted to the packed food. For this purpose, various packaging can be applied: active packaging, intelligent and smart packaging, modified atmosphere packaging, and packaging with enhanced material.

Active packaging, apart from the primary function of packaging such as barrier against external factors, has some additional capabilities that delay ripening (Kraśniewska et al., 2020). Active packaging slows down respiration, inhibits microbial development, and reduces moisture migration to maintain food's freshness, quality, and safety. Some modern food packaging features moisture, ethylene, oxygen, and carbon dioxide scavengers, as well as antioxidants and antibacterial agents, making them active (Dutta and Sit, 2024). Antimicrobial chemicals can preserve food by lengthening the shelf life while guaranteeing food safety and nutrition by inhibiting microbial growth. Gas-absorbing systems control the atmosphere or absorb moisture in the packaging, conserving the food flavor. All those systems incorporated in the packaging make it active to extend the food's shelf life.

Another type of packaging is smart or intelligent packaging. This packaging differs from active packaging in that it provides information about the condition of the product inside. Smart packaging does not directly slow the spoilage process but plays a crucial role in monitoring and communicating the condition of the food contained. This can be achieved in several ways. By including indicators or sensors, it is capable of detecting changes in temperature, gas composition, or microbial activity and communicating this in real-time to consumers in terms of freshness, safety, and potential spoilage. For instance, smart packaging can contain a label that irreversibly changes color while spoiling or if the product has been stored in inappropriate conditions according to the manufacturer's recommendations. Another solution is to implement it into the packaging RFID device or tag, which allows to trace the packaging in the supply chain. The RFID device can also be equipped with several electronic sensors detecting changes in food products, such as pH, conductivity, humidity, temperature, gas, etc (Zuo et al., 2022).

The last group of packaging is modified atmosphere packaging (MAP), in which specific gas composition, mainly mixtures of carbon dioxide, nitrogen, and oxygen, is used to prevent microbial growth or food oxidation. However, in MAP, conventional films based on raw fossil materials are used mainly, but there are some reports confirming the use of more sustainable materials for this type of packaging (Czerwiński et al., 2021).

5.3 Starch-based packaging materials for shelf-life enhancement

Starch-based materials can be successfully applied for packaging or coating purposes due to several advantages, for instance, biodegradability, barrier properties, and compatibility with food (Majeed et al., 2023). At the same time, starch-based material can be easily adjusted by chemical modification of starch or additions of various additives, nano- and microparticles, essential oils, extracts, etc., to achieve desired properties, that is, shelf-life extension. Active compounds are essential for improving the functionality and effectiveness of starch-based films.

Starch-based film or coating formulation is well-known and has been discussed in several studies. Fig. 5.1 illustrates the scheme of development of packaging or coating with active agents for extending shelf-life purposes. Generally, the coating is formed from starch, water, and plasticizer heated above the gelatinization temperature of starch. The active compounds are mainly added during the starch solution forming. Then, the solution can be cast on Petri dishes or Teflon plates to develop a film for packing or coating by spraying, dipping, or spreading on the food, for example, fruit or vegetables.

In this chapter, emphasis is made on various additives to starch-based films. Active additives are categorized into three groups: nanoparticles, NPs, essential oils, and antioxidation agents. It should be emphasized that active agents can also be combined, or the starch film itself may have properties that allow extending food freshness.

5.3.1 Starch coating or packaging

Starch-based films or coating do not demonstrate antimicrobial or antioxidant properties by themselves (Moghadas et al., 2025) but might extend the shelf-life of the packed product due to their specific barrier properties. Packaging or coating prepared without active additives, combining starch, cellulose, chitosan, or another biodegradable or bio-friendly polymer, can also be utilized for a shelf-life extension. Increasing the hydrophobicity of the material and its affinity to water allows the development of an excellent water barrier (Yousuf et al., 2018). The starch-based coating acts as a barrier to the transition of gases, such as oxygen, carbon dioxide, or water vapor,

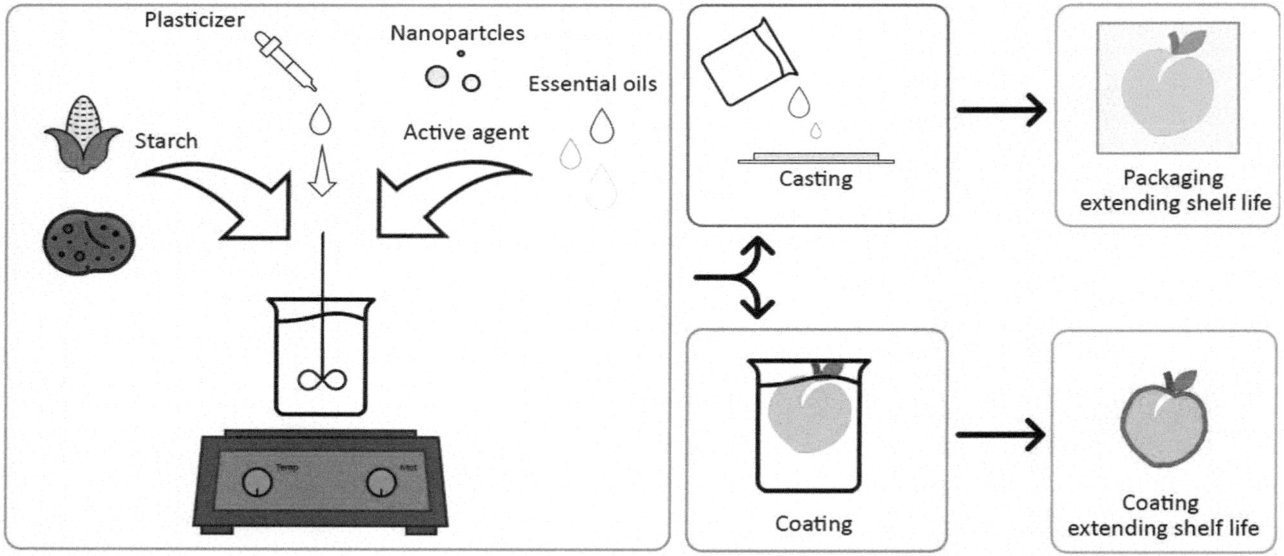

Figure 5.1 Summary of the development of starch-based films with active agents for packaging or coating. *Based on Z. Żołek-Tryznowska, A. Jeznach, E. Bednarczyk, T. Murawski, K. Piłczyńska, S. Sikora, M. Tryznowski, Effect of hydrophobic nano-silica content on the surface properties of corn-starch films. Ind. Crop. Prod. 214 (2024), 118582. https://doi.org/10.1016/j.indcrop.2024.118582; Júnior, L. M., Vieira, R. P., & Anjos, C. A. R. (2020). Kefiran-based films: fundamental concepts, formulation strategies and properties. Carbohydr. Polym. 246, 116609. https://doi.org/10.1016/j.carbpol.2020.116609.*

resulting in delaying the ripening process and reducing microbial growth (Basiak et al., 2022).

Therefore, several reports confirm the effect of extending food freshness through starch coating or packaging. The contribution of natural-based polymers can combine the properties of both components. For instance, Fakhouri et al. reported the combination of corn starch with waxy starch and gelatin using sorbitol or glycerol as a plasticizer as a composite coating for Red Crimson grapes (Fakhouri et al., 2015). Due to the lower water vapor permeability of the developed coating, delaying ripening for 21 days was observed. Moreover, the combination of acetylated cassava starch and hydroxyethyl cellulose can also be found in food preservation and successfully used to increase the shelf life of guavas fruits (Francisco et al., 2020).

Starch coating enriched in starch-whey protein protects plum fruit from ripening (Basiak et al., 2022). This work also compared the affectivity of various ways of applying starch-based materials. The starch-based films were applied as coating and wrapping as packaging, and the wrapping was more effective, resulting in a lower weight loss. Furthermore, potato starches of various potato origins did not have a positive effect on the shelf-life of blueberries, but Median-Jaramillo reported a decrease of respiration rate and greater bright and firmness of the coated fruits (Medina-Jaramillo et al., 2019).

The starch-based coating can also be a greener alternative to plastic wrapping based on fossil raw materials for green cucumbers. Patel and Panigrahi reported the influence of starch glucose coating on its beneficial effect in delaying the ripening process and simultaneously increasing the shelf-life of cucumbers during long storage (Patel and Panigrahi, 2019). Cucumbers are suitable for consumption for up to five to seven days after harvesting and, if covered with starch glucose coating, even for up to 30 days.

The coating with SC resulted in dehydrated products with better color and higher retention of trans-α-carotene and trans-β-carotene, increasing the pro-vitamin A content.

The application of SC for walnuts could help to reduce storage space, once removing the husk and shell of walnuts results in less space requirements. In addition, SC could reduce the cost of storage by reducing the need for refrigeration.

5.3.2 Nanoparticles

Nanoparticles, NPs, according to the simplest definition, are tiny particles of size up to 100 nm in any direction (Strambeanu et al., 2015). NPs have many chemical and physical properties that are different from bulk materials. For example, gold in the form of bulk metal has a characteristic golden shine, and in the form of 20 nm NPs has a wine-red color; silver, platinum, or palladium NPs are yellowish gray and black, respectively (Horikoshi and Serpone, 2013). According to their origin, NPs can be divided into three different categories: (i) organic, (ii) inorganic, and (iii) carbon-based (Anu Mary Ealia and Saravanakumar, 2017). Organic NPs are composed of organic compounds such as lipids or polymers and include a group of various polymers such as dendrimers, micelles, etc (Shatrohan Lal, 2014). They are mainly applied in the biomedical or pharmaceutical sectors, for instance, in drug delivery systems (Peters et al., 2011).

NPs, due to their remarkable properties such as optical properties, surface area, uniformity, and possibility of functionalization, have attracted interest in various applications. One of them is the enhancement of starch-based materials to extend the shelf-life of food products. For this purpose, various nanomaterials, including cellulose nanofibers, Montmorillonite, and various metal or metal oxide nanoparticles (Ag, ZnO, CuO, etc.), can be used for mechanical properties improvement, changes barrier properties, and due to the antimicrobial activity of the additives delaying ripening of the food. Table 5.1 summarizes the application of various NPs for enhancing the shelf life of food products, including fruits, vegetables, dairy, meat, etc. It is well known that food deteriorates due to microorganism contamination. At the same time, metal and metal oxide are characterized by antimicrobial activity, which prevents the growth of microorganisms (He and Hwang, 2016). As shown in Table 5.1, zinc oxide, titanium oxide, copper oxide, silver, and selenium NPs were incorporated into the starch matrix film to extend the shelf life of food or inhibit various bacteria's growth. It should be highlighted that, the usage of AgNPs is restricted by EU. Also, the NPs properties can be reinforced with multiple essential oils or extracts, such as thyme oil or elderberry extract. Moreover, besides using metal or metal oxide NPs, some researchers report the functionalization of natural-based NPs. Li et al. proposed a synthesis of multifunctional organic NPs by combining epigallocatechin-3-gallate (phenolic compound extracted from tea) with cysteine and cinnamyl aldehyde (Li et al., 2024). The developed coating was used to extend the shelf-life of strawberries and oranges and was characterized by excellent biocompatibility. Finally, the development of carbon dots to extend shelf life was also reported. Carbon dots synthesized from radish residue were successfully applied to starch-chitosan films. Due to functional groups on carbon dots,

Table 5.1 Application of various NPs for enhancing shelf-life of food products or inhibition effect against microorganisms.

Starch source	Nanoparticles	Food	Conclusion	Reference
Hydroxypropyl cassava starch	Zinc oxide	Passion fruit	Coating slowed ripening and reduced nutrient loss of passion fruits at 42 days of storage.	Congying et al. (2024)
Rice starch	Silver Zinc oxide	Peach fruits	Decrease the overall surface microbial load and enhance the shelf life of packed peaches compared to control films.	Kaur et al. (2017)
Cassava starch	TiO_2 ZnO	Pisang mas Bananas Tomatoes	The addition of 0.01%w/w of TiO_2 to the starch films extended the shelf life of banana up to 14 days and tomatoes up to 21 days.	Tunma (2018)
Pearl millet grains starch	Cellulosic nanocrystals	Red grapes	The films as packaging material improved the shelf life of the grapes for 15 days at 5°C.	Punia Bangar et al. (2022)
Corn starch	Selenium nanoparticles	Litchi	Prevent litchis from spoilage and extend its shelf life by antioxidant, antibacterial, and UV blocking, properties.	Yu et al. (2024)
Arrowroot starch	Cellulose nanocrystals	Strawberry	The coatings increased the concentration of esters in the strawberries volatile profile.	(Oliveira Filho et al., 2022)
Starch[a]	Combination of Ag, ZnO and CuO	-	Strangest effect of retarding S. aureus and E. coli growth respectively, by CuO and Ag nanoparticles.	Peighambardoust et al. (2019)

Corn starch	Green silver nanoparticles (G-AgNPs)	Grapes, mushrooms and tomatoes	The presence of starch And G-AgNPs significantly reduced weight loss after 6 days, with cherry tomatoes decreasing by 8.59% and green grapes by 6.77%.	Trotta et al. (2024)
Corn starch	AgPNs	Cheese	Nanocomposite films extended the shelf-life of fresh cheese samples for 21 days.	Ortega et al. (2017)
Starch[a]	ECC_{NPs} (epigallocatechin gallate and cysteine and cinnamyl aldehyde organic NPs)	Strawberries Oranges	SC-ECCNPs coating film can prolong the shelf life of fruits at room temperature.	Li et al. (2024)
Cassava starch	AgNPs enriched in oregano essential oil	-	Inhibition of *Escherichia coli* (*E. coli*) and *Staphylococcus aureus* (*S. aureus*) growth.	Srikhao et al. (2021)
Starch[a]	TiO_2 and elderberry extract	Cherry tomato	Reduced weight loss and no signs of microorganism activity up to 22 days of storage.	Jayakumar et al. (2022)
Tapioca starch	ZnO nanospheres, nanorods, NPs	-	Significant antibacterial activity against *Escherichia coli* bacteria of films enriched in ZnO	Tamimi et al. (2021)

(Continued)

Table 5.1 Application of various NPs for enhancing shelf-life of food products or inhibition effect against microorganisms.—cont'd

Starch source	Nanoparticles	Food	Conclusion	Reference
Corn starch	CuO NPs	-	Fungistatic effect of CuO enriched starch films against *Alternaria alternata* fungus	Malafatti et al. (2023)
Corn starch	Carbon dots based on radish residue	Salmon fillets	Delaying the salmon fillets' spoilage and increasing their shelf life by approximately 4 days	Wang et al. (2024a)

[a]The starch source was not provided.

composite material exhibits high antioxidant and antibacterial delaying the salmon fillets' spoilage and increasing their shelf life by approximately 4 days (Wang et al., 2024a).

5.3.3 Essential oils

Essential oils, EOs, are a liquid mixture derived from plants containing natural-based bioactive organic compounds (Valdivieso-Ugarte et al., 2019; Hossen et al., 2024). Due to their antimicrobial and antioxidant properties against microorganisms causing food spoilage and deterioration, they are promising active agents in the food industry that can prolong shelf life by developing coatings or packaging containing EOs or blending them during food processing.

In general, EOs are a mixture of various volatile organic compounds obtained by distillation from various plants (Sadgrove et al., 2021). The chemical composition of EOs is complex, including various chemical components, among which the following organic compounds can be distinguished: terpenes, phenolics, and alcohols (Li et al., 2015). The presence of phenolic and terpenoid chemical compounds is responsible for the antibacterial and antioxidant properties of EOs. The chemical composition of essential oils is the subject of intensive research, and their quantity can range from a few to several hundred and is related to plant origin. Some of the primary active compounds and their contents are summarized in Table 5.2. For example, Walasek-Janusz et al. investigated the chemical composition of Oregano EO from various origins and have marked from 50 up to 90 various organic compounds in Oregano EO depending on the oil origin (Walasek-Janusz et al., 2024). The component with the highest content is carvacrol, a phenolic monoterpenoid responsible for the characteristic odor of the oregano plant. Carvacrol is also present in different plants; for instance, it is a component of thyme EOs. Besides carvacrol, thyme EO contains thymol as a major compound (Gavaric et al., 2015). Furthermore, cinnamaldehyde is one of the 19 chemical compounds of natural cinnamon EO, and the content ranges from 55% to 90% depending on Cinnamomum species (Guo et al., 2024).

The molecules of EOs are small with an odor typical of a given oil, which might affect food's sensory quality and sensory properties (Valdivieso-Ugarte et al., 2019). Another problem is their volatility, and poor stability, which is related to the small particle size. A solution for this problem is the encapsulation of EOs, which might be necessary if

Table 5.2 List of some EOs successfully used in starch films or coatings for food preservation and shelf-life enhancement.

EOs	Active compound	Content in %	Chemical structure	Ref.
Oregano EO	Carvacrol	76.64–85.70	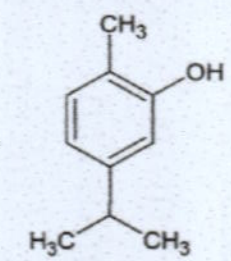	Walasek-Janusz et al. (2024)
Thyme EO	Thymol	41.6	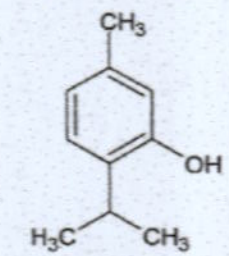	Gavaric et al. (2015)
Cinnamomum EO	Cinnamaldehyde	55–90	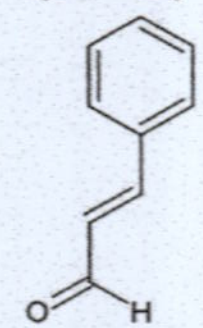	Guo et al. (2024)
Orange EO	*d*-Limoneno	95.96	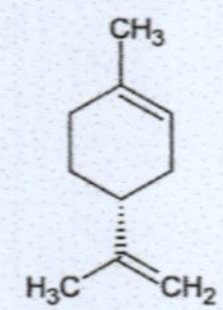	do Evangelho et al. (2019)
Clove EO	Eugenol	66.9–88.8	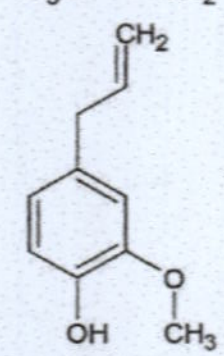	Fan et al. (2024)

they are to be used in food preservation. At the same time, encapsulation allows for the gradual release of active EOs during storage, even upt a 1 month. Last but not least, the chemical composition of EOs is strongly related to their origin and the environment in which the plant grows, such as sunlight or humidity, and recreating the chemical composition of EOs is challenging. Despite those issues, EOs are successfully used in starch-based packaging or coatings to delay ripening. Preventing food spoilage is mainly related to the antimicrobial effects of active compounds in EOs, inhibiting the growth of various bacteria and fungi on food surfaces. The inhibitory effect of starch-based films with EOs is generally related to the increasing EOs content. However, it must be noted that greater EO content may affect other properties, such as the mechanical properties of the films (Souza et al., 2013) and the sensory quality of food. Therefore, a specific optimal composition that positively affects all important parameters must be selected.

Table 5.3 lists various EOs successfully used for the enhancement of the shelf-life of various foods, that is, fruits and vegetables, meat, and dairy. According to the literature reports, the starch-based coating enriched in EOs inhibits bacteria and fungi growth, preventing food deterioration and spoilage. As it was mentioned before, the EOs could be encapsulated, whereas the capsule is synthesized from natural-based or synthetic polymers. Ferreira et al. reported cinnamon and Ho wood EOs encapsulating in chitosan nanocapsules (Ferreira et al., 2021). The film's antimicrobial activity and low water permeability allow the extended shelf life of strawberries up to 15 days. Furthermore, the EOs can also be encapsulated in starch nanofibers or microspheres (Glenn et al., 2010; Fonseca et al., 2020) or nanosilica (Yan et al., 2022). In this case, the application of natural-based capsules incorporated in starch films allows the development of fully biodegradable and environmentally friendly packaging. On the other side, the EOs can be encapsulated in a synthetic polyurethane based on toluene diisocyanate and xylitol and then added to a potato starch film to prevent blueberries and raspberries decay (Wang et al., 2023).

5.3.4 Antioxidation compounds

Due to oxidation, food products lose their quality and nutritional value during storage and transportation. This process is the primary reason for food spoilage, apart from microorganisms. In the oxidation process, the reactive oxygen in oxygen-based radicals, such as oxygen radicals, hydroxyl radicals,

Table 5.3 Essential oils used for the enhancement of the shelf-life of various foods.

Starch origin	Essential oil	Food	Improved properties	Reference
Sweet potato starch	Thyme essential oil + montmorillonite (MMT) nanoclays	Baby spinach leaves	Reduced the population of *E. coli* and *S. Typhi* on fresh baby spinach leaves to below detectable levels within 5 days.	Issa et al. (2017)
Corn starch	Cinnamon oil or Ho wood encapsulated in chitosan nanocapsules	Strawberry	Films with 1 wt% loaded with EOs extend the strawberries' shelf life for 15 days without the appearance of fungi contamination and lower weight loss.	Ferreira et al. (2021)
Purple sweet potato starch	Oregano	-	Excellent antimicrobial activity and the inhibition rate against *Escherichia coli* and *Listeria monocytogenes*.	Chen et al. (2020)
Potato starch	Green tea extract or BHT	Beef meat	These active films appear promising for the development of active antioxidant packaging for use with fresh beef.	u Nisa et al. (2015)
Corn starch	Clove and cinnamon	Beef meat	The incorporation of spice EOs in CS films may provide antimicrobial and antioxidant activities that could improve the stability of raw meat.	Radha krishnan et al. (2015)
Amylose corn starch	Thymus kotschyanus Pomegranate peel extract	Beef meat	Combination of PPE and TEO to increase the meat shelf life.	Mehdizadeh et al. (2020)

Potato starch	Thyme oil microemulsion	Pork meat	The shelf life of chilled meat coated with 5% (v/v) of thyme oil microcapsules increased to 14 days.	Yuan et al. (2021)
Cassava starch	*Lippia gracilis* Schauer essential oil	Guava	The shelf life of guavas coated with chitosan-cassava starch with 3.0% of *L. gracilis* Schauer EO mixture during storage at room temperature increased. The developed coating demonstrated microbiological effect against Gram-positive and Gram-negative bacteria.	De Aquino, Blank and De Aquino Santana (2015)
Corn starch	Thyme	Mango	The corn starch film enriched with thyme EO extended the shelf life of mangos up to 10 days and inhibited the growth of *Botryodiplodia theobromae Pat* and *Colletotrichum gloeosporioides Penz* on fruit surface.	Cai et al. (2020)
Corn starch	Orange	-	The film with addition of orange EO exhibit higher antibacterial activity against Staphylococcus aureus and Listeria monocytogenes.	do Evangelho et al. (2019)
Corn starch	Clove	Bread	Composite film with incorporation of clove EO extended the shelf life up to 9 days of bread.	Fan et al. (2024)

(*Continued*)

Table 5.3 Essential oils used for the enhancement of the shelf-life of various foods.—cont'd

Starch origin	Essential oil	Food	Improved properties	Reference
Cassava starch	Cinnamon	Fish fillets	Films based on cassava starch and fish gelatin with cinnamon EO addition extended the shelf life of fish filets to 10 days.	Arul Raj et al. (2024)
Chayotextle starch	Rosemary or cinnamon EO	Bread	Films based on chayotextle starch enhanced with rosemary oil or cinnamon oil extended the shelf life of bread to 15 days.	Luna-Niño et al. (2025)
Corn starch	Eucalyptus citriodora and melaleuca bracteata	Plum	The composite films based on starch and chitosan hydrogel incorporated with encapsulated EOs enhanced the shelf-life of plums for 15 days without any fungal contamination.	Sharma et al. (2023)
Potato starch	Tea tree, lavender, and perilla leaf EOs in polyurethane nanocapsules	Raspberries, blueberries	Effective fruit preservation and prolonged of shelf life up to 7 days.	Wang et al. (2023)
Corn starch	Lemongrass	Strawberry	Films enriched in lemongrass EO, and montmorillonite or starch nanocrystals extended the quality of strawberries up to 8 and 6 days, respectively.	Singh et al. (2023)

Table 5.4 Chemical compounds used as antioxidants are listed with their E number used in the EU.

Compound	E number
Ascorbic acid, vitamin C	E300
Vitamin E (tocopherols)	E306-E309
Potassium sorbate	E202[a]
Sodium benzoate	E211[a]
Butylated hydroxy anisole	E320
Butylated hydroxytoluene	E321
Citric acid	E330

[a]Classified as preservatives.

or superoxide anions, react with lipids, proteins, or vitamins (Leyva-Porras et al., 2021). These reactions can cause undesirable changes, including color loss, flavor breakdown, or even the formation of toxic compounds. Antioxidant compounds or agents can be added to the film or coating to prevent the oxidation process. They act as scavengers reacting with free radicals to inhibit their further reaction and restrict oxidative destruction. Table 5.4 provides an overview of antioxidant compounds in this chapter. All the additives approved in the EU for usage as food additives have their own E number. The group of food additives with E numbers in the range of 300–399 are listed as antioxidants and acidity regulators. Some of them were successfully used for starch-based active packaging with antioxidant properties.

Synthetic antioxidants such as butylated hydroxy anisole or butylated hydroxytoluene are cheap and highly effective. However, their usage in food preservation is limited according to EU regulations. Furthermore, some chemical compounds, such as potassium sorbate and sodium benzoate, are known and widely used in the food industry as preservatives.

Natural antioxidants can be used as an alternative to synthetic antioxidants. EOs and extracts from herbs, pulp, or fruit by-products, such as peels, husks, leaves, or seeds, are also known for their antioxidant properties. EOs or plant extracts are rich in phenolic compounds, which can react with free radicals, giving an antioxidant effect. Menzel et al. reported the usage of extract from sunflower hull to potato starch films to produce renewable and based on biodegradable resources active food packaging material (Menzel et al., 2019). Finally, some of the by-products of plant production at harvest can be utilized as a source of natural oxidants. The plant by-products can be a valuable source of various bioactive compounds. Menzel et al. reported the incorporation of antioxidants extracted from postharvest rice straw into starch films. The extract of rice straw was rich in phenolic compounds, and starch film exhibited antioxidant activity to prolong the shelf life of selected foods, such as nuts or fresh fruit salads (Menzel, 2020). Last but not least, rosemary extract can be also incorporated into cassava starch film, resulting in active films with a significant antioxidant activity for packaging purposes (Piñeros-Hernandez et al., 2017). It should be noted that natural EOs and plant extracts are not approved for use in food additives apart from rosemary EO, which is marked with the E392 symbol.

Despite EOs or plant extracts, some natural proteins, such as lactoferrin and lysozyme, are recognized as natural antioxidants. Moreno et al. have incorporated lactoferrin and lysozyme for potato starch films and confirmed their antimicrobial activity and reduction of fat oxidation in minced pork meat (Moreno et al., 2015). Vitamins are also known as commonly used compounds as antioxidants. Ascorbic acid, also known as vitamin C, can be encapsulated in nanoliposomes and show antioxidant activity and is used as a functional agent for starch-based packaging film (Khuntia et al., 2022). The film enriched in vitamin C was used as packaging for apples, and the gradual release of vitamin C prevented the browning of the apples. Furthermore, vitamin E is also known for its antioxidant properties. Unfortunately, vitamin E is more sensitive to oxygen and light and is poorly soluble in water. Therefore, vitamin E must be either encapsulated or closed in a polymer matrix as nanofibers (Hategekimana et al., 2015; Celebioglu and Uyar, 2017).

There are several works dealing with the antioxidant capacity of various organic acid. Citric and ascorbic acids have been added to corn starch to extend the shelf life of mangoes (Chiumarelli et al., 2010) and guavas (Martínez-Ortiz et al., 2019). The acids retarded the ripening process and reduced changes in surface color when compared with fruits treated with starch without organic acids. Starch enriched with ascorbic acid has also been used to prevent microbial growth and increase the specific volume of coated bread (Araujo Galvão et al., 2018). Bonilla et al. have compared the incorporation of basil EO, thyme EO, citric acid, and α-tocopherol (vitamin E) to the wheat-chitosan films (Bonilla et al., 2013). Apart from those antioxidants, α-tocopherol exhibited the greatest antioxidant activity in dry films. This effect may be related to the highly volatile properties of

EOs. The solution to this problem, as was mentioned before, might be an encapsulation of EOs.

Nevertheless, incorporating antioxidants decreases food spoilage and increases preservation, enhancing shelf-life.

5.3.5 Starch modification

Besides adding various active additives to the starch film matrix, other methods of modification result in shelf-life enhancement. As it was mentioned, starch-based films are mainly developed by gelatinization of native starch and subsequently casting or melting. Various plasticizers and active compounds are added to develop films with the required properties. However, the starch might be modified in chemical, physical, or enzymatic ways before preparing the film. Physical and chemical modification of starch is used primarily now to enhance the properties and appearance of the starch film.

Chemical modification of starch films is the key to unshackling their use in food packaging to enhance the shelf life of food. Chemical modification influences several properties, such as tensile, barrier properties, and functionality, to improve native starch drawbacks. Starch modification is performed using various techniques, such as cross-linking, esterification, etherification or acetylation. Chemical modification of starch-based films leads to lower oxygen or water vapor permeability (Punia Bangar et al., 2022), which may result in extended storage time and delayed ripening.

Starch's most common physical modification is the thermal or sonification treatment of native starch and starch solution, respectively. Thermal treatment of native starch results in structural changes, degradation of starch macromolecules, and a decrease in crystallinity followed by changes in permeability (Zhu et al., 2016). This effect might be utilized for film development with controlled permeability and migration of active compounds.

Enzymatically modified starch is mainly utilized in non-food applications such as the pharmaceutical industry for drug delivery systems with gradual drug release. However, it might also be used to enhance the freshness and shelf-life of bread products. In this case, the starch modified by the amylase enzyme improved the shelf life of starch-based bread (Kweon et al., 1994).

The influence of the physical or chemical modification of starch on the shelf-life of food should be further analyzed, and to the best of our knowledge, no reports confirmed the effect on food quality. However, the modified starch is mainly used with other active additives, such as nanoparticles

EOs, plant extract, etc., and it is impossible to clearly state whether the effect is the modification itself or the action of active particles.

5.4 Challenges and limitations

Due to growing environmental concerns, consumers seek products with less natural impact and a lower carbon footprint based on natural or renewable resources, which are "eco-friendly" or "green" and are in a sustainability trend. Starch-based films follow this course. Starch-based films are edible, biodegradable, and, at the same time, raw material for production is cheap. Unfortunately, despite all the benefits, there are some drawbacks and limitations to introducing starch-based films into the wide market.

Firstly, some of the properties limit the usage of starch-based films. Starch-based films have poorer mechanical properties than conventional fossil-based plastics used in packaging. This is a problem for printing these materials. Another problem is affinity to water, which results in changes in their properties with humidity changes. This issue might be solved by incorporating various nanoparticles, resulting in tensile and hydrophobicity enhancement.

Most starch-based packaging products have bio-based additives such as essential oils, polyphenols, nanoparticles, and antimicrobial agents to enhance functional properties. Incorporating them into the starch matrix is related to some challenges. However, long-term effects on human health and food safety of these compounds remain under investigation. Migration of active ingredients into foods, potential toxicity at different levels, and compatibility with packaging conditions are some of the concerns to be thoroughly investigated. Some of the nanoparticles and EOs are not allowed to be used in food. Only rosemary EO is approved for usage as a food additive from all known EOs, like basil, oregano, cinnamon, cumin, etc., which found application as additives for coatings or packaging from starch.

New packaging technologies, including chemically treated starch and incorporating various nanomaterials or active compounds, require regulatory approval and adherence to international safety standards. Manufacturers must conduct extensive testing to meet standards set by authorities like the U.S. FDA or EFSA, and local food safety agencies. This includes demonstrating biodegradability, food safety, migration levels, and environmental testing.

References

Amit, S.K., Mezbah Uddin, Md, Rahman, R., Rezwanul Islam, S.M., 2017. Mohidus Samad Khan, A review on mechanisms and commercial aspects of food preservation and processing. Agric. Food Secur. 6 (1). https://doi.org/10.1186/s40066-017-0130-8.

Anu Mary Ealia, S., Saravanakumar, M.P., 2017. A review on the classification, characterisation, synthesis of nanoparticles and their application. IOP Conf. Ser. Mater. Sci. Eng. 263 (3), 032019. https://doi.org/10.1088/1757-899x/263/3/032019.

De Aquino, A.B., Blank, A.F., De Aquino Santana, L.C.L., 2015. Impact of edible chitosan-cassava starch coatings enriched with Lippia gracilis Schauer genotype mixtures on the shelf life of guavas (Psidium guajava L.) during storage at room temperature. Food Chem. 171, 108–116. https://doi.org/10.1016/j.foodchem.2014.08.077.

Araujo Galvão, G.D., De Nadae, J., Clemente, D.H., Chinen, G., De Carvalho, M.M., 2018. Circular economy: overview of barriers. Proced. CIRP 73, 79–85. https://doi.org/10.1016/j.procir.2018.04.011.

Arul Raj, J.S., Kasi, M., Karuppiah, P., Hirad, A.H., 2024. Green packaging solutions for extending the shelf life of fish fillet: development and evaluation of cinnamon essential oil-infused cassava starch and fish gelatin edible films. ACS Omega 9 (46), 45898–45910. https://doi.org/10.1021/acsomega.4c05249.

Basiak, E., Linke, M., Debeaufort, F., Lenart, A., Geyer, M., 2022. Impact of biodegradable materials on the quality of plums. Coatings 12 (2), 226. https://doi.org/10.3390/coatings12020226.

Bonilla, J., Fortunati, E., Vargas, M., Chiralt, A., Kenny, J.M., 2013. Effects of chitosan on the physicochemical and antimicrobial properties of PLA films. J. Food Eng. 119 (2), 236–243. https://doi.org/10.1016/j.jfoodeng.2013.05.026.

Bradford, K.J., Dahal, P., Van Asbrouck, J., Kunusoth, K., Bello, P., Thompson, J., Wu, F., 2018. The dry chain: reducing postharvest losses and improving food safety in humid climates. Trends Food Sci. Technol. 71, 84–93. https://doi.org/10.1016/j.tifs.2017.11.002.

Cai, C., Ma, R., Duan, M., Deng, Y., Liu, T., Lu, D., 2020. Effect of starch film containing thyme essential oil microcapsules on physicochemical activity of mango. LWT 131, 109700. https://doi.org/10.1016/j.lwt.2020.109700.

Celebioglu, A., Uyar, T., 2017. Antioxidant vitamin E/Cyclodextrin inclusion complex electrospun nanofibers: enhanced water solubility, prolonged shelf life, and photostability of vitamin E. J. Agric. Food Chem. 65 (26), 5404–5412. https://doi.org/10.1021/acs.jafc.7b01562.

Chen, S., Wu, M., Lu, P., Gao, L., Yan, S., Wang, S., 2020. Development of pH indicator and antimicrobial cellulose nanofibre packaging film based on purple sweet potato anthocyanin and oregano essential oil. Int. J. Biol. Macromol. 149, 271–280. https://doi.org/10.1016/j.ijbiomac.2020.01.231.

Chiumarelli, M., Pereira, L.M., Ferrari, C.C., Sarantópoulos, C.I.G.L., Hubinger, M.D., 2010. Cassava starch coating and citric acid to preserve quality parameters of fresh-cut "Tommy atkins" mango. J. Food Sci. 75 (5), E297. https://doi.org/10.1111/j.1750-3841.2010.01636.x.

Congying, H., Meifang, W., Islam, M.N., Cancan, S., Shengli, G., Hossain, A., Xiaohuang, C., 2024. Cassava starch-based multifunctional coating incorporated with zinc oxide nanoparticle to enhance the shelf life of passion fruit. J. Food Process. Preserv. 2024. https://doi.org/10.1155/2024/3026836.

Czerwiński, K., Rydzkowski, T., Wróblewska-Krepsztul, J., Kumar Thakur, V., 2021. Towards impact of modified atmosphere packaging (MAP) on shelf-life of polymer-film-packed food products: challenges and sustainable developments. Coatings 11 (12), 1504. https://doi.org/10.3390/coatings11121504.

Dutta, D., Sit, N., 2024. Comprehensive review on developments in starch-based films along with active ingredients for sustainable food packaging. Sustain. Chem. Pharm. 39, 101534. https://doi.org/10.1016/j.scp.2024.101534.

Edo, G.I., Ugbune, U., Ezekiel, G.O., Onoharigho, F.O., Agbo, J.J., 2024. Cyperus esculentus (tiger nut): its application in agriculture, food, health and nutrition. A review. Vegetos 37 (3), 757–766. https://doi.org/10.1007/s42535-023-00672-8.

Eurostat (2022) Food waste and food waste prevention - estimates - Statistics Explained. Available at: https://ec.europa.eu/eurostat/statistics-explained/index.php?title=Food_waste_and_food_waste_prevention_-_estimates (Accessed: 1 February 2025).

do Evangelho, J.A., da Silva Dannenberg, G., Biduski, B., el Halal, S.L.M., Kringel, D.H., Gularte, M.A., Fiorentini, A.M., da Rosa Zavareze, E., 2019. Antibacterial activity, optical, mechanical, and barrier properties of corn starch films containing orange essential oil. Carbohydr. Polym. 222. https://doi.org/10.1016/j.carbpol.2019.114981.

Fakhouri, F.M., Martelli, S.M., Caon, T., Velasco, J.I., Mei, L.H.I., 2015. Edible films and coatings based on starch/gelatin: film properties and effect of coatings on quality of refrigerated Red Crimson grapes. Postharvest Biol. Technol. 109, 57–64. https://doi.org/10.1016/j.postharvbio.2015.05.015.

Fan, X., Zhang, B., Zhang, X., Ma, Z., Feng, X., 2023. Incorporating Portulaca oleracea extract endows the chitosan-starch film with antioxidant capacity for chilled meat preservation. Food Chem. X 18, 100662. https://doi.org/10.1016/j.fochx.2023.100662.

Fan, S., Yin, X., Liu, X., Wang, G., Qiu, W., 2024. Enhancing bread preservation through non-contact application of starch-based composite film infused with clove essential oil nanoemulsion. Int. J. Biol. Macromol. 263 (Pt 1), 130297. https://doi.org/10.1016/j.ijbiomac.2024.130297.

Ferreira, R.R., Souza, A.G., Quispe, Y.M., Rosa, D.S., 2021. Essential oils loaded-chitosan nanocapsules incorporation in biodegradable starch films: a strategy to improve fruits shelf life. Int. J. Biol. Macromol. 188, 628–638. https://doi.org/10.1016/j.ijbiomac.2021.08.046.

Fonseca, L.M., Radünz, M., dos Santos Hackbart, H.C., da Silva, F.T., Camargo, T.M., Bruni, G.P., Monks, J.L.F., da Rosa Zavareze, E., Dias, A.R.G., 2020. Electrospun potato starch nanofibers for thyme essential oil encapsulation: antioxidant activity and thermal resistance. J. Sci. Food Agric. 100 (11), 4263–4271. https://doi.org/10.1002/jsfa.10468.

Francisco, C.B., Pellá, M.G., Silva, O.A., Raimundo, K.F., Caetano, J., Linde, G.A., Colauto, N.B., Dragunski, D.C., 2020. Shelf-life of guavas coated with biodegradable starch and cellulose-based films. Int. J. Biol. Macromol. 152, 272–279. https://doi.org/10.1016/j.ijbiomac.2020.02.249.

Gavaric, N., Mozina, S.S., Kladar, N., Bozin, B., 2015. Chemical profile, antioxidant and antibacterial activity of thyme and oregano essential oils, thymol and carvacrol and their possible synergism. J. Essent. Oil Bear. Plants 18 (4), 1013–1021. https://doi.org/10.1080/0972060X.2014.971069.

Glenn, G.M., Klamczynski, A.P., Woods, D.F., Chiou, B., Orts, W.J., Imam, S.H., 2010. Encapsulation of plant oils in porous starch microspheres. J. Agric. Food Chem. 58 (7), 4180–4184. https://doi.org/10.1021/jf9037826.

Gram, L., Ravn, L., Rasch, M., Bruhn, J.B., Christensen, A.B., Givskov, M., 2002. Food spoilage - interactions between food spoilage bacteria. Int. J. Food Microbiol. 78 (1–2), 79–97. https://doi.org/10.1016/S0168-1605(02)00233-7.

Guo, J., Jiang, X., Tian, Y., Yan, S., Liu, J., Xie, J., Zhang, F., Yao, C., Erwei, H., 2024. Therapeutic potential of cinnamon oil: chemical composition, pharmacological actions, and applications. Pharmaceuticals 17 (12), 1700. https://doi.org/10.3390/ph17121700.

Hammond, S.T., Brown, J.H., Burger, J.R., Flanagan, T.P., Fristoe, T.S., Mercado-Silva, N., Nekola, J.C., Okie, J.G., 2015. Food spoilage, storage, and transport:

implications for a sustainable future. Bioscience 65 (8), 758–768. https://doi.org/10.1093/biosci/biv081.

Hategekimana, J., Masamba, K.G., Ma, J., Zhong, F., 2015. Encapsulation of vitamin E: effect of physicochemical properties of wall material on retention and stability. Carbohydr. Polym. 124, 172–179. https://doi.org/10.1016/j.carbpol.2015.01.060.

He, X., Hwang, H.M., 2016. Nanotechnology in food science: functionality, applicability, and safety assessment. J. Food Drug Anal. 24 (4), 671–681. https://doi.org/10.1016/j.jfda.2016.06.001.

Horikoshi, S., Serpone, N., 2013. Introduction to Nanoparticles Microwaves in Nanoparticle Synthesis: Fundamentals and Applications. Wiley, pp. 1–24. https://doi.org/10.1002/9783527648122.ch1.

Hossen, M.A., Shimul, I.M., Sameen, D.E., Rasheed, Z., Dai, J., Li, S., Qin, W., Tang, W., Chen, M., Liu, Y., 2024. Essential oil–loaded biopolymeric particles on food industry and packaging: a review. Int. J. Biol. Macromol. 265. https://doi.org/10.1016/j.ijbiomac.2024.130765.

Huis in't Veld, J.H.J., 1996. Microbial and biochemical spoilage of foods: an overview. Int. J. Food Microbiol. 33 (1), 1–18. https://doi.org/10.1016/0168-1605(96)01139-7.

Illikoud, N., Rossero, A., Chauvet, R., Courcoux, P., Pilet, M.F., Charrier, T., Jaffrès, E., Zagorec, M., 2019. Genotypic and phenotypic characterization of the food spoilage bacterium Brochothrix thermosphacta. Food Microbiol. 81, 22–31. https://doi.org/10.1016/j.fm.2018.01.015.

Ioannidis, A.G., Kerckhof, F.M., Riahi Drif, Y., Vanderroost, M., Boon, N., Ragaert, P., De Meulenaer, B., Devlieghere, F., 2018. Characterization of spoilage markers in modified atmosphere packaged iceberg lettuce. Int. J. Food Microbiol. 279, 1–13. https://doi.org/10.1016/j.ijfoodmicro.2018.04.034.

Issa, A., Ibrahim, S.A., Tahergorabi, R., 2017. Impact of sweet potato starch-based nanocomposite films activated with thyme essential oil on the shelf-life of baby spinach leaves. Foods 6 (6), 1–13. https://doi.org/10.3390/foods6060043.

Jaya, S., Sudhagar, M., Das, H., 2002. Stickiness of food powders and related physicochemical properties of food components. J. Food Sci. Technol. 39 (1), 1–7.

Jayakumar, A., Radoor, S., Kim, J.T., Rhim, J.W., Parameswaranpillai, J., Nandi, D., Srisuk, R., Siengchin, S., 2022. Titanium dioxide nanoparticles and elderberry extract incorporated starch based polyvinyl alcohol films as active and intelligent food packaging wraps. Food Packag. Shelf Life 34. https://doi.org/10.1016/j.fpsl.2022.100967.

Ju, J., Xie, Y., Guo, Y., Cheng, Y., Qian, H., Yao, W., 2019. The inhibitory effect of plant essential oils on foodborne pathogenic bacteria in food. Crit. Rev. Food Sci. Nutr. 59 (20), 3281–3292. https://doi.org/10.1080/10408398.2018.1488159.

Júnior, L.M., Vieira, R.P., Anjos, C.A.R., 2020. Kefiran-based films: fundamental concepts, formulation strategies and properties. Carbohydr. Polym. 246, 116609. https://doi.org/10.1016/j.carbpol.2020.116609.

Karanth, S., Feng, S., Patra, D., Pradhan, A.K., 2023. Linking microbial contamination to food spoilage and food waste: the role of smart packaging, spoilage risk assessments, and date labeling. Front. Microbiol. 14. https://doi.org/10.3389/fmicb.2023.1198124.

Kathuria, D., Hamid, S.G., Thakur, A., 2023. Maillard reaction in different food products: effect on product quality, human health and mitigation strategies. Food Control 153, 109911. https://doi.org/10.1016/j.foodcont.2023.109911.

Kaur, M., Kalia, A., Thakur, A., 2017. Effect of biodegradable chitosan–rice-starch nanocomposite films on post-harvest quality of stored peach fruit. Starch - Stärke 69 (1–2). https://doi.org/10.1002/star.201600208.

Khuntia, A., Kumar, R., Premjit, Y., Mitra, J., 2022. Release behavior of vitamin C nanoliposomes from starch–vitamin C active packaging films. J. Food Process. Eng. 45 (9). https://doi.org/10.1111/jfpe.14075.

Kraśniewska, K., Galus, S., Gniewosz, M., 2020. Biopolymers-based materials containing silver nanoparticles as active packaging for food applications–A review. Int. J. Mol. Sci. 21 (3), 698. https://doi.org/10.3390/ijms21030698.

Kristott, J., 2000. Fats and oils. In: The Stability and Shelf-Life of Food. Woodhead publishing, pp. 279–309.

Kurzyna-Szklarek, M., Cybulska, J., Zdunek, A., 2022. Analysis of the chemical composition of natural carbohydrates – an overview of methods. Food Chem. 394, 133466. https://doi.org/10.1016/j.foodchem.2022.133466.

Kweon, M.R., Park, C.S., Auh, J.H., Cho, B.M., Yang, N.S., Park, K.H., 1994. Phospholipid hydrolysate and antistaling amylase effects on retrogradation of starch in bread. J. Food Sci. 59 (5), 1072–1076. https://doi.org/10.1111/j.1365-2621.1994.tb08193.x.

Leyva-Porras, C., Román-Aguirre, M., Cruz-Alcantar, P., Pérez-Urizar, J.T., Saavedra-Leos, M.Z., 2021. Application of antioxidants as an alternative improving of shelf life in foods. Polysaccharides 2 (3), 594–607. https://doi.org/10.3390/polysaccharides2030036.

Li, H., Chen, C., Cao, X., 2015. Essential oils-oriented chiral esters as potential pesticides: asymmetric syntheses, characterization and bio-evaluation. Ind. Crop. Prod. 76, 432–436. https://doi.org/10.1016/j.indcrop.2015.07.027.

Li, W., Zhang, J., Chen, X., Xiao, Z., Zhou, J., Sun, H., Wang, S., Liu, Y., 2024. Organic nanoparticles incorporated starch/carboxymethylcellulose multifunctional coating film for efficient preservation of perishable products. Int. J. Biol. Macromol. 275 (Pt 1), 133357. https://doi.org/10.1016/j.ijbiomac.2024.133357.

Lianou, A., Panagou, E.Z., Nychas, G.J.E., 2016. Microbiological spoilage of foods and beverages. In: Subramaniam, P. (Ed.), The Stability and Shelf Life of Food, second ed. Woodhead Publishing, Duxford, UK, pp. 3–42.

Luna-Niño, M.F., Beristain-Guevara, C.I., Pascual-Pineda, L.A., Rascón-Díaz, M.P., Luna-Solano, G., Jiménez-Fernández, M., 2025. Characterization of chayotextle starch films supplemented with essential oils and their effect as a coating on the shelf life of bread. Starch/Staerke 77 (3). https://doi.org/10.1002/star.202400057.

Mafe, A.N., Edo, G.I., Makia, R.S., Joshua, O.A., Akpoghelie, P.O., Gaaz, T.S., Jikah, A. N., Yousif, E., Isoje, E.F., Igbuku, U.A., Ahmed, D.S., Essaghah, A.E.A., Umar, H., 2024. A review on food spoilage mechanisms, food borne diseases and commercial aspects of food preservation and processing. Food Chemistry Advances 5. https://doi.org/10.1016/j.focha.2024.100852.

Mageswari, A., Subramanian, P., Srinivasan, R., Karthikeyan, S., Gothandam, K.M., 2015. Astaxanthin from psychrotrophic Sphingomonas faeni exhibits antagonism against food-spoilage bacteria at low temperatures. Microbiol. Res. 179, 38–44. https://doi.org/10.1016/j.micres.2015.06.010.

Majeed, T., Dar, A.H., Pandey, V.K., Dash, K.K., Srivastava, S., Shams, R., Jeevarathinam, G., Singh, P., Echegaray, N., Pandiselvam, R., 2023. Role of additives in starch-based edible films and coating: a review with current knowledge. Prog. Org. Coating 181. https://doi.org/10.1016/j.porgcoat.2023.107597.

Malafatti, J.O.D., Domingues, M.Á., Meirelles, M.R., Peres, L.G.S., Bresolin, J.D., Paris, E. C., 2023. Antifungal and ultraviolet–visible barrier properties in starch films reinforced with CuO nanoparticles. J. Appl. Polym. Sci. 140 (33). https://doi.org/10.1002/app.54290.

Man, C.M.D., 2015. Introduction to Shelf Life of Foods – Frequently Asked Questions Shelf Life. Wiley, pp. 1–40. https://doi.org/10.1002/9781118346235.ch1.

Martínez-Ortiz, M.A., Palma-Rodríguez, H.M., Montalvo-González, E., Sáyago-Ayerdi, S.G., Utrilla-Coello, R., Vargas-Torres, A., 2019. Effect of using microencapsulated ascorbic acid in coatings based on resistant starch chayotextle on the quality of guava fruit. Sci. Hortic. 256. https://doi.org/10.1016/j.scienta.2019.108604.

Medina-Jaramillo, C., Estevez-Areco, S., Goyanes, S., López-Córdoba, A., 2019. Characterization of starches isolated from Colombian native potatoes and their application as novel edible coatings for wild andean blueberries (vaccinium meridionale swartz). Polymers 11 (12), 1937. https://doi.org/10.3390/polym11121937.
Mehdizadeh, T., Tajik, H., Langroodi, A.M., Molaei, R., Mahmoudian, A., 2020. Chitosan-starch film containing pomegranate peel extract and Thymus kotschyanus essential oil can prolong the shelf life of beef. Meat Sci. 163, 108073. https://doi.org/10.1016/j.meatsci.2020.108073.
Menzel, C., 2020. Improvement of starch films for food packaging through a three-principle approach: antioxidants, cross-linking and reinforcement. Carbohydr. Polym. 250, 116828. https://doi.org/10.1016/j.carbpol.2020.116828.
Menzel, C., González-Martínez, C., Chiralt, A., Vilaplana, F., 2019. Antioxidant starch films containing sunflower hull extracts. Carbohydr. Polym. 214, 142–151. https://doi.org/10.1016/j.carbpol.2019.03.022.
Moghadas, H.C., Smith, J.S., Tahergorabi, R., 2025. Recent advances in the application of edible coatings for shelf-life extension of strawberries: a review. Food Bioprocess Technol. 18 (2), 1079–1103. https://doi.org/10.1007/s11947-024-03517-7.
Moreno, O., Atarés, L., Chiralt, A., 2015. Effect of the incorporation of antimicrobial/antioxidant proteins on the properties of potato starch films. Carbohydr. Polym. 133, 353–364. https://doi.org/10.1016/j.carbpol.2015.07.047.
Moschopoulou, E., Moatsou, G., Syrokou, M.K., Paramithiotis, S., Drosinos, E.H., 2019. Food Quality Changes during Shelf Life Food Quality and Shelf Life. Elsevier, Greece, pp. 1–31. https://doi.org/10.1016/B978-0-12-817190-5.00001-X.
Murata, M., 2021. Browning and pigmentation in food through the Maillard reaction. Glycoconj. J. 38 (3), 283–292. https://doi.org/10.1007/s10719-020-09943-x.
Navina, B., Keshav Huthaash, K., Velmurugan, N.K., Korumilli, T., 2023. Insights into recent innovations in anti browning strategies for fruit and vegetable preservation. Trends Food Sci. Technol. 139. https://doi.org/10.1016/j.tifs.2023.104128.
Ndraha, N., Hsiao, H.I., Vlajic, J., Yang, M.F., Lin, H.T.V., 2018. Time-temperature abuse in the food cold chain: review of issues, challenges, and recommendations. Food Control 89, 12–21. https://doi.org/10.1016/j.foodcont.2018.01.027.
von Neubeck, M., Baur, C., Krewinkel, M., Stoeckel, M., Kranz, B., Stressler, T., Fischer, L., Hinrichs, J., Scherer, S., Wenning, M., 2015. Biodiversity of refrigerated raw milk microbiota and their enzymatic spoilage potential. Int. J. Food Microbiol. 211, 57–65. https://doi.org/10.1016/j.ijfoodmicro.2015.07.001.
Nwosu, L.C., Edo, G.I., Özgör, E., 2022. The phytochemical, proximate, pharmacological, GC-MS analysis of Cyperus esculentus (Tiger nut): a fully validated approach in health, food and nutrition. Food Biosci. 46. https://doi.org/10.1016/j.fbio.2022.101551.
Odeyemi, O.A., Burke, C.M., Bolch, C.J.S., Stanley, R., 2018. Evaluation of spoilage potential and volatile metabolites production by Shewanella baltica isolated from modified atmosphere packaged live mussels. Food Res. Int. 103, 415–425. https://doi.org/10.1016/j.foodres.2017.10.068.
Odeyemi, O.A., Alegbeleye, O.O., Strateva, M., Stratev, D., 2020. Understanding spoilage microbial community and spoilage mechanisms in foods of animal origin. Compr. Rev. Food Sci. Food Saf. 19 (2), 311–331. https://doi.org/10.1111/1541-4337.12526.
Oliveira Filho, J.G.d., Albiero, B.R., Calisto, Í.H., Bertolo, M.R.V., Oldoni, F.C.A., Egea, M.B., Bogusz Junior, S., de Azeredo, H.M.C., Ferreira, M.D., 2022. Bionanocomposite edible coatings based on arrowroot starch/cellulose nanocrystals/carnauba wax nanoemulsion containing essential oils to preserve quality and improve shelf life of strawberry. Int. J. Biol. Macromol. 219, 812–823. https://doi.org/10.1016/j.ijbiomac.2022.08.049.

Ortega, F., Giannuzzi, L., Arce, V.B., García, M.A., 2017. Active composite starch films containing green synthetized silver nanoparticles. Food Hydrocoll. 70, 152–162. https://doi.org/10.1016/j.foodhyd.2017.03.036.

Ozmen, L., Langrish, T.A.G., 2002. Comparison of glass transition temperature and sticky point temperature for skim milk powder. Dry. Technol. 20 (6), 1177–1192. https://doi.org/10.1081/DRT-120004046.

Patel, C., Panigrahi, J., 2019. Starch glucose coating-induced postharvest shelf-life extension of cucumber. Food Chem. 288, 208–214. https://doi.org/10.1016/j.foodchem.2019.02.123.

Peighambardoust, S.J., Peighambardoust, S.H., Mohammadzadeh Pournasir, N., Pakdel, P., 2019. Properties of active starch-based films incorporating a combination of Ag, ZnO and CuO nanoparticles for potential use in food packaging applications. Food Packag. Shelf Life 22. https://doi.org/10.1016/j.fpsl.2019.100420.

Peters, R., Dam, G.t., Bouwmeester, H., Helsper, H., Allmaier, G., Kammer, F.v., Ramsch, R., Solans, C., Tomaniová, M., Hajslova, J., Weigel, S., 2011. Identification and characterization of organic nanoparticles in food. TrAC, Trends Anal. Chem. 30 (1), 100–112. https://doi.org/10.1016/j.trac.2010.10.004.

Petruzzi, L., Corbo, M.R., Sinigaglia, M., Bevilacqua, A., 2017. Microbial spoilage of foods: fundamentals. In: Bevilacqua, A., Corbo, M.R., Sinigaglia, M. (Eds.), The Microbiological Quality of Food. Woodhead Publishing, Duxford, UK, pp. 1–21.

Piñeros-Hernandez, D., Medina-Jaramillo, C., López-Córdoba, A., Goyanes, S., 2017. Edible cassava starch films carrying rosemary antioxidant extracts for potential use as active food packaging. Food Hydrocoll. 63, 488–495. https://doi.org/10.1016/j.foodhyd.2016.09.034.

Pothakos, V., Devlieghere, F., Villani, F., Björkroth, J., Ercolini, D., 2015. Lactic acid bacteria and their controversial role in fresh meat spoilage. Meat Sci. 109, 66–74. https://doi.org/10.1016/j.meatsci.2015.04.014.

Programme, U.N.E. (2024) Food Waste Index Report 2024. Think Eat Save: Tracking Progress to Halve Global Food Waste. Available at: https://wedocs.unep.org/20.500.11822/45230.

Prokopov, T., Tanchev, S., 2007. Methods of food preservation BT - food safety. In: McElhatton, A., Marshall, R.J. (Eds.), Food Safety: A Practical and Case Study Approach. Springer US, Boston, MA, pp. 3–25. https://doi.org/10.1007/978-0-387-33957-3_1.

Punia Bangar, S., Whiteside, W.S., Ozogul, F., Dunno, K.D., Cavender, G.A., Dawson, P., 2022. Development of starch-based films reinforced with cellulosic nanocrystals and essential oil to extend the shelf life of red grapes. Food Biosci. 47. https://doi.org/10.1016/j.fbio.2022.101621.

Radha krishnan, K., Babuskin, S., Rakhavan, K.R., Tharavin, R., Azhagu Saravana Babu, P., Sivarajan, M., Sukumar, M., 2015. Potential application of corn starch edible films with spice essential oils for the shelf life extension of red meat. J. Appl. Microbiol. 119 (6), 1613–1623. https://doi.org/10.1111/jam.12932.

Raposo, A., Pérez, E., de Faria, C.T., Ferrús, M.A., Carrascosa, C., 2017. Food Spoilage by pseudomonas spp.-An Overview Foodborne Pathogens and Antibiotic Resistance. wiley, Portugal, pp. 41–71. https://doi.org/10.1002/9781119139188.ch3.

Reyes, V., Cahill, E., Mis Solval, K.E., 2024. The potential for reducing food waste through shelf-life extension: actionable insights from data digitization. Sustainability 16 (7), 2986. https://doi.org/10.3390/su16072986.

Rodríguez-Rojas, A., Arango Ospina, A., Rodríguez-Vélez, P., Arana-Florez, R., 2019. ¿What is the new about food packaging material? A bibliometric review during 1996–2016. Trends Food Sci. Technol. 85, 252–261. https://doi.org/10.1016/j.tifs.2019.01.016.

Sadgrove, N.J., Padilla-González, G.F., Leuner, O., Melnikovova, I., Fernandez-Cusimamani, E., 2021. Pharmacology of natural volatiles and essential oils in food, therapy, and disease prophylaxis. Front. Pharmacol. 12. https://doi.org/10.3389/fphar.2021.740302.

Sharma, K., Kaur, M., Lalhall, A., Gupta, I., Batish, D.R., Sharma, A., 2023. Eucalyptus citriodora and melaleuca bracteata essential oils encapsulated chitosan hydrogels embedded in corn starch films for improving the shelf life of plums. Trends in Carbohydrate Research 14 (4), 1–11.

Shatrohan Lal, R.K., 2014. Synthesis of organic nanoparticles and their applications in drug delivery and food nanotechnology: a review. Journal of Nanomaterials & Molecular Nanotechnology 03 (04). https://doi.org/10.4172/2324-8777.1000150.

Siddiqui, S.A., Erol, Z., Rugji, J., Taşçı, F., Kahraman, H.A., Toppi, V., Musa, L., Di Giacinto, G., Bahmid, N.A., Mehdizadeh, M., Castro-Muñoz, R., 2023. An overview of fermentation in the food industry - looking back from a new perspective. Bioresources and Bioprocessing 10 (1). https://doi.org/10.1186/s40643-023-00702-y.

Singh, R.P., Anderson, B.A., 2004. The Major Types of Food Spoilage: An Overview Understanding and Measuring the Shelf-Life of Food. Elsevier Inc., United States, pp. 1–23. https://doi.org/10.1016/B978-1-85573-732-7.50005-6.

Singh, P., Kaur, G., Singh, A., Dar, B.N., 2023. Effect of montmorillonite and starch nanocrystals based biodegradable films loaded with lemongrass oil nanoemulsion on quality, enzymatic activity and shelf life of strawberry. Food Chemistry Advances 3, 100343. https://doi.org/10.1016/j.focha.2023.100343.

Skibsted, L.H., Risbo, J., Andersen, M.L., 2010. Chemical Deterioration and Physical Instability of Food and Beverages Chemical Deterioration and Physical Instability of Food and Beverages. Elsevier Inc., Denmark, pp. 1–789. https://doi.org/10.1533/9781845699260.

Snyder, A.B., Worobo, R.W., 2018. The incidence and impact of microbial spoilage in the production of fruit and vegetable juices as reported by juice manufacturers. Food Control 85, 144–150. https://doi.org/10.1016/j.foodcont.2017.09.025.

Snyder, A.B., Churey, J.J., Worobo, R.W., 2019. Association of fungal genera from spoiled processed foods with physicochemical food properties and processing conditions. Food Microbiol. 83, 211–218. https://doi.org/10.1016/j.fm.2019.05.012.

Snyder, A.B., Martin, N., Wiedmann, M., 2024. Microbial food spoilage: impact, causative agents and control strategies. Nat. Rev. Microbiol. 22 (9), 528–542. https://doi.org/10.1038/s41579-024-01037-x.

Sofi, S.A., Singh, J., Rafiq, S., Ashraf, U., Dar, B.N., Nayik, G.A., 2018. A comprehensive review on antimicrobial packaging and its use in food packaging. Curr. Nutr. Food Sci. 14 (4), 305–312. https://doi.org/10.2174/1573401313666170609095732.

Soro, A.B., Noore, S., Hannon, S., Whyte, P., Bolton, D.J., O'Donnell, C., Tiwari, B.K., 2021. Current sustainable solutions for extending the shelf life of meat and marine products in the packaging process. Food Packag. Shelf Life 29. https://doi.org/10.1016/j.fpsl.2021.100722.

Souza, A.C., Goto, G.E.O., Mainardi, J.A., Coelho, A.C.V., Tadini, C.C., 2013. Cassava starch composite films incorporated with cinnamon essential oil: antimicrobial activity, microstructure, mechanical and barrier properties. LWT Food Sci. Technol. (Lebensm.-Wiss.-Technol.) 54 (2), 346–352. https://doi.org/10.1016/j.lwt.2013.06.017.

Srikhao, N., Kasemsiri, P., Ounkaew, A., Lorwanishpaisarn, N., Okhawilai, M., Pongsa, U., Hiziroglu, S., Chindaprasirt, P., 2021. Bioactive nanocomposite film based on cassava starch/polyvinyl alcohol containing green synthesized silver nanoparticles. J. Polym. Environ. 29 (2), 672–684. https://doi.org/10.1007/s10924-020-01909-2.

Strambeanu, N., Demetrovici, L., Dragos, D., Lungu, M., 2015. Nanoparticles: Definition, Classification and General Physical Properties Nanoparticles' Promises and Risks:

Characterization, Manipulation, and Potential Hazards to Humanity and the Environment. Springer International Publishing, Romania, pp. 3–8. https://doi.org/10.1007/978-3-319-11728-7_1.

Subramaniam, P., Wareing, P. (Eds.), 2016. The Stability and Shelf Life of Food. Woodhead Publishing.

Suppakul, P., Miltz, J., Sonneveld, K., Bigger, S.W., 2003. Active packaging technologies with an emphasis on antimicrobial packaging and its applications. J. Food Sci. 68 (2), 408–420. https://doi.org/10.1111/j.1365-2621.2003.tb05687.x.

Tamimi, N., Mohammadi Nafchi, A., Moghaddam, H.H.-, Baghaie, H., 2021. The effects of nano-zinc oxide morphology on functional and antibacterial properties of tapioca starch bionanocomposite. Food Sci. Nutr. 9 (8), 4497–4508. https://doi.org/10.1002/fsn3.2426.

Teixeira-Costa, B.E., Andrade, C.T., 2022. Natural polymers used in edible food packaging—history, function and application trends as a sustainable alternative to synthetic plastic. Polysaccharides 3 (1), 32–58. https://doi.org/10.3390/polysaccharides3010002.

Trotta, F., Da Silva, S., Massironi, A., Mirpoor, S.F., Lignou, S., Ghawi, S.K., Charalampopoulos, D., 2024. Advancing food preservation: sustainable green-AgNPs bionanocomposites in paper-starch flexible packaging for prolonged shelf life. Polymers 16 (7). https://doi.org/10.3390/polym16070941.

Tunma, S., 2018. Starch based nanocomposites in active packaging for extended shelf life of fresh fruits. Walailak J. Sci. Technol. 15 (4), 273–281. https://doi.org/10.48048/wjst.2018.3849.

u Nisa, I., Ashwar, B.A., Shah, A., Gani, A., Gani, A., Masoodi, F.A., 2015. Development of potato starch based active packaging films loaded with antioxidants and its effect on shelf life of beef. J. Food Sci. Technol. 52 (11), 7245–7253. https://doi.org/10.1007/s13197-015-1859-3.

Urugo, M.M., Teka, T.A., Gemede, H.F., Mersha, S., Tessema, A., Woldemariam, H.W., Admassu, H., 2024. A comprehensive review of current approaches on food waste reduction strategies. Compr. Rev. Food Sci. Food Saf. 23 (5). https://doi.org/10.1111/1541-4337.70011.

Valdivieso-Ugarte, M., Gomez-Llorente, C., Plaza-Díaz, J., Gil, Á., 2019. Antimicrobial, antioxidant, and immunomodulatory properties of essential oils: a systematic review. Nutrients 11 (11), 2786. https://doi.org/10.3390/nu11112786.

Vigar, V., Myers, S., Oliver, C., Arellano, J., Robinson, S., Leifert, C., 2020. A systematic review of organic versus conventional food consumption: is there a measurable benefit on human health? Nutrients 12 (1), 7. https://doi.org/10.3390/nu12010007.

Walasek-Janusz, M., Grzegorczyk, A., Malm, A., Nurzyńska-Wierdak, R., Zalewski, D., 2024. Chemical composition, and antioxidant and antimicrobial activity of oregano essential oil. Molecules 29 (2), 435. https://doi.org/10.3390/molecules29020435.

Wang, H., He, A., Yang, X., 2018. Dynamics of microflora on conveyor belts in a beef fabrication facility during sanitation. Food Control 85, 42–47. https://doi.org/10.1016/j.foodcont.2017.09.017.

Wang, W., Zhang, W., Lin, L., Deng, W., Liu, M., Hu, J., 2023. Biodegradable starch-based packaging films incorporated with polyurethane-encapsulated essential-oil microcapsules for sustained food preservation. Int. J. Biol. Macromol. 235, 123889. https://doi.org/10.1016/j.ijbiomac.2023.123889.

Wang, D., Wang, X., Zhou, S., Ren, L., Meng, Y., Ma, R., Wang, S., Liu, Z., Alamri, A.S., Alhomrani, M., Zhang, Z., Cui, F., Li, T., Li, J., 2024a. Radish residue carbon dots-based novel starch/chitosan film with high antioxidant, biocompatibility, and antibacterial activities for salmon fillets' active packaging. Int. J. Biol. Macromol. 273 (Pt 2), 133107. https://doi.org/10.1016/j.ijbiomac.2024.133107.

Wang, X., Guo, J., Ren, H., Jin, J., He, H., Jin, P., Wu, Z., Zheng, Y., 2024b. Research progress of nanocellulose-based food packaging. Trends Food Sci. Technol. 143, 104289. https://doi.org/10.1016/j.tifs.2023.104289.

Yan, X., Cheng, M., Zhao, P., Wang, Y., Chen, M., Wang, X., Wang, J., 2022. Fabrication and characterization of oxidized esterified tapioca starch films encapsulating oregano essential oil with mesoporous nanosilica. Ind. Crop. Prod. 184, 115033. https://doi.org/10.1016/j.indcrop.2022.115033.

Yousuf, B., Qadri, O.S., Srivastava, A.K., 2018. Recent developments in shelf-life extension of fresh-cut fruits and vegetables by application of different edible coatings: a review. LWT 89, 198–209. https://doi.org/10.1016/j.lwt.2017.10.051.

Yu, Y., Zhou, J., Chen, Q., Xie, F., Zhang, D., He, Z., Cheng, S., Cai, J., 2024. Self-reinforced multifunctional starch nanocomposite film for litchi fruit postharvest preservation. Chem. Eng. J. 486, 150262. https://doi.org/10.1016/j.cej.2024.150262.

Yuan, L., Feng, W., Zhang, Z., Peng, Y., Xiao, Y., Chen, J., 2021. Effect of potato starch-based antibacterial composite films with thyme oil microemulsion or microcapsule on shelf life of chilled meat. LWT 139, 110462. https://doi.org/10.1016/j.lwt.2020.110462.

Zhu, Y., Romain, C., Williams, C.K., 2016. Sustainable polymers from renewable resources. Nature 540 (7633), 354–362. https://doi.org/10.1038/nature21001.

Żołek-Tryznowska, Z., Jeznach, A., Bednarczyk, E., Murawski, T., Piłczyńska, K., Sikora, S., Tryznowski, M., 2024. Effect of hydrophobic nano-silica content on the surface properties of corn-starch films. Ind. Crop. Prod. 214, 118582. https://doi.org/10.1016/j.indcrop.2024.118582.

Zuo, J., Feng, J., Gameiro, M.G., Tian, Y., Liang, J., Wang, Y., Ding, J., He, Q., 2022. RFID-based sensing in smart packaging for food applications: a review. Future Foods 6, 100198. https://doi.org/10.1016/j.fufo.2022.100198.

CHAPTER SIX

Active and intelligent for food packaging

Arif Ozcan[1,6], **Fayçal Dergal**[2,3], **Ramzi Khiari**[4,5], **Bilge Nazli Altay**[6]
[1]Marmara University, Faculty of Applied Sciences, Department of Printing Technologies, Istanbul, Turkey
[2]Scientific and Technical Research Center in Physico-Chemical Analysis (CRAPC), Tipaza, Algeria
[3]Laboratory of Catalysis and Synthesis in Organic Chemistry (LCSCO), Faculty of Sciences, University of Tlemcen, Tlemcen, Algeria
[4]University of Grenoble Alpes, CNRS, Grenoble INP, LGP2, Grenoble, France
[5]Department of Textile, Higher Institute of Technological Studies of Ksar Hellal, Ksar Hellal, Tunisia
[6]Rochester Institute of Technology, College of Engineering Technology, Department of Packaging and Graphic Media Science, Rochester, NY, United States

6.1 Introduction

In the food industry, the combined effects of global population growth, emerging pandemics, and increasing consumer awareness have accelerated the introduction of new regulations and quality-control practices. At the same time, the widespread use of digital media has increased consumer demand for food products that are safe, fresh, transparent, and sustainably produced. These pressures have driven the rapid adoption of active and intelligent packaging technologies, which not only extend shelf life and enhance food safety but also provide real-time information on product quality and storage conditions, thereby improving traceability, consumer trust, and supply-chain efficiency (Jeantet et al., 2016; Rehman et al., 2020; Li et al., 2022; Abdullahi et al., 2025; Akkemik and Güner, 2020). Alongside regulatory pressures and rising consumer expectations, the rapid evolution of modern food production technologies, shifts in marketing and retail strategies, and the globalization of food supply chains have further intensified the need for extended product shelf life. Today's food products are often manufactured at centralized facilities and distributed across long distances to reach diverse markets, exposing them to varying environmental and handling conditions. These logistical and commercial realities make conventional packaging increasingly inadequate, thereby reinforcing the importance of advanced active and intelligent packaging systems that can preserve product quality, ensure safety, and provide real-time monitoring throughout storage, transportation, and retail display (Janjarasskul and Suppakul, 2018; Caon et al., 2016).

Starch-Based Sustainable Food Packaging
ISBN: 978-0-443-36346-7
https://doi.org/10.1016/B978-0-443-36346-7.00007-2

Food packaging aims to meet customer expectations by preserving product quality, providing information about the product, and ensuring food safety during distribution and throughout the shelf life (Salgado et al., 2021; Young et al., 2020). A ready-to-eat food product is considered safe if it does not pose a risk to consumer health and therefore does not threaten human health at any stage of the "farm-to-fork" chain. Food safety is defined within the framework of national and international regulations, as well as established standards and control measures. Its main purpose is to ensure the transport of food, protect it from environmental factors, and provide consumers with information about its contents. Distributing products without packaging is impossible today. Consequently, packaging is a critical source of information that can significantly influence consumers' perceptions of a product and their purchasing decisions (Ahmed et al., 2022).

In developed countries, food packaging accounts for nearly half of the entire packaging industry (Pereira et al., 2015). On a global scale, to ensure freshness, safety, increasing product diversity, and authenticity throughout the supply chain, establishing traceability systems for food and similar products has become a priority for authorities, the food industry, and retailers. These systems aim to enable real-time tracking and evaluation of products, reduce the carbon footprint, and provide reliable, accurate, fast, cost-effective solutions that pose no risk to the environment or human health (Hassan et al., 2023; Pal et al., 2019). Within this new concept, packaging manufacturers have faced numerous challenges, which has motivated researchers to develop next-generation solutions that go far beyond traditional packaging approaches. As a result, efforts to replace conventional passive packaging systems with active and intelligent packaging solutions have increased. While these new systems retain the basic functions of traditional passive packaging, they also enable product monitoring and evaluation, as well as spoilage-delaying mechanisms that help extend shelf life (Nopwinyuwong et al., 2010; Rukchon et al., 2014). These systems can be defined as next-generation technologies that use smart sensors and labels integrated into packaging to help monitor the condition of the product (Mustafa and Andreescu, 2018). In general, smart packaging can be classified into two main categories: intelligent packaging and active packaging (Loucanova et al., 2017; Schaefer and Cheung, 2018).

Active packaging refers to a class of packaging systems that are designed to actively interact with the food product or its surrounding headspace to maintain or improve food quality, safety, and shelf life. Unlike conventional passive packaging, which only provides a physical barrier, active packaging incorporates functional components such as oxygen and ethylene scavengers, carbon dioxide regulators, moisture absorbers, antimicrobial agents, and antioxidants that either remove undesirable substances or release beneficial compounds into the package environment. Through these controlled interactions, active packaging slows down microbial growth, oxidation, and physiological degradation processes, thereby preserving freshness, sensory attributes, and nutritional value while reducing the need for chemical preservatives. Active packaging designed to preserve or improve the sensory, safety, and quality attributes of food, whereas intelligent packaging is designed to monitor the condition of the food and its surrounding environment and to communicate information about product quality, safety, and storage history. Rather than altering the package atmosphere, intelligent packaging relies on indicators, sensors, data carriers, and sometimes RFID or digital technologies to detect changes such as temperature change, gas composition, pH shifts, microbial activity, or package integrity. These changes are typically displayed through visible color variations or digital signals that allow producers, retailers, and consumers to assess the freshness and safety of the product in real time, enabling better decision-making, reducing food waste, and improving supply-chain transparency (Soltani Firouz et al., 2021).

Active and intelligent packaging systems have attracted strong interest and adoption, particularly in the food and pharmaceutical industries, because they enable the extension of product shelf life, optimization and shortening of production processes, and simplification of complex distribution logistics. They also make it possible to minimize—ideally eliminate—the use of chemical preservatives in food formulations, to develop more sales and marketing strategies and options in food packaging applications, and to differentiate products through higher quality, greater product diversity, and more advanced marketing tools, all while ensuring consumer safety (Amin et al., 2022; Liu et al., 2025).

All active and intelligent packaging technologies that are currently commercially available or under development are designed and produced to perform a specific function. Today, virtually all the food products we consume are packaged, and each product is unique. Each product has distinct intrinsic properties—such as pH, nutrient content, water activity, antimicrobial compounds, biological structure, and the type and load of microorganisms—as well as differing external conditions, including storage environment, humidity, and headspace gas composition. Therefore, the active or intelligent packaging systems developed for these products must be designed with great precision. A schematic representation of active and intelligent packaging systems is shown in Fig. 6.1.

Despite the many functional and economic benefits offered by plastic packaging materials, they also bring numerous disadvantages, which have today become a source of concern for manufacturers, consumers, and governments. For this reason, global efforts are being undertaken to address issues related to their production, use, and disposal (Evans et al., 2020; Asgher et al., 2020; Li et al., 2021).

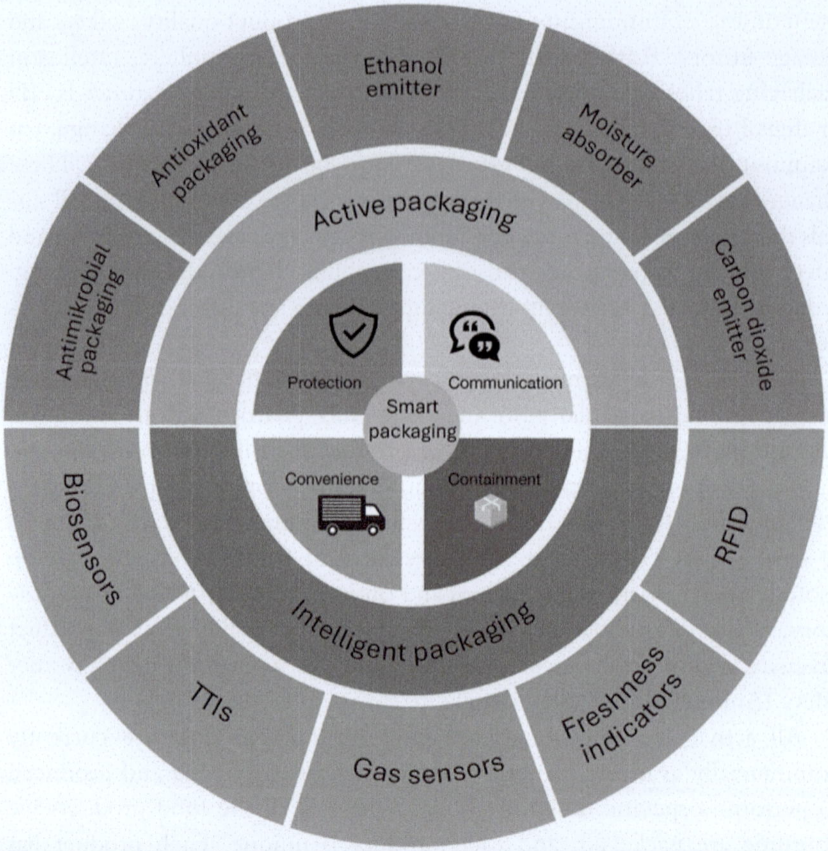

Figure 6.1 Functional model of smart packaging, combining active technologies that modify the package environment with intelligent systems that sense, record, or communicate product status. *Ozcan, A., 2020. New approaches in smart packaging technologies, in* International Symposium on Graphic Engineering and Design, University of Novi Sad - Faculty of Technical Sciences, Department of Graphic Engineering and Design, Turkey, *pp. 21–34. https://doi.org/10.24867/GRID-2020-p1.*

The increasing widespread use of plastic packaging and the resulting growth in plastic waste have forced all stakeholders to take action (Sundqvist-Andberg and Åkerman, 2021). These actions are generally aimed at policies and regulations that reduce or ban single-use plastics, as well as at promoting the circular economy of existing plastics (Barnes, 2024).

In the EU, the Circular Economy Action Plan was adopted by the European Commission in 2015, followed by the European Strategy for Plastics in 2018, both aimed at reducing plastic waste (Moment, E. S. I., 2014; Union, I., 2014). In addition, in October 2023 the proposal for the Packaging and Packaging Waste Regulation (PPWR) was voted on, emphasizing the need to define reusable packaging. In the United States, several states have now incorporated extended producer responsibility (EPR) packaging legislation into state law to reduce plastic waste and promote the reuse of packaging. EPR laws are legal instruments that place end-of-life responsibility for products on producers. These laws establish requirements to reduce plastic packaging use and plastic waste generation to meet state recycling and climate targets. To encourage the reuse of packaging, the legislation also provides incentives by exempting reusable packaging from EPR program fees. In the United States, the first Plastics Strategy to combat plastic pollution was released in 2024. The Plastics Strategy aims to phase out single-use plastics in federal procurement by 2027 and across all federal operations by 2035 (Farrell et al., 2024).

As the industry shifts toward these advanced, sustainable, and intelligent packaging systems, the need for robust, non-destructive analytical techniques has become paramount to validate their efficacy and safety. Raman spectroscopy has emerged as a highly versatile and powerful tool in this domain, offering rapid, label-free chemical characterization of both the food product and the packaging matrix. Advanced modalities of this technique, such as Spatially Offset Raman Spectroscopy (SORS) and hyperspectral Raman chemical imaging, enable the through-package detection of food adulterants, the monitoring of microbial degradation, and the mapping of active antimicrobial agents without ever breaching the package integrity (Lohumi et al., 2018; Clemente et al., 2016). By providing a direct window into the internal biochemical environment, these spectroscopic methods bridge the gap between active preservation mechanisms and real-time intelligent monitoring.

Furthermore, in the context of the global circular economy and the increasingly stringent regulations surrounding plastic waste management,

Raman spectroscopy plays a critical role in the end-of-life processing and forensic auditing of packaging materials. Its ability to generate highly specific molecular fingerprints allows for the rapid, high-accuracy identification and sorting of complex post-consumer polymers, such as PET, PE, and PVC, which is essential for optimizing industrial recycling streams (Moreira et al., 2025). Additionally, it serves as a formidable forensic instrument for verifying supply-chain authenticity; by analyzing the microscopic chemical origins of packaging colorants and binders, Raman microscopy can unequivocally distinguish genuine materials from illicit counterfeit packaging, thereby ensuring both environmental compliance and consumer safety across globalized markets (Kwok and Taylor, 2012).

6.2 Active packaging systems

Active packaging represents a major shift from conventional passive barrier systems toward functional materials that deliberately interact with the packaged food and its surrounding environment to preserve quality, enhance safety, and extend shelf life. Rather than simply isolating food from external conditions, active packaging systems are designed to regulate the internal package atmosphere through controlled release and scavenging mechanisms, such as the removal of oxygen or moisture and the release of antimicrobial or antioxidant agents. In starch-based active packaging, this functionality is achieved by incorporating bioactive ingredients, such as essential oils, plant extracts, organic acids, phenolic compounds, and antimicrobial nanoparticles, into the polymer matrix, enabling targeted control over microbial growth, oxidation, and moisture migration. Depending on their dominant function, active packaging systems can be classified into antimicrobial, antioxidant, humidity-regulating, and gas-scavenging or gas-releasing systems, each addressing a specific deterioration pathway in foods. However, despite their promising performance and sustainability advantages, these systems also face important challenges, including stability of active compounds, controlled release behavior, interactions with food matrices, regulatory constraints, and scalability for industrial applications.

6.2.1 Active packaging working mechanism

Active packaging systems work with the food that is packaged or the air around it to keep the food fresh for a time. These systems are different from packaging because they have special parts that do things like release helpful substances or remove bad stuff that can make the food go bad.

We can put these systems into two groups: release systems and scavenging systems. Release systems slowly let out things like stuff that kills germs or antioxidants or carbon dioxide or ethanol into the air around the food or right onto the food. These things help stop germs from growing slow down reactions that can make the food go bad and keep the food fresh. This often happens because the packaging material lets these helpful things move out slowly. For example, in packaging made from starch, the starch can act like a carrier that lets helpful substances move out over time. On the hand scavenging systems work by absorbing or removing bad things from the air around the food like oxygen, moisture, ethylene or bad smells that can make the food go bad. Systems that remove oxygen help stop the food from getting old and rotten. They also stop germs that need oxygen from growing. Systems that absorb moisture help stop water from collecting and germs from growing in foods that have a lot of moisture. Systems that remove ethylene are especially helpful for fruits and vegetables because they slow down the process of getting old. Both release and scavenging systems can be put into packaging films, coatings, sachets or pads depending on the type of food and how we want to keep it fresh.

6.2.2 Functional ingredients used in starch-based active packaging

Starch-based packaging offers an adaptable platform for incorporating functional additives that enhance both food packaging and shelf-life stability (Arruda et al., 2025). These active components can be added directly into the biopolymer matrix or applied as coating to impart antimicrobial, antioxidant, moisture-regulating, or gas-scavenging properties to help stop germs from growing and slow down reactions that can make the food go bad control moisture and/or remove gases. Essential oils and plant extracts derived from herbs and spices such as oregano, thyme, cinnamon, clove, and rosemary are commonly incorporated due to their strong natural antimicrobial activity. Their active constituents, primarily phenolic compounds, terpenoids, and flavonoids, disrupt microbial cell membranes and inhibit enzymatic processes, making them highly effective against bacteria, yeasts, and molds. Beyond antimicrobial functions, many of these botanicals also provide antioxidant protection by delaying lipid oxidation and pigment degradation in foods sensitive to oxidative spoilage. Natural components like polyphenols, flavonoids, tocopherols and ascorbic acid can slow down these reactions. Organic acids such as citric, lactic, acetic, and sorbic acids further strengthen antimicrobial performance. These acids reduce pH locally, interfere with microbial metabolic pathways, and can destabilize cell membranes, creating an unfavorable environment for bacterial and fungal growth. Their compatibility with starch matrices allows for controlled release, enabling sustained antimicrobial activity throughout storage.

Antimicrobial nanoparticles such as silver (Ag), zinc oxide (ZnO), and titanium dioxide (TiO_2) represent another class of functional additives that can be integrated into starch-based systems. These nanoparticles exhibit strong, broad-spectrum antimicrobial activity through mechanisms such as reactive oxygen species generation, membrane disruption, and interference with microbial DNA replication. When distributed within starch matrices, they provide long-lasting antibacterial performance while maintaining film transparency and mechanical stability. In addition, nanoscale fillers like nanocellulose, nano-clays, or layered silicates can significantly reinforce starch-based films by enhancing tensile strength, thermal stability, and barrier resistance to oxygen and moisture (Muñoz-Gimena et al., 2023; Altay et al., 2017, 2021, 2024).

Through the combined use of these natural bioactive compounds, organic acids, antioxidants, and antimicrobial nanomaterials, starch-based packaging can deliver improved freshness retention, spoilage prevention, and functional performance. At the same time, such materials remain more sustainable and environmentally compatible than conventional petroleum-based plastics, aligning with growing demand for biodegradable and renewable packaging solutions.

6.2.3 Active packaging types

Packaging includes many systems that help control things that can make food go bad. One of the used systems is the one that removes oxygen because oxygen can make fats go bad colors fade and germs grow. These systems remove the oxygen that is left in the packaging, which slows down reactions that can make the food go bad and helps the food longer. We also use systems that control moisture, for fresh foods like meat, poultry and produce because controlling moisture helps stop water from collecting and germs from growing. Packaging that stops germs from growing includes substances that can stop these germs either by touching the food or by releasing helpful substances slowly. Packaging that slows down reactions that can make the food go bad includes substances that stop these reactions from happening which helps prevent the food from getting old changing color or losing nutrients. Systems that remove ethylene are very important for fruits and vegetables because ethylene is a hormone that helps plants ripen and get old. Removing ethylene from the packaging helps the fruits and vegetables longer. There are also systems being developed like ones that release carbon dioxide to stop germs from growing in certain foods ones that release ethanol to prevent mold from growing in bakery products and ones that absorb bad smells that the food can make during storage.

6.2.4 Limitations and challenges

Even though active packaging is very helpful in keeping food fresh and safe there are still challenges that stop it from being used more widely. One of the problems is making sure that the helpful substances are released slowly and stay effective for a long time. This is especially hard for substances that can be affected by the environment or that can break down during processing and storage. Another big problem is that the materials used for packaging like starch-based polymers can be sensitive to moisture, not very strong and not very good at keeping air out compared to regular plastics. Adding ingredients can also change the properties of the packaging like how flexible it is, how see-through it is, how it lets air through and how strong it is. It is also important to make sure that the helpful substances do not move into the food much which could affect the taste, smell or make it unsafe. Getting approval from agencies is also a challenge because active packaging systems must follow strict rules about what can come into contact with food and these rules can be different in different countries. The cost of materials tiny particles and helpful ingredients can also be higher than regular packaging materials, which can stop it from being used widely. In terms of consumer, what they think and whether they accept packaging technologies, especially ones that use tiny particles or chemical additives is very important and needs to be addressed with clear communication, about safety and benefits.

6.3 Intelligent packaging systems

Intelligent packaging systems play a crucial role in enhancing food safety and quality by enabling the continuous, real-time monitoring of products throughout the entire supply chain, from processing and storage to transportation and retail (Mirza Alizadeh et al., 2021). These systems can perform multiple intelligent functions—including sensing, data recording, condition monitoring, and information transmission—which together support shelf-life extension, safety assurance, quality control, early warning of deterioration, and improved decision-making by producers, retailers, and consumers. One of the most defining features of intelligent packaging is its ability to communicate product status to users, often by translating chemical or biological changes associated with food spoilage, such as oxidation or reduction reactions, into visible or digitally readable signals (Abekoon et al., 2024; Janjarasskul and Suppakul, 2018).

Conventional expiration dates do not always accurately reflect the true condition of food products, as spoilage can occur earlier depending on temperature fluctuations, handling, and storage conditions. Intelligent packaging overcomes this limitation by directly monitoring key variables such as temperature history, microbial activity, package integrity, gas composition, and freshness, thereby reducing food waste and lowering the risk of foodborne illness. Driven by the rapid growth of the global food industry and rising consumer demand for fresher, safer, and more nutritious foods, these advanced systems are increasingly replacing traditional static packaging approaches.

Its most distinctive feature is that it typically informs consumers about product conditions during storage or transportation by assessing food spoilage through oxidation or reduction levels (Mkhari et al., 2025; Kuswandi and Jumina, 2020). Expiration dates on food products may not always be reliable, as products can spoil earlier depending on environmental conditions, creating risks for consumer safety and industrial efficiency. With intelligent packaging, factors such as temperature changes, microbial spoilage, package integrity, and freshness can be monitored, thereby minimizing food waste and reducing the risk of foodborne illnesses (Müller and Schmid, 2019).

The growing food industry and the increasing demand for fresher and more nutritious products have driven a shift away from traditional methods and toward the search for new systems (Fadiji et al., 2022; Stoma and Dudziak, 2022). Some of the new techniques that have emerged in recent years include coatings, antimicrobial packaging, antioxidant packaging, and modified atmosphere packaging. Intelligent packaging systems are closely linked to active packaging systems. Active packaging interacts directly with the food contents to enhance shelf life and improve food quality (Ahmed et al., 2022). Here, the role of preserving and protecting food is carried out not passively, as in traditional packaging, but actively through additional functions. On the other hand, intelligent packaging focuses on the ability to detect or monitor specific characteristics of the packaged product, the internal packaging environment, or surrounding conditions (Sohail et al., 2018). The core technologies of intelligent packaging include indicators, data carriers, and sensors.

6.3.1 Intelligent packaging types

In packaged products, factors such as moisture content, packaging temperature, and the levels of ethylene, oxygen, and carbon dioxide can affect the freshness of fruits and vegetables (Eom et al., 2012). Indicators can essentially measure the concentration of specific components, determine their presence or absence, and even evaluate the degree of reactivity between different components (Soltani Firouz et al., 2021). These indicators manifest themselves as directly observable characteristic changes (Müller and Schmid, 2019). Depending on the desired function, indicators can be placed either inside or outside the package and, unlike sensors, do not require any receiver, transmitter, or transducer (Sohail et al., 2018). In intelligent packaging systems, commonly used indicators include time–temperature indicators, gas indicators, freshness indicators, pH indicators, and pathogen indicators.

6.3.1.1 Freshness colorimetric indicators

Freshness refers to a food product that shows no signs of aging or wilting and whose cellular structure remains intact (Baietto and Wilson, 2015). In recent years, numerous studies have been conducted on freshness indicators applicable to intelligent food packaging to detect rotting and spoilage in food systems (Zhao et al., 2025; Zhai et al., 2025; Bhargava et al., 2020; Kalpana et al., 2019; Mohammadalinejhad et al., 2020; Liu et al., 2019). Today, freshness sensors that detect microbial changes, gas composition, and pH variations in food are very popular, and they are also highly practical for consumers to read or interpret (Zhang et al., 2019). Natural or synthetic colorants can be used in freshness sensors; however, the use of natural materials has increased in recent years and has become more important (Hailu et al., 2025). For this purpose, anthocyanins, carotenoids, chlorophyll, and similar natural compounds are commonly used (Liang et al., 2019; Yun et al., 2019; Chen et al., 2020). Potential microbial contamination in foods increases the risk of foodborne illness and consequently shortens shelf life. Therefore, the food industry requires accurate, cost-effective, fast, and reliable methods to monitor food freshness in real time. In this context, freshness indicators play a crucial role in tracking food products during storage and transportation (Abekoon et al., 2024; Zhang et al., 2022). Freshness indicators determine the microbiological status of a product by interacting with metabolites produced during microbial growth or by responding to freshness deterioration. These metabolites include CO_2, oxygen, glucose, ethanol, lactic acid, total volatile basic nitrogen (TVB-N) compounds, and H_2S. To enable contact with these chemicals, freshness indicators must be placed inside the package (Lee et al., 2019; Park et al., 2015).

In terms of starch, a wide range of starch-based indicator films has been developed for intelligent food packaging, each incorporating natural pigments or bioactive extracts to enable visible, pH-dependent color changes in response to spoilage compounds such as ammonia and volatile basic nitrogen. Potato starch combined with betacyanin from paperflower has been used to create labels that shift from pink to yellow as fish deteriorates under chilled storage, providing a clear real-time freshness signal (Naghdi et al., 2021). Other formulations, such as potato starch blended with blueberry anthocyanins and phycocyanin, fabricated through 3D-printing, have demonstrated strong ammonia sensitivity and can be paired with smartphone color-analysis apps to track the freshness of salmon and beef (Fig. 6.2) (Bao et al., 2024). Starch–chitosan matrices enriched with Lonicera caerulea anthocyanins have also shown reliable differentiation between fresh, sub-fresh, and spoiled shrimp based on their distinct color transitions (Li et al., 2022). Pea-starch and κ-carrageenan films containing black raspberry extract function effectively as intelligent labels for pork, exhibiting pronounced pH and ammonia responsiveness as the meat ages (Jiang et al., 2024). Similarly, CMC/starch films with purple sweet potato anthocyanin shift from red toward blue-green as fish spoilage progresses, mirroring changes in pH and TVB-N levels (Jiang et al., 2020). Waste-derived starch from Artocarpus lakoocha combined with curcumin has produced multifunctional films capable of signaling chicken freshness while also offering antioxidant and antimicrobial benefits (Mali and Pandey, 2024). Collectively, these systems illustrate the versatility of starch as a biodegradable carrier for natural dyes, enabling cost-effective, responsive, and environmentally friendly smart packaging solutions.

6.3.1.2 Time-temperature indicators (TTIs)

Temperature can directly affect a product's shelf life, safety, and the associated chemical degradation due to time-dependent physical kinetics (Ghaani et al., 2016; Biji et al., 2015; Pereira et al., 2015). Temperature fluctuations in the packaged-product environment can promote the growth of undesirable microorganisms and lead to product spoilage (Wang et al., 2015). Temperature indicators show whether a food product has been exposed to

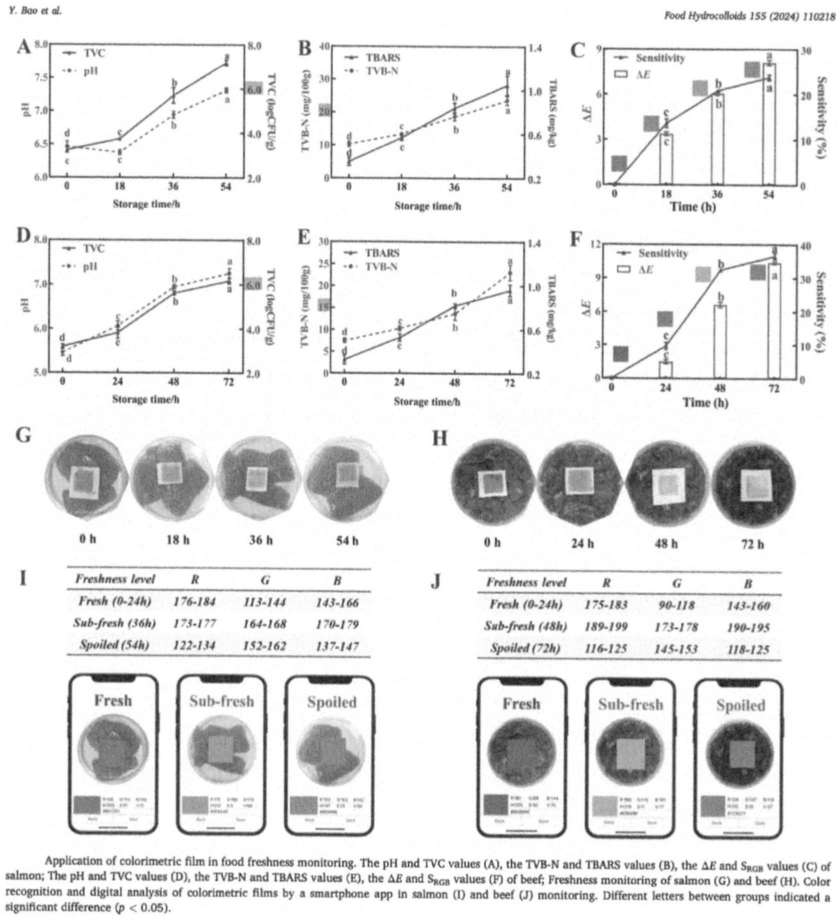

Figure 6.2 Overview of the colorimetric film indicator's performance in monitoring food freshness: For salmon, changes in pH and total viable count (A), TVB-N and TBARS values (B), and ΔE and SRGB color metrics (C) were tracked. For beef, corresponding shifts in pH and microbial load (D), TVB-N and TBARS (E), and ΔE and SRGB values (F) were evaluated. Freshness progression for salmon (G) and beef (H) was visually documented, while smartphone-based color extraction enabled digital analysis of the films in both products (I, J). Statistical differences between groups are indicated by distinct letter labels ($p < .05$). *Bao, Y., Wang, M., Li, J., Yang, X., Wen, B., Wang, L., He, Y., Si, X., Li, B., 2024. A starch-based 3D printed intelligent colorimetric film co-loaded natural pigments for visualizing food freshness: effect of nozzle size on gel structure formation.* Food Hydrocoll. *155, 110218. https://doi.org/10.1016/j.foodhyd.2024.110218.*

excessive heat or excessive cold outside a defined reference temperature range. Their primary function was originally defined as informing consumers (Han et al., 2005); today, however, TTIs are used at various control points throughout the entire food supply chain to ensure that required checks are carried out at defined reference times (Müller and Schmid, 2019). TTIs record the thermal history of a product, creating a "fingerprint" (Wang et al., 2018; Chen et al., 2021) and facilitating the identification of periods when the product has been outside the specified reference temperature range (Abekoon et al., 2024).

TTIs are classified according to their operating principles as physical, chemical, biological, and enzymatic systems (Gregor-Svetec, 2018; Pandian et al., 2021). Physical TTIs operate through different color-change mechanisms based on physical properties such as nanoparticle-based systems, diffusion-based systems, and electronic systems (Girardeau et al., 2020). Chemical TTIs function through chemical reactions that produce a specific color and include photochemical, polymerization-based, and oxidation-reaction-based TTIs (Girardeau et al., 2020). Biological TTIs include yeast- and lactic acid bacteria-based TTIs, Streptococcus-based TTIs, TTIs based on photonic lattice changes, and TTIs based on thermochromic polymer/dye mixtures. Enzymatic TTIs operate via a hydrolysis reaction involving an enzyme and its substrate, resulting in a color change (Gao et al., 2020). The color change can be tuned by considering various parameters such as enzyme type and concentration, the presence of activators or inhibitors, pH level, and buffer composition (Siddiqui et al., 2022). TTIs provide measurable outputs through color development, color movement, or mechanical deformation. Because they are user-friendly, easy to apply, and relatively low-cost, their commercial use has become widespread (Ahmed et al., 2018). Consumer satisfaction levels are high. They are generally used in the form of labels attached to packages or as direct labels (Kim et al., 2012; Sohail et al., 2018; Poyatos-Racionero et al., 2018) and are especially preferred for food products stored under refrigerated or frozen conditions (Pereira et al., 2015).

Starch is increasingly used as the functional platform for TTIs and thermal history indicators (THIs) in food packaging, with several open-access demonstrations spanning different sensing principles and products. In starch–iodine enzyme TTIs, the starch–iodine complex provides an initial blue hue that progressively fades as amylase degrades starch, creating a temperature- and time-dependent discoloration that records a product's thermal exposure and can be adapted to various perishables (Figs. 6.3 and 6.4) (Brizio and Prentice, 2015). Starch hydrogel THIs made from readily available starch track thermal accumulation via Maillard/browning chemistry; the resulting changes in absorbance, fluorescence, viscoelastic moduli,

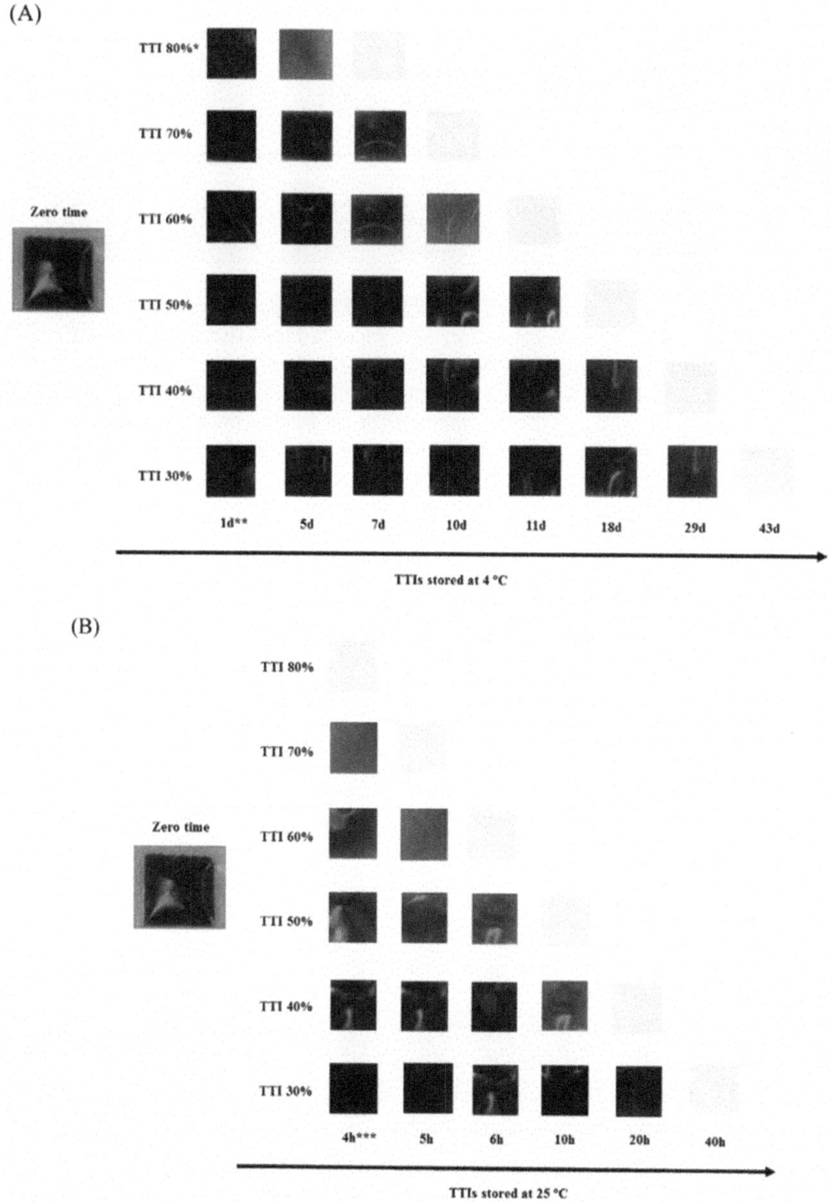

Figure 6.3 Discoloration patterns displayed by the TTI prototypes during storage: (A) held at 4°C and (B) at 25°C. *Indicates amylase concentration; **time in days; **time in hours. *Brizio, A.P.D.R., Prentice, C., 2015. Development of an intelligent enzyme indicator for dynamic monitoring of the shelf-life of food products.* Innov. Food Sci. Emerg. Technol. *30, 208–217. https://doi.org/10.1016/j.ifset.2015.04.001.*

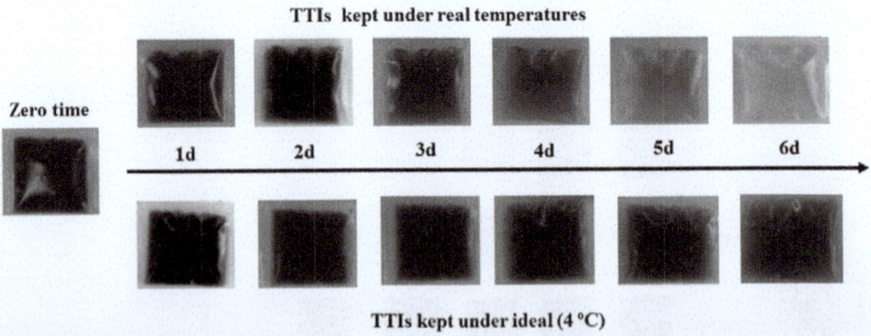

Figure 6.4 Day-to-day shifts in coloration were recorded for TTIs formulated with 60% amylase, revealing clear differences between indicators kept at the ideal storage temperature of 4°C and those subjected to realistic, fluctuating temperature conditions. *Brizio, A.P.D.R., Prentice, C., 2015. Development of an intelligent enzyme indicator for dynamic monitoring of the shelf-life of food products.* Innov. Food Sci. Emerg. Technol. *30, 208–217. https://doi.org/10.1016/j.ifset.2015.04.001.*

and RGB color values enable quantitative time–temperature history and have been demonstrated for milk, with the optical/viscoelastic signal closely aligned with microbial growth during storage (Fig. 6.5) (Sun et al., 2025). Cassava thermoplastic starch films can also serve as carriers for myoglobin/nitrite chemistries, where the protein's thermally driven color transition provides a natural, biodegradable TTI alternative to electronic sensors (Liu et al., 2022b). Finally, starch-based edible films embedded with blackberry particles act as temperature-abuse indicators: a heating/sterilization step (≈127°C) triggers an irreversible red to brown shift in the berry pigments within the starch matrix, producing a clear on-pack record of excessive thermal exposure (Liu et al., 2022b).

6.3.1.3 Gas indicators

Packaged products, especially fruits and vegetables, undergo changes over time due to variations in gas generation and microbial or microbiological activity (He et al., 2023; Alamdari et al., 2019). Gas indicators are in direct contact with the gases in the packaged food environment and provide information about food quality and safety by monitoring spoilage in real time (Ma et al., 2022; Mohammadian et al., 2020). Moisture release and respiration of fresh produce, microbial spoilage, material permeability, and environmental factors such as temperature directly affect the quality, safety, integrity, and shelf life of packaged foods (Fang et al., 2017). Gas sensors monitor changes in the package environment and are triggered in

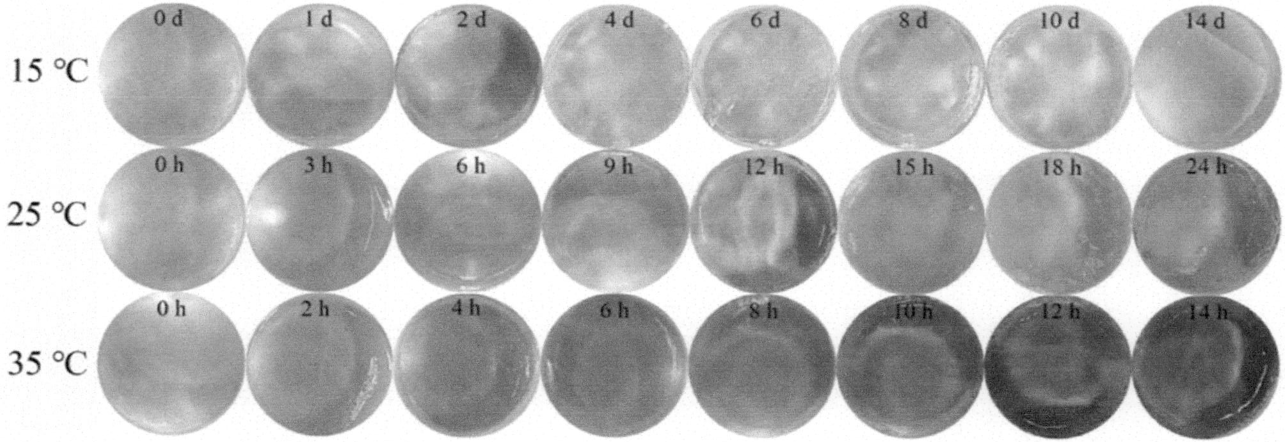

Figure 6.5 Color changes of the THI hydrogel based on starch at different temperatures for perishable products. *Sun, K., Liang, X., Zhou, S., Wang, J., Lu, F., Deng, Q., 2025. Thermal history indicator hydrogel based on starch with multiple signal channels for visualized monitoring thermal accumulation of perishable products.* China J. Food Sci. Technol. *62 (8), 1569–1579. https://doi.org/10.1007/s13197-024-06130-w.*

response to these changes, indicating the condition of the product. In short, gas indicators placed inside the package operate by responding to changes in gas composition. These responses are irreversible color changes. The indicators track levels of gases associated with spoilage, including oxygen, carbon dioxide (CO_2), hydrogen sulfide, ethanol, and water vapor (Heo and Lim, 2024).

Multiple starch-based gas indicators have been developed, particularly those responsive to ammonia/TVB-N and CO_2, and have been effectively applied to shrimp, pork, fish, and mushrooms for real-time freshness monitoring. Corn starch/polyvinyl alcohol (PVA) films loaded with UiO-66 MOF and anthocyanin nanocomplexes act as ammonia sensors, where the MOF boosts adsorption and accelerates a strong purple to gray to green color shift during shrimp or pork spoilage (Fig. 6.6) (Wu et al., 2024b). UiO-66 is a zirconium (Zr)-based metal–organic framework (MOF) constructed from Zr^{4+} ions. It has attracted significant interest because of its distinctive physical and chemical characteristics, including permanent porosity, exceptionally high surface area, and catalytic functionality, making it highly relevant for applications in gas storage and adsorption (Wu et al., 2024b). Another study presented corn starch/PVA films carrying ovalbumin-carboxymethylcellulose nanocomplexes respond to CO_2 and basic gases in sealed packs, registering mushroom freshness via a purple to pink change (Liu et al., 2022b). For higher sensitivity to basic volatiles in seafood, corn starch/chitosan films are reported incorporating rose-anthocyanin–loaded amylopyctin nanoparticles and a silica airgel use the airgel's high porosity to increase capture and deliver distinct purple-red to orange–gray to gray color transitions for shrimp (Wu et al., 2024a). Potato starch/PVA blended with ZnCu-BTC MOF nanorods yields a colorimetric ammonia film with strong response and stable long-term color, suitable for real-time prawn monitoring (Wang et al., 2024).

Simpler labels reported to be made by dipping starch/PVA films in Syzygium oleana pigment, which turns blue upon ammonia exposure and has been proposed as a smart freshness marker (Listyarini et al., 2020). Beyond ammonia alone, pea starch/κ-carrageenan films with black raspberry extract demonstrated to respond to both pH and NH_3, with color changes that align to rising volatile nitrogen in pork and to NH_3 vapor tests (Jiang et al., 2024). Likewise, potato starch combined with betacyanin delivers a pH/NH_3-sensitive label capable of detecting low ammonia levels and tracking fish spoilage in real time (Naghdi et al., 2021). These formulations show how starch systems, paired with bio-pigments and porous or

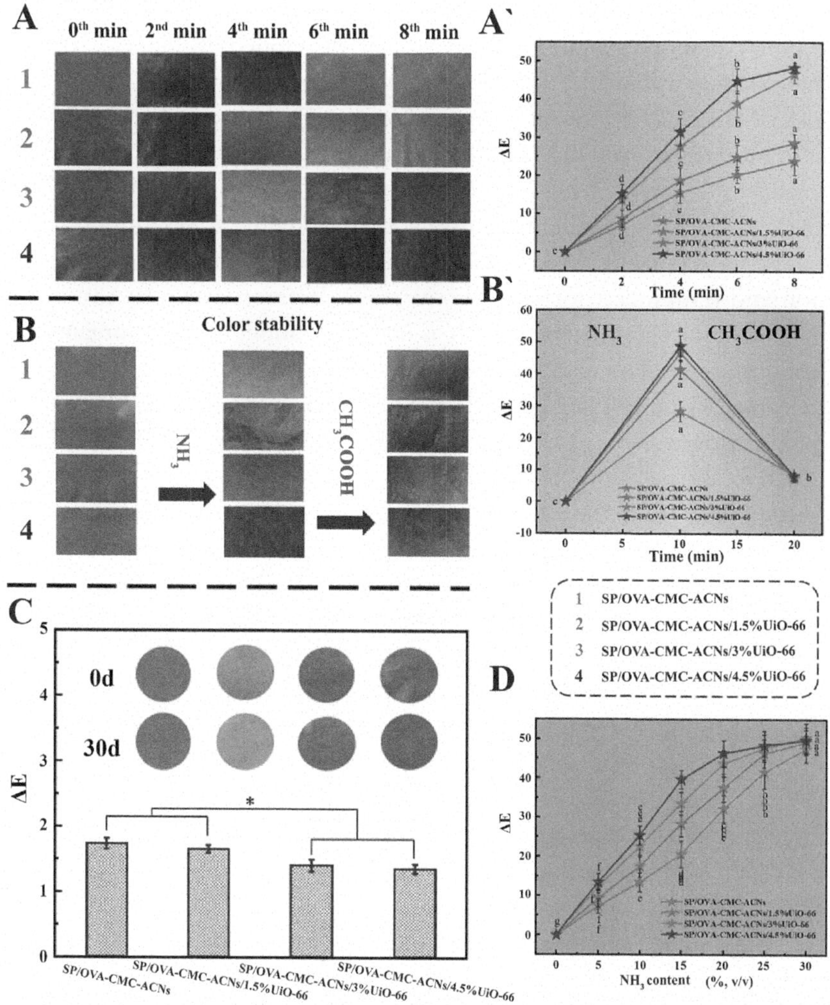

Figure 6.6 Composite films with varying UiO-66 contents respond under different conditions. Panel A shows the films' color changes after exposure to 20% NH3, with the corresponding ΔE values presented in A′. Panel B displays the films' ability to reverse color, alongside the associated ΔE color difference measurements in B′. Panel C highlights the films' color stability during storage, while panel D reports the ΔE values obtained when the films were subjected to different ammonia concentrations. *Wu, W., Liu, L., Zhou, Y., Shao, P., 2024. Highly ammonia-responsive starch/PVA film with gas absorption system as the 'bridge' for visually spoilage monitoring of animal-derived food.* Food Chem. *430, 137032. https://doi.org/10.1016/j.foodchem.2023.137032.*

MOF additives, can deliver rapid, visible, and application-specific freshness readouts across shrimp, pork, fish, and mushrooms.

6.3.1.4 Pathogen indicators

Recently, pathogen indicators have been developed to directly detect contamination caused by microbiological changes, particularly in highly perishable foods such as meat, fish, and poultry. These pathogens are important to monitor and identify within the food chain because they pose risks to human health and can cause various diseases. The indicators function by detecting, recording, and transmitting the biochemical reactions caused by pathogens. For this purpose, they employ bioreceptors for analyte recognition and transducers for signal conversion. Bioreceptors may consist of organic or biological materials such as enzymes, antigens, microorganisms, hormones, or nucleic acids, while transducers may be optical, colorimetric, or electrochemical. Like other types of sensors, these sensors typically display changes through color variation (Adeyemi and Fawole, 2023; Bedi et al., 2022; George et al., 2023).

Studies can be found in the literature where starch is utilized to create antimicrobial or pathogen-targeting films designed to inhibit or signal the presence of foodborne pathogens. These systems often incorporate functional additives to enhance their activity. For example, hydroxypropyl starch and PVA composite films incorporating biosynthesized ZnO nanoparticles demonstrate strong antibacterial and antifungal activity against *E. coli*, *S. aureus*, and several spoilage fungi, while remaining biodegradable, non-cytotoxic and achieving approximately 85% biodegradation in 4 weeks (Hashem et al., 2023). In addition, sago starch–PVA–citric acid films (starch-citrate) found to inhibit 98%–99% of *Salmonella Typhimurium*, *E. coli* O157:H7, *Listeria monocytogenes*, and *Aspergillus/Rhizopus* species, extending bread shelf life up to 40 days (Fig. 6.7) (Romainor et al., 2022). Starch films enhanced with essential oils such as clove, basil, and cinnamon significantly inhibit *S. aureus*, *Bacillus cereus*, and *E. coli*, and have been proposed as smart antimicrobial packaging (Mallick et al., 2020). Starch/chitosan/alginate composite reinforced with plant extracts or nanofillers show enhanced activity against *S. epidermidis, C. albicans*, and other microorganisms, with chestnut extract noted as especially potent (Wardejn et al., 2024). Moreover, corn starch combined with carboxymethyl chitosan is reported to achieve about 99.7% kill of *E. coli* and 99.8% of *S. aureus*, along with favorable mechanical and barrier properties (Dang et al., 2025).

6.3.1.5 pH indicators

Sensors are systems that detect, record, and transmit changes in product quality that may occur within the packaged environment or during

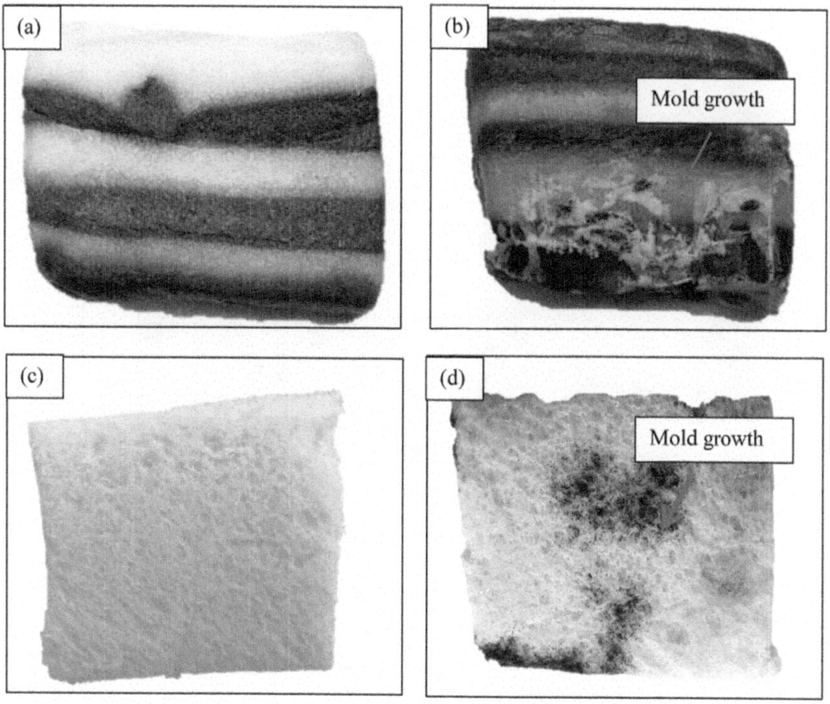

Figure 6.7 Photographs show (a) a slice of cake wrapped in a starch–citrate film and (b) a slice of cake packaged in a commercial food wrapper after 10 days, as well as (c) a piece of bread wrapped in the starch–citrate film and (d) a piece of bread stored in a commercial wrapper after 40 days. *Romainor, A.N., Chin, S.F., Lihan, S., 2022. Antimicrobial starch-based film for food packaging application.* Starch - Stärke. *74 (3–4), 2100207. https://doi.org/10.1002/star.202100207.*

transportation (Kuswandi et al., 2012; Pacquit et al., 2007). Protein-rich foods (meat, poultry, seafood) generate total volatile basic nitrogen (TVBN), primarily ammonia, trimethylamine, and dimethylamine, as endogenous enzymes and psychrotrophic microbiota metabolize amino acids during storage. The progressive increase of TVBN alkalizes the package headspace, providing a reliable proxy for freshness loss that can be detected by pH-responsive indicator systems embedded in or affixed to packages. Colorimetric pH indicators translate this chemical shift into a visible, instrument-free signal suitable for real-time monitoring along the cold chain (Fig. 6.8) (Alamdari et al., 2021). A typical on-package pH indicator (sometimes also called a spoilage or freshness sensor) comprises three functional elements: (1) pH-sensitive dye (natural or synthetic) that undergoes a reversible or quasi-irreversible spectral change across a defined

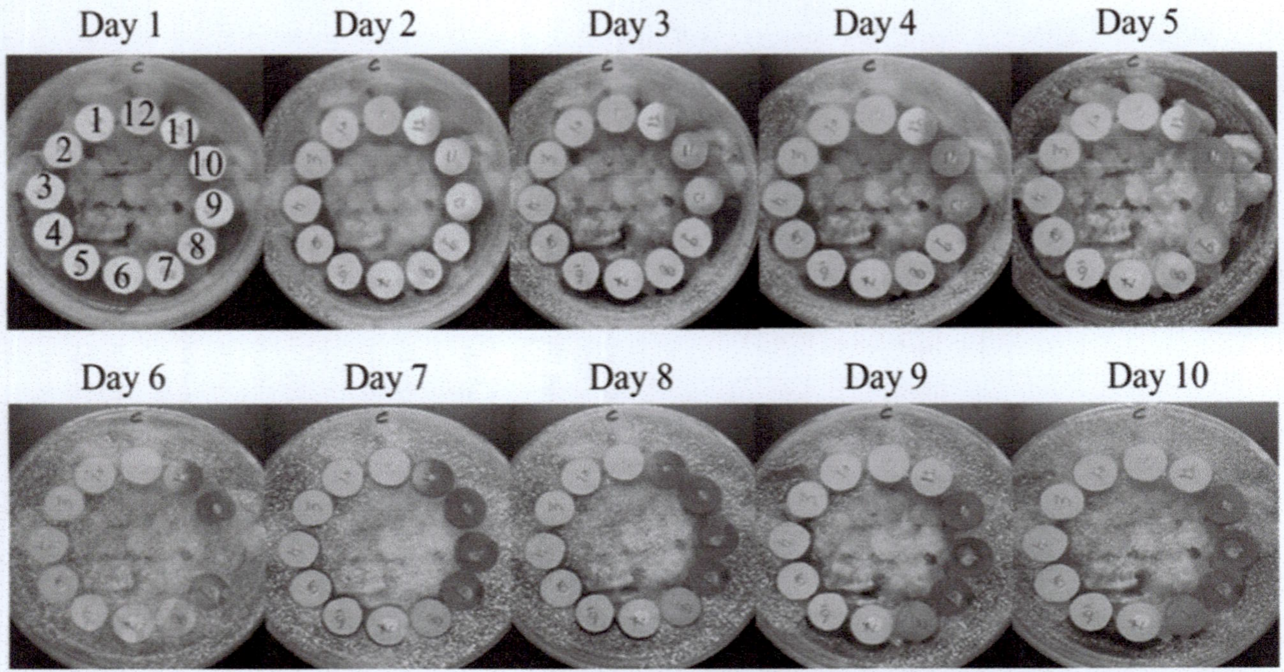

Figure 6.8 Progressive daily photographs illustrating the color response of NFSBH detectors with starting pH values ranging from #1 (0.25) to #12 (5.20). *Alamdari, N.E., Aksoy, B., Aksoy, M., Beck, B.H., Jiang, Z., 2021. A novel paper-based and pH-sensitive intelligent detector in meat and seafood packaging.* Talanta *224, 1211913. https://doi.org/10.1016/j.talanta.2020.121913.*

pH window; (2) solid matrix (film or paper) to immobilize and retain the dye while allowing diffusion of alkaline volatiles; (3) formulation aids (binders and pigments/fillers) that tune porosity, gas adsorption, mechanical integrity, and coating processability (Alamdari et al., 2021). Matching the dye transition range to the expected TVBN trajectory of a target product enables tiered signaling (e.g., "fresh," "approaching limit," "spoiled"). Practical systems also manage mass-transfer (gas access to the indicator), optics (contrast, readability), and adhesion/printability for conversion on paper or polymer substrates.

Recent work demonstrates a paper-based, pH-sensitive freshness detector manufactured by coating a functional formulation onto filter paper, an approach that is scalable to conventional paper-mill equipment (rod/knife coating, hot-air drying). The coating combines Bromocresol purple (BCP) as the pH dye (from yellow to purple across pH $\approx$ 5.2–6.8) tailored to respond to amine release; Bentonite (Na-montmorillonite) as a layered platelet pigment that increases gas adsorption and creates tortuous diffusion paths; and soybean-hull (SBH) binder, optionally nanofibrillated (NFSBH), providing abundant carboxyl and protein functionalities that strengthen coating cohesion, dye immobilization, and bonding to cellulose fibers (Alamdari et al., 2021). To map indicator response to microbiological limits, the study fabricated indicator sets with different initial coating pH values (0.25–5.20) and co-stored them with catfish fillets at 4°C (Fig. 6.8). Daily color change was compared with total viable count (TVC). Indicators with lower initial pH triggered later (requiring more TVBN), creating a ladder of labels from early warning to end-of-shelf-life—with the pH $\approx$ 3.0 label aligning to the spoilage day near the accepted seven log CFU/g threshold. This provides a practical tuning strategy: select higher-pH labels for early freshness loss and lower-pH labels for end-point spoilage. The coated paper disks produced high-contrast yellow to purple transitions that were easily photographed/visually read. Activated labels exhibited 12 h/24 h (4°C) visible persistence in a controlled ammonia challenge, and in practice the color change was effectively irreversible, preserving a history record of a temperature change or extended storage event that is useful for supply-chain accountability. Similarly, starch-based indicators reliably change color in response to pH changes to detect spoilage volatiles. A study present cassava starch film for pH-sensitive intelligent film application where onion peel, a polyphenol-rich agro-industrial byproduct, valorized by incorporating in into cassava starch-based films (Torche et al., 2025). The films exhibited selective antibacterial action (notably against *P. aeruginosa* and

E. coli) and broad pH-indicator behavior (light to dark brown across pH 1–13). Applied as on-pack labels for minced beef stored at 4°C, the films tracked freshness over 13 days, and the color changes correlated strongly with both microbial counts and pH shifts that signals the onset of spoilage (Fig. 6.9) (Torche et al., 2025).

Intelligent pH sensitive starch films with butterfly pea (BPE) anthocyanins also developed to track a visible shift from light pink to green across pH 1 to 12 to alert spoilage over 6 days (Mary et al., 2020). This research work evaluated the influence of BPE and TiO_2 on the physical and structural properties of starch film. Several studies utilized similar labels successfully to pork, tuna, poultry, fish, shrimp, blueberries, cheese, chicken, milk and pastry packaging that provides visible responses during storage (Versino et al., 2023).

6.3.1.6 RFID, NFC, QR code-based smart systems

Starch-based materials are increasingly incorporated into smart food-packaging technologies, where they function mainly as biodegradable matrices for active or intelligent films rather than as direct platforms for complex electronics such as Radio-Frequency Identification (RFID) or Near-Field Communication (NFC). Recent literature presents starch and other biopolymers as essential components of sustainable packaging systems, often paired with sensors, indicators, or data carriers, including barcodes, Quick Response (QR) codes, and RFID tags, to enhance safety, traceability, and waste reduction, though these reviews rarely indicate that the electronic tags themselves are fabricated on starch films (D'Almeida and de Albuquerque, 2024; Versino et al., 2023; Zuo et al., 2022).

Intelligent starch-based films typically incorporate pH- or gas-responsive dyes, such as anthocyanins, to enable color-changing freshness indicators, which are sometimes linked to smartphone imaging as a low-cost analytical tool, effectively creating a simple app-supported information system (Xie et al., 2025; Kılıç and Kahyaoglu, 2025; Cui et al., 2021). Broader smart-packaging research emphasizes RFID, NFC, and QR technologies for real-time product tracking, quality monitoring, and supply-chain authentication, but these systems are generally presented as attachable or printable modules compatible with a range of packaging substrates rather than inherently starch-based solutions (Ghaani et al., 2016; Zuo et al., 2022). Accordingly, major reviews describe two parallel yet increasingly intersecting trends of bio-derived matrices such as starch that provide biodegradability and integrated chemical sensing, and

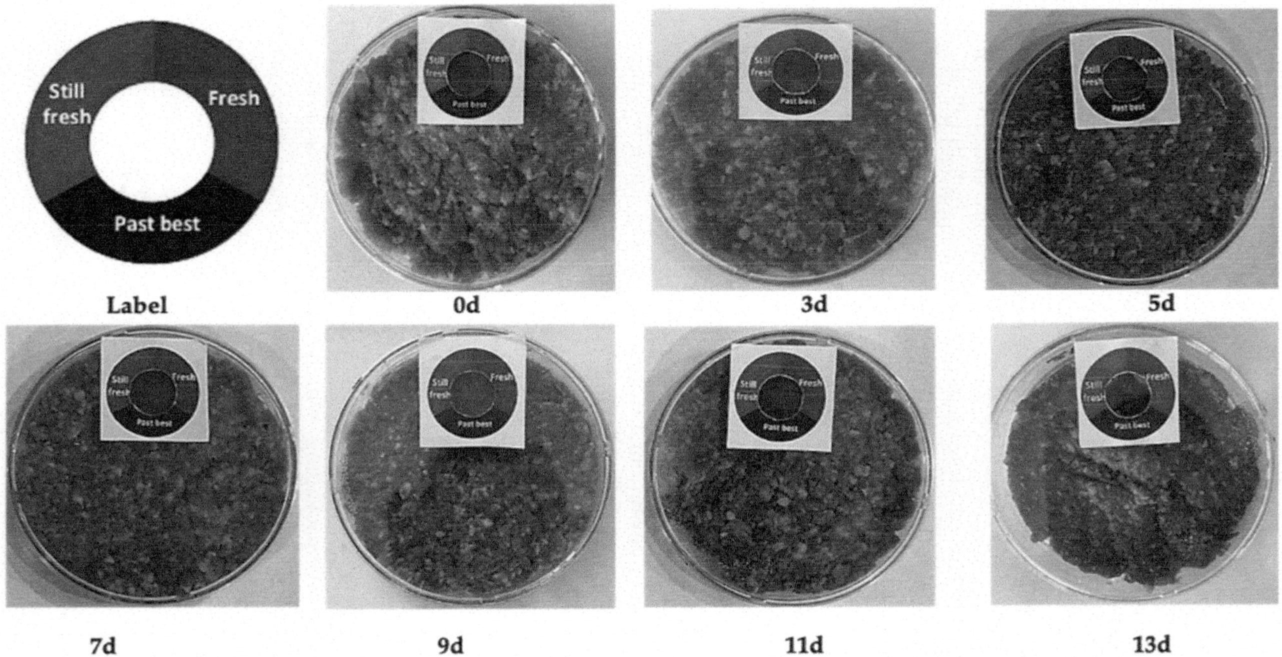

Figure 6.9 Color reference chart for the pH-responsive label for minced beef held at 4°C, showing transition through fresh, still fresh and past best at days 0–13. *Torche, A., Chouana, T., Sanah, I., Djeghim, F., Khodja, E.A.T., Mezreb, K., Elboutachfaiti, R., Delattre, C., D'Elia, M., Rastrelli, L., 2025. pH-sensitive cassava starch/onion peel powder films as colorimetric indicators for minced beef freshness monitoring.* Algeria Foods. *14 (17), 2974. https://doi.org/10.3390/foods14172974.*

electronic or optical data-carrier systems that enable digital connectivity. Full convergence of these capabilities into a single starch-based platform is highlighted as an emerging but still developing area of research (Soltani Firouz et al., 2021).

6.3.2 Advanced analytical and sensing technologies in intelligent packaging

In the modern food industry, the convergence of exponential global population growth, heightened consumer awareness, and the emergence of global health crises has necessitated a fundamental transformation in quality-control practices and food safety regulations (Lohumi et al., 2018; Liu et al., 2021). Concurrently, the ubiquity of digital media has amplified consumer expectations, driving an unprecedented demand for food products that are demonstrably safe, exceptionally fresh, fully transparent in their origins, and sustainably manufactured. To meet these multifaceted demands, the industry is rapidly moving away from traditional packaging paradigms toward advanced active and intelligent packaging technologies. These next-generation systems not only rigorously enhance food safety and extend shelf life but also offer real-time, dynamic information regarding product quality and environmental storage conditions, thereby optimizing supply-chain efficiency and bolstering consumer trust (Lohumi et al., 2018; Liu et al., 2021).

The globalization of modern food supply chains has severely exposed the limitations of conventional passive packaging. Today, food products are frequently manufactured in massive, centralized facilities before being distributed across vast geographical distances to reach diverse international markets. Throughout this journey, products are subjected to highly variable handling and environmental conditions that can compromise their integrity. Traditional passive packaging, which functions merely as a static physical barrier against external contaminants, is increasingly inadequate for mitigating these complex logistical realities. Consequently, there is an imperative to deploy advanced packaging systems capable of actively preserving sensory attributes and monitoring potential hazards from the point of production to the consumer's table.

Simultaneously, the environmental devastation associated with the ubiquitous use and disposal of petroleum-derived plastic packaging has triggered urgent global legislative and regulatory responses. In the European Union, comprehensive frameworks such as the 2015 Circular Economy

Action Plan and the 2018 European Strategy for Plastics were established to drastically reduce plastic waste. This was further reinforced by (He et al., 2023) proposal for the Packaging and Packaging Waste Regulation (PPWR), which emphasizes the critical need for reusable and sustainable packaging. In parallel, various states within the United States have enacted extended producer responsibility (EPR) laws, which shift the financial and logistical burden of end-of-life product management onto the producers, while the 2024 Plastics Strategy aims to phase out single-use plastics in federal procurement by 2027.

Driven by these stringent environmental policies, the packaging sector is experiencing a massive shift toward biodegradable biopolymers, with starch emerging as a preeminent candidate. Starch is highly favored because it is a low-cost, fully biodegradable, natural polysaccharide derived from highly renewable resources such as corn, potatoes, cassava, and rice. However, native starch possesses a granular structure characterized by strong intermolecular hydrogen bonds, rendering it non-thermoplastic and exceedingly difficult to process into continuous films. To overcome this, the starch must be destructured using heat and shear forces in the presence of plasticizers, such as glycerol or sorbitol, converting it into thermoplastic starch (TPS). While TPS can be processed similarly to synthetic polymers and provides high flexibility, its inherent hydrophilic nature remains a significant challenge, causing it to lose mechanical strength and vital barrier properties when exposed to high-humidity environments. Despite these processing hurdles, when properly functionalized, starch serves as an excellent matrix for "smart" packaging, effectively bridging the gap between active preservation and intelligent communication.

6.3.2.1 Spatially offset Raman spectroscopy for non-destructive through-packaging analysis

Spatially Offset Raman Spectroscopy (SORS) has emerged as a revolutionary, non-destructive analytical technique capable of penetrating opaque, highly diffusely scattering barriers, such as plastic containers and multilayered paperboard, to analyze the chemical composition of enclosed food products. By overcoming the profound limitations of conventional backscattering Raman spectroscopy, SORS represents a massive leap forward for auditing food supply chains, detecting fraudulent adulteration, and identifying toxic contaminants.

6.3.2.1.1 The spatially offset mechanism

In conventional Raman spectroscopy, the excitation laser and the signal collection optics are aligned at the exact same point (a 0 mm offset). While effective for surface analysis, this backscattering geometry is severely limited when analyzing packaged foods, as the resulting spectrum is overwhelmingly dominated by the intense Raman scattering and fluorescence of the superficial packaging layer, completely masking the internal food's signature.

SORS solves this problem by physically separating the laser illumination point from the signal collection zone. When laser photons penetrate a sample, they undergo multiple random scattering events. Photons that scatter near the surface tend to re-emerge very close to the incident laser point. Conversely, photons that migrate deeper into the internal subsurface layer (the food) travel laterally before eventually resurfacing. Because the attenuation speed of the subsurface Raman signal over a lateral distance is much slower than that of the surface layer signal, moving the collection point away from the laser incidence point selectively filters out the packaging noise, as illustrated by the photo paths (Fig. 6.10a) and the offset-dependent spectra (Fig. 6.10b).

6.3.2.1.2 Chemometric quantification

Random Forest and PLS Models The SORS approach produce very complex multidimensional spectra that contain thousands of variables and require some form of chemometric or machine learning treatment in order to generate quantitative concentration information.

- **Random Forest (RF):** Random Forest is an ensemble classifier for supervised learning. They are used because of their ability to handle high dimensional data with multicollinearity and less overfitting. They work by averaging the prediction from many decision trees on different subsets of the data.
- **Partial Least Squares (PLS):** Partial Least Squares (PLS) is a set of statistical methods that simplifies the data in the spectral vectors by reducing them to latent variables which contain the most important information from the original spectra. Common examples of PLS methods include Partial Least Squares Regression (PLSR) and Partial Least Squares Discriminant Analysis (PLS-DA). PLSR is used to determine the continuous quantitative variables such as percentage of adulterants in a sample. PLS-DA is used to classify samples in certain categories such as fresh meat and spoilt meat.

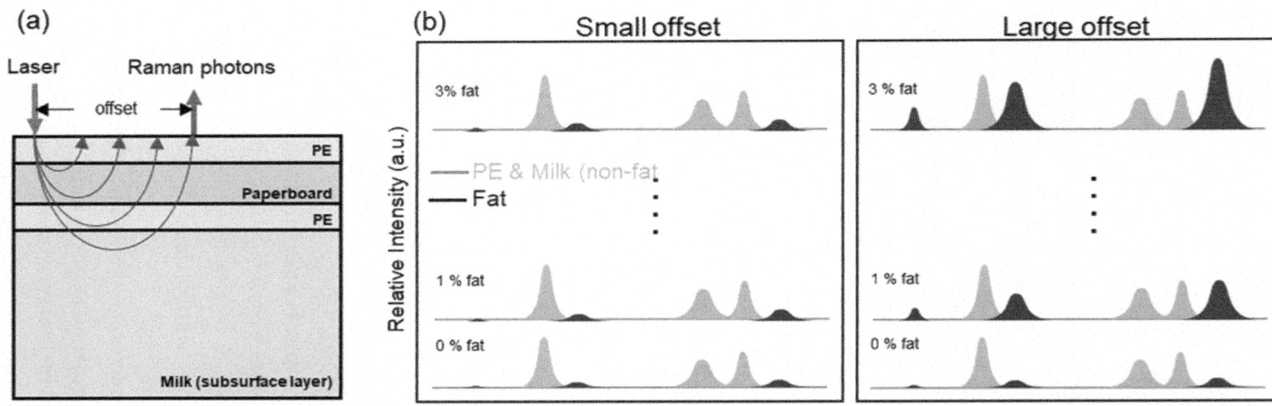

Figure 6.10 (a) Optical paths of Raman photons in deep Raman measurements of milk contained in a carton. (b) Simulated Raman spectra of carton-contained milk with different fat concentrations at small and large offsets. *Song, S.W., Jeong, Y.C., Park, C.R., Kim, H.M., 2023. Quantitative fat analysis of milk using a line-illumination spatially offset Raman probe through carton packaging.* Analyst. *148 (14), 3321–3329. https://doi.org/10.1039/d3an00416c.*

6.3.2.1.3 Detection of adulteration through packaging

- **Glucose Adulteration in Sugar:** We have previously shown that the adulteration of glucose in soft sugar can be detected using a cheap 0.5 mm thick red plastic filter with a line-scan Raman imaging system and optimal 3 mm spatial offset to remove the strong plastic bands at 1062, 1095 and 1294 cm^{-1}. An RF model produced an R^2 of 0.957 and RPD of 4.846.
- **Margarine Adulteration in Butter:** A macro-scale line-scan SORS system (utilizing a 14 cm wide laser line) was engineered to quantify margarine adulteration in butter. Butter and margarine share a prominent CH2 scissoring peak at 1442 cm^{-1}, which is typically obscured by plastic covers. By establishing optimal spatial offsets of 4 and 6 mm, the overlapping packaging peaks virtually disappeared. Coupling this SORS data with a PLSR model caused the R^2 value to soar to 0.922 and the Standard Error of Prediction (SEP) to plummet to 3.9% (Fig. 6.11).
- **Milk Fat Analysis and Melamine Detection:** In Liquid Milk Cartons The liquid milk cartons that consumers pour milk into every day are made of highly opaque multi-layer structure composed of polyethylene (PE)/paperboard/PE. As a result, the milk fat peaks at 1302 cm^{-1}, 1440 cm^{-1} are obscured. Using a multiple-depth line-mapping fiber optic probe with a maximum probe offset of 7 mm, the milk fat content in liquid milk cartons can be profiled out. Melamine, a known toxic compound, can be detected at concentration levels as low as 0.24 wt.% through its ring-breathing peak at 674 cm^{-1}, confirming that SORS can quantify nutritional components and detect hazardous contaminants in one scan through highly complex barriers (Fig. 6.12).

6.3.2.2 Raman chemical imaging and mapping for active packaging characterization

Raman Mapping Spectroscopy: Raman mapping spectroscopy is a nondestructive analytical technique for the in-depth investigation of the molecular structure and morphology of a wide range of biological samples and packaging materials at the micrometer and nanometer scale.

6.3.2.2.1 Investigating antimicrobial effects and visualizing cell damage

Clemente et al. utilized high-spatial-resolution Raman imaging to investigate the cell damage caused by benzyl isothiocyanate (BITC), an antimicrobial active packaging agent, on *Aspergillus ochraceus* (an ochratoxin A-producing mold). Raman maps revealed that BITC exposure led to

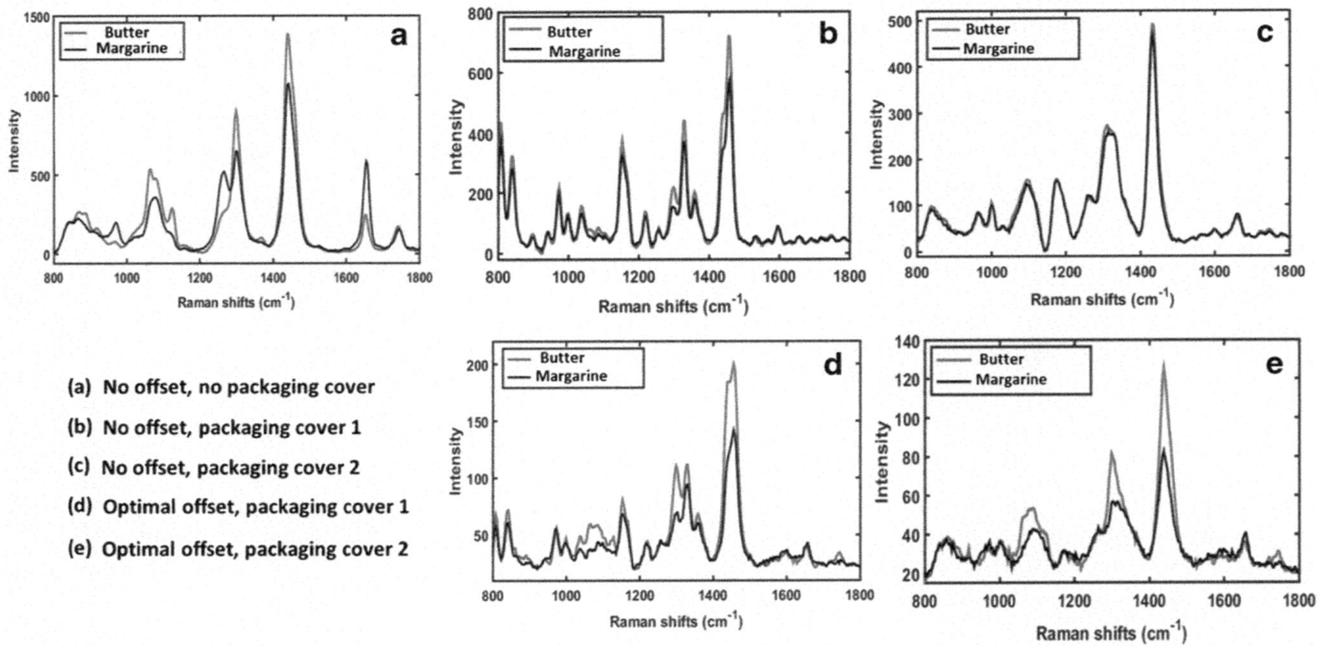

Figure 6.11 Preprocessed Raman spectra for butter and margarine acquired with conventional backscattering geometry and SORS spectra obtained at optimal offsets. *Lohumi, S., Lee, H., Kim, M.S., Qin, J., Cho, B.K., 2018. Through-packaging analysis of butter adulteration using line-scan spatially offset Raman spectroscopy.* Anal. Bioanal. Chem. *410 (22), 5663–5673. https://doi.org/10.1007/s00216-018-1189-1.*

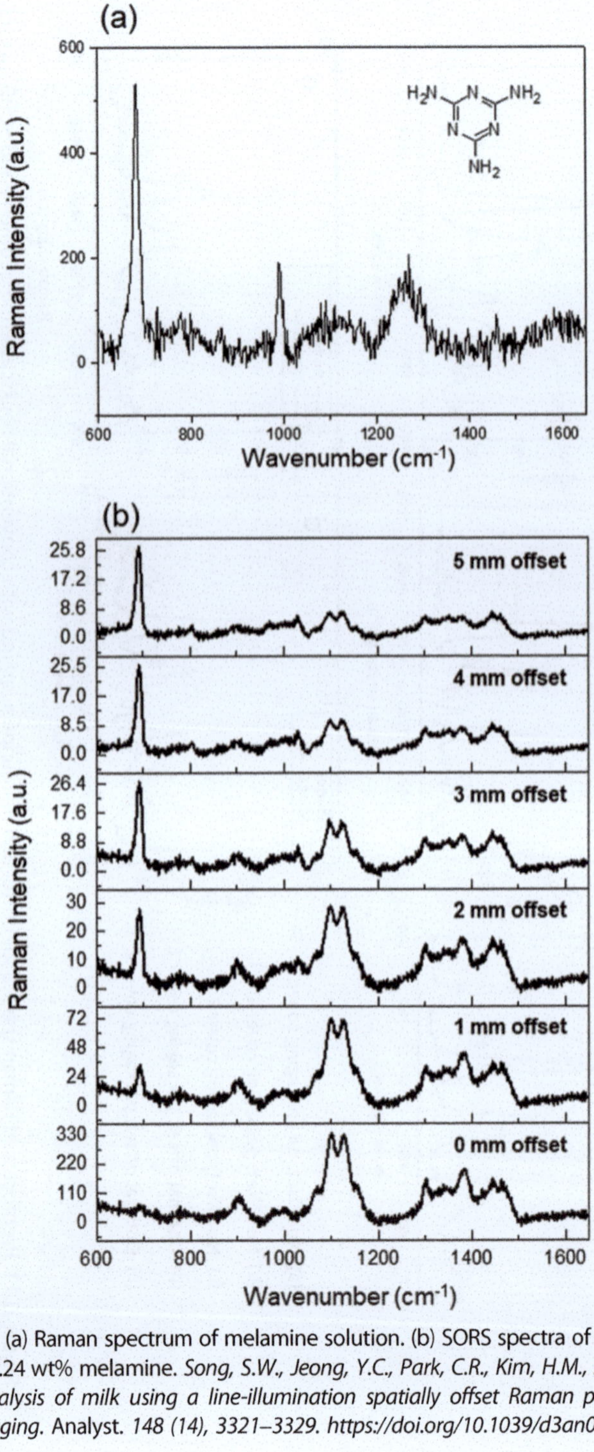

Figure 6.12 (a) Raman spectrum of melamine solution. (b) SORS spectra of milk samples containing 0.24 wt% melamine. *Song, S.W., Jeong, Y.C., Park, C.R., Kim, H.M., 2023. Quantitative fat analysis of milk using a line-illumination spatially offset Raman probe through carton packaging.* Analyst. *148 (14), 3321–3329. https://doi.org/10.1039/d3an00416c.*

structural disruptions such as cell wall deformities and clump formations inside the hyphae (Fig. 6.13).

Spectra showed drastic reductions in bands corresponding to proteins (e.g., 747, 1002, and 1448 cm^{-1}), indicating protein aggregation caused by BITC conjugating with thiol groups. Similarly, reductions in lipid and phospholipid bands (e.g., 1309, 1448, and 1657 cm^{-1}) suggested reduced cell fluidity. Utilizing Principal Component Analysis (PCA), researchers mapped the exact areas of vulnerability, proving severe impairment of essential cellular functions (Fig. 6.14). Similarly, combining dielectrophoresis (DEP) with Raman imaging allows for rapid, single-cell characterization of foodborne pathogens (*E. coli*, *S. aureus*) in planktonic suspension, aiding the development of antimicrobial nano systems.

6.3.2.2.2 Verifying the distribution of active agents

Beyond biological interactions, Raman mapping is essential for verifying the distribution of active agents within packaging films, such as starch-based matrices. Direct addition of active components can lead to uncontrolled release or instability. Raman mapping allows researchers to observe how agents are distributed. For instance, FT-Raman mapping coupled with Multivariate Curve Resolution (MCR) can visualize the spatial distribution of active ingredients (like salicylic acid) within solid formulations. Furthermore, Band-Target Entropy Minimization (BTEM) algorithms can be used to reverse-engineer multi-layered commercial laminates, successfully mapping the spatial distribution and thickness of distinct layers like cellulose, talc, and polyethylene (Moreira et al., 2025).

6.3.2.3 Material identification for circular economy and forensic applications

Raman spectroscopy is an analytical technique used to measure the inelastic scattering of laser light by molecules. By exploiting differences in molecular vibrations, it provides a highly specific molecular signature. Therefore, Raman spectroscopy is one of the preferred analytical techniques for characterization of materials and plastics, for example in starch-based matrices (Table 6.1), plastic recycling or forensic applications (Moreira et al., 2025; Lohumi et al., 2018; Liu et al., 2021; Song et al., 2023; Clemente et al., 2016).

6.3.2.3.1 Polymer identification for post-consumer plastic recycling

Successful recycling requires accurate identification of polymers, as misclassification degrades the recycled end-product. Raman spectroscopy directly measures structural composition regardless of dirt or biofilms:

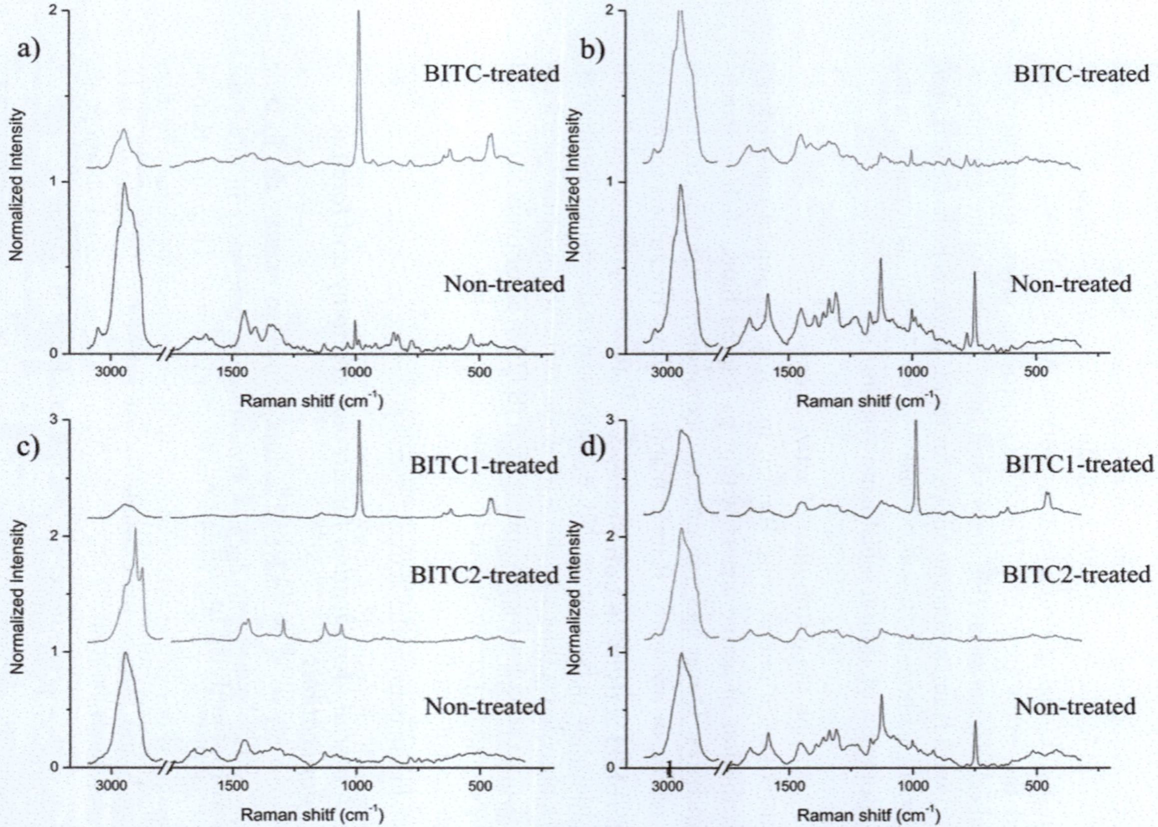

Figure 6.13 Normalized Raman imaging spectra so that an average spectrum of all the maps for each same sample ($n = 5-8$) was taken. Antifungal activity in liquid medium: (a) nontreated and BITC-treated spores and (b) nontreated and BITC-treated hyphae. Antifungal activity in vapor phase: (c) nontreated, BITC1-treated, and BITC2-treated spores and (d) nontreated, BITC1-treated, and BITC2-treated hyphae. *Clemente, I., Aznar, M., Nerín, C., 2016. Raman imaging spectroscopy as a tool to investigate the cell damage on aspergillus ochraceus caused by an antimicrobial packaging containing benzyl isothiocyanate.* Anal. Chem. *88 (9), 4772–4779. https://doi.org/10.1021/acs.analchem.6b00116.*

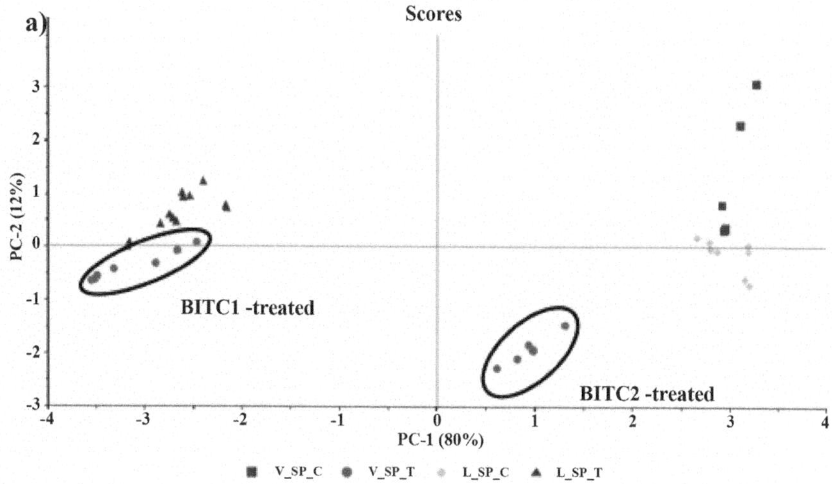

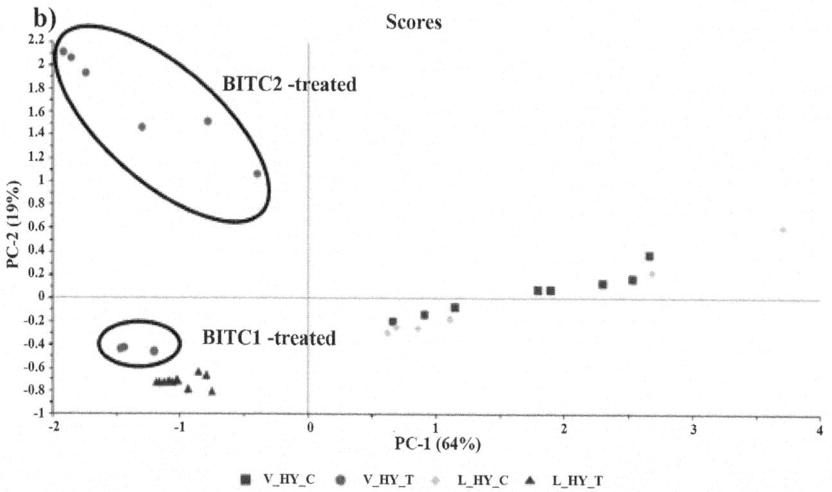

Figure 6.14 Principal component analysis obtained with Raman spectra of spores (a) and hyphae (b) in different assays: vapor condition, nontreated (■) and BITC-treated (●) and in a liquid condition; nontreated (◆) and BITC-treated (▲). *Clemente, I., Aznar, M., Nerín, C., 2016. Raman imaging spectroscopy as a tool to investigate the cell damage on aspergillus ochraceus caused by an antimicrobial packaging containing benzyl isothiocyanate.* Anal. Chem. *88 (9), 4772–4779. https://doi.org/10.1021/acs.analchem.6b00116.*

- **PET:** Identifiable by prominent peaks at 645, 871, 1623, and 1737 cm^{-1}.
- **PE-HD and PE-LD:** Raman is sensitive enough to differentiate these structural variations of polyethylene based on the degree of polymer

Table 6.1 Integrative summary of starch-based matrices, active agents, and advanced analytical monitoring technologies.

Sample material	Analysis technique	Primary goal	Key components identified	Optimal offset distance	Statistical performance (R^2/RMSEP)	Source
Adulterated sugar with plastic packaging	Line-scan spatially offset Raman imaging (SORS)	Detection of adulterated sugar (castor sugar and glucose)	Soft sugar, glucose	3 mm	$R^2 = 0.957$, RMSEP = 0.413	Kwok and Taylor (2012)
Butter adulterated with margarine	Line-scan spatially offset Raman spectroscopy (SORS)	Quantitative analysis through plastic packaging covers	Butter, margarine	6 mm (Cover 1), 4 mm (Cover 2)	$R^2 = 0.948$, (SEP = 3.2%) for Cover 1; $R^2 = 0.948$ (SEP = 3.9%) for Cover 2	Widjaja and Garland (2012)
Milk in carton packaging	Line-illumination spatially offset Raman probe (SORS)	Quantitative fat analysis and toxic chemical detection	Milk fat, melamine	5 mm (table-top), 7 mm (fiber probe)	$R^2 = 0.987$ (blue carton), $R^2 = 0.991$ (green carton)	Widjaja and Garland (2012)
Model pharmaceutical tablets	Raman mapping with BTEM	Detection of trace components and impurities	Acetaminophen, lactose, avicel, magnesium stearate	–	Trace detection down to 0.2%	Liu et al. (2021)
Salicylic acid film tablets	FT-Raman mapping with Multivariate Curve Resolution (MCR)	Quantification and spatial distribution of API	Salicylic acid, lactose, starch	–	$R^2 = 0.99$; LOD = 0.35%	Lohumi et al. (2018)
Lamb meat	Hand-held Raman spectroscopy (in-pack)	Prediction of total viable microbial count (TVC)	Microbial load (TVC)	–	$R^2 = 0.29$; RMSE = 1.34	Holman et al. (2025)

Postconsumer packaging plastics	Raman spectroscopy combined with PLS	Identification for recycling purposes	PET, PE-HD, PVC, PE-LD, PP, PS, dyes (phthalocyanine blue, titanium dioxide)	–	100% classification accuracy	Moreira et al. (2025)
Pharmaceutical packaging (Cialis®)	Raman microscopy and 2D-correlation spectroscopy (2DCOS)	Authentication of genuine and counterfeit packaging	Cellulose, calcium carbonate, styrene-based polymer, yellow azo-type pigments, unknown colorants	–	–	Kwok and Taylor (2012)
Commercial laminate packaging	Raman mapping with BTEM	Reverse engineering of layered materials	Cellulose form I, cellulose form II, talc, polyethylene, anatase, calcite, polystyrene co-polymer	–	–	Liu et al. (2021)
Singapore postage stamp	Raman mapping with BTEM	Authentication of consumer goods	Phthalocyanine blue, calcite, organic yellow dye	–	–	Liu et al. (2021)
Ancient archeological shards (Qing dynasty)	Raman mapping with BTEM	Characterization of excavated artifacts	Silica quartz, carbon, hematite, glassy silicate, lanthanide	–	–	Liu et al. (2021)

(Continued)

Table 6.1 Integrative summary of starch-based matrices, active agents, and advanced analytical monitoring technologies.—cont'd

Sample material	Analysis technique	Primary goal	Key components identified	Optimal offset distance	Statistical performance (R^2/RMSEP)	Source
Human nails (finger and toe)	Raman points measurements with BTEM	Identification of bio-constituents and disease biomarkers	Keratins, proteins, polypeptides, fatty acid, carotenoids, ferulic acid	–	–	Liu et al. (2021)
Bio-active glass	Raman mapping with BTEM	Monitoring growth of hydroxyapatite/bone apatite formation	Amorphous calcium phosphate, octa-calcium phosphate, matured carbonated hydroxyapatite	–	–	Liu et al. (2021)
Biodegradable polymeric double-walled microspheres	Raman mapping with BTEM	Spatial distribution of polymers for drug delivery	Poly(L-lactide), poly (D,L-lactide-co-glycolide), poly (-caprolactone)	–	–	Liu et al. (2021)
Cheese (Edam, Kefalotyri, Parmesan) in LDPE packaging	Raman and ATR-FTIR spectroscopy	Detection of polyethylene migration	Low-density polyethylene (LDPE)	–	–	Katsara et al. (2021)
Aspergillus ochraceus (mold)	Raman imaging spectroscopy	Investigation of cell damage caused by antimicrobial packaging	Lipids, proteins, saccharides, amino acids, NADH, phosphate	–	–	Clemente et al. (2016)

chain branching. Specific peaks at 1141, 1425, and 1470 cm^{-1} in PE-HD are subtly shifted in PE-LD.

- **PVC:** Characterized by peaks at 636, 696, 1429, and 1437 cm^{-1}.

Coupling this rich spectral data with PLS regression models allows automated systems to achieve 100% accuracy in classifying post-consumer plastics, as well as identifying specific inorganic dyes (e.g., titanium dioxide) (Lohumi et al., 2018; Liu et al., 2021; Song et al., 2023; Clemente et al., 2016).

6.3.2.3.2 Detecting counterfeit pharmaceutical packaging using 2D-COS

Beyond recycling, Raman spectroscopy is a formidable weapon against counterfeit medicines. A pioneering study by Kwok and Taylor demonstrated that analyzing the external carton's chemical composition is a robust method for detecting forgeries. They paired Raman microscopy with Two-Dimensional Correlation Spectroscopy (2D-COS), an advanced technique that converts 1D spectra into 2D "dynamic spectra" maps:

- **Synchronous Maps:** Display similarities in spectral intensity changes. Positive cross-peaks indicate materials are homogeneously mixed, while negative cross-peaks prove they belong to distinct materials.
- **Asynchronous Maps:** Highlight dissimilarities, helping resolve overlapping peaks.

Applying 2D-COS made it possible to visualize in a spectacular way the composition differences between the authentic and the counterfeited packaging. The authentic packaging is characterized by the presence of a specific grade of calcium carbonate used as white printed pigment (auto-peak at 1087 cm^{-1}) and synthetic azo-type yellow pigments.

6.3.3 Smart packaging technologies and production techniques used in starch-based active and intelligent packaging systems

In the food industry, packaging has traditionally been limited to passive functions such as transporting, preserving, and marketing the product. However, modern consumer demands and food safety concerns have necessitated the development of dynamic systems where packaging interacts with or reports on the status of the food. At this point, the concepts of active packaging and intelligent packaging, although often used interchangeably, clearly differentiate in terms of their fundamental functions. Active packaging is based on the principle of incorporating components (e.g., antimicrobial agents or oxygen scavengers) into the system to alter the internal environment of the packaging to extend the shelf life or improve the safety of the food; that is, the system performs a protective action on the food. In contrast, intelligent packaging is a communication-oriented technology that monitors parameters such as the freshness, integrity, or temperature history of the food, and transmits information to the consumer or supply chain actors. In the literature, the combination of these two technologies is generally grouped under the umbrella of "smart" packaging (Ahari and Soufiani, 2021; Yildirim et al., 2018; Soltani Firouz et al., 2021).

In the development of these advanced packaging systems, interest in biodegradable polymers is increasing due to the environmental pollution pressure created by petroleum-derived plastics. Starch, a natural polysaccharide, stands out as one of the strongest candidates due to its availability from renewable sources (corn, potatoes, cassava, rice, etc.), low cost, and complete biodegradability. However, natural starch does not exhibit thermoplastic properties due to its granular structure and strong intermolecular hydrogen bonds, making it difficult to process. To convert this material into packaging film, it needs to be destructured under heat and shear forces, in the presence of plasticizers such as glycerol or sorbitol and transformed into thermoplastic starch. Although thermoplastic starch offers the advantage of being processable similarly to synthetic polymers, its hydrophilic structure tends to lose its mechanical strength and barrier properties in high-humidity environments (Jiang et al., 2020; Khan et al., 2017; Onyeaka et al., 2022).

When evaluating the performance of starch-based materials as food packaging, a comparative analysis with other biopolymers is critical to understanding the material's strengths and areas for improvement. For example, chitosan is known for its natural antimicrobial activity and better mechanical properties, but it is more expensive than starch and may need to be dissolved in acidic solvents. Cellulose, especially in its nanocrystalline form, is an excellent reinforcing agent and offers high thermal stability, although its film-forming ability is limited, and it is difficult to process on its own. Polylactic acid (PLA), on the other hand, is notable for its high strength and commercial availability, but it is more brittle than starch, and its biodegradability requires specific composting conditions. Although starch films offer higher flexibility and gas barrier potential (especially oxygen) compared to these polymers, they are often blended with other biopolymers or nanofillers to reduce their water vapor permeability, which is their biggest drawback (Kumari et al., 2022; Niranjana Prabhu and Prashantha, 2018; Halonen et al., 2020; Karnwal et al., 2025).

The active and smart components used in the functionalization of the starch matrix enable the packaging to transition from a "passive preservative" role to an "active preservative and communicator" role. The most common approach in smart packaging applications is the use of color-changing indicators that react to volatile amines or pH changes that occur during food spoilage. Natural pigments, such as anthocyanins (like red cabbage, rose, and blueberry extracts) or curcumin, are integrated into starch films to function as sensors that visualize the freshness of meat and fish products through a pH-dependent color scale. On the active packaging side, systems are designed to extend the shelf life of food by adding essential oils (thyme, clove, etc.), antimicrobial peptides, or metal oxide nanoparticles (Ag, ZnO, TiO2) to the starch matrix. The hydrophilic nature of starch can provide an advantage in the controlled release of these active components, especially in foods with high water activity (Bangar et al., 2021; Roy and Rhim, 2021; Sani et al., 2021; Liu et al., 2022a).

The use of starch-based biopolymers in food packaging presents various processing challenges due to the hydrophilic character and brittleness inherent in the material's natural structure. However, equipping these materials with active (protective) and smart (tracking) functions has become possible through innovative approaches in production technologies. Below, the basic production techniques used in converting starch matrices into functional packaging and the effects of these processes on material properties are examined in light of the literature.

6.3.3.1 Encapsulation (micro and nano encapsulation)

Direct addition of active and smart components to the starch matrix can compromise the stability of these components during processing or lead to uncontrolled release. Encapsulation technologies play a critical role in protecting sensitive components such as essential oils or antioxidants and ensuring their controlled release into the target environment. Especially in starch-based systems, encapsulating the active component within a wall material or emulsion system increases the bioavailability of the component.

A prominent approach in the literature is the use of Pickering emulsions. For example, Pickering emulsions containing citrus essential oil (CEO) stabilized with pectin and nanolignin have been used to activate starch films. These structures have improved mechanical properties by filling the gaps between starch molecules and have ensured the long-term release of the essential oil (Zhang et al., 2025). Similarly, in the integration of low-solubility bioactive components such as quercetin into a starch matrix, microencapsulation improves the dispersion of the component within the matrix while increasing its stability to light and heat (Shafi et al., 2025). As a more advanced technology, the addition of tea tree oil to the starch film by loading it into covalent organic frameworks (COFs) has enabled the development of a pH-sensitive release mechanism (Cao et al., 2025).

6.3.3.2 Nanofiber film production by electrospinning

Electrospinning is a heat-free technique that allows the production of ultrafine fibers from polymer solutions under high voltage. This method is particularly advantageous in terms of preserving heat-sensitive active ingredients (anthocyanins, enzymes, etc.). However, since the low molecular entanglement of natural starch makes it difficult to form fibers on its own, it usually requires processing with auxiliary polymers such as polyvinyl alcohol (PVA) (García-Guzmán et al., 2022). In recent studies, core-shell fibers have been produced using electrospinning techniques. For example, fat-soluble curcumin and clove essential oil are placed in the core part of the fiber, and water-soluble anthocyanins are placed in the shell part. This structure combines both active protection (antioxidant/antimicrobial) and intelligent monitoring (color change with pH change) functions in a single fiber and has been successfully applied in shrimp packaging (Lv et al., 2025). Since starch-based nanofibers produced by electrospinning have a much higher surface area and porosity compared to cast films, they can function as more sensitive sensors by responding to pH changes within seconds (Lv et al., 2025).

6.3.3.3 Sensor printing with bioink

The use of starch-based bioinks in the integration of sensor technologies into packaging is noteworthy in terms of sustainability and biocompatibility. In these systems, starch acts as a flexible and transparent substrate on which active indicators (e.g., pH-sensitive dyes) are immobilized. In systems called "optoelectronic noses," acid-based indicators printed or impregnated onto starch films interact with volatile components (VOCs) released during food spoilage, changing color.

For example, in a study where 10 different pH indicators were integrated into a starch-based film, volatile gases released during the spoilage of meat products were detected by the sensor array on the film, creating a color map that could be analyzed with the help of a smartphone. In this system, starch offered better optical transparency and gas permeability compared to paper or synthetic polymers (Roveda et al., 2024). In addition, low-cost sensors developed using combinations of cellulose fibers and starch can detect specific degradation gases such as ammonia (Cui et al., 2024).

6.3.3.4 3D and 4D printing applications

Additive manufacturing (3D printing) allows starch-based materials to be produced in complex geometries and customized functions. Starch is widely used in 3D printing inks because it can form gels with suitable rheological properties. However, in recent years, the concept of "4D printing," which refers to materials that change shape or color over time in response to environmental stimuli (pH, humidity, temperature), has come to the forefront (Montoya-Yepes et al., 2024).

In 4D printing applications, starch gels containing natural pigments such as anthocyanins can change color when exposed to external factors such as microwave heat or pH changes, providing information about the condition of the product (Montoya et al., 2021). For example, smart labels printed using potato starch and red cabbage extract can be used to visualize the freshness of food. This technology has great potential, especially for personalized nutrition and interactive packaging designs (Karnwal et al., 2025).

6.3.3.5 Solvent casting

Solvent casting, the most commonly used method at the laboratory scale, is based on the principle of gelatinization of starch followed by evaporation of the solvent (usually water). This method is ideal for ensuring homogeneous distribution of active ingredients within the polymer matrix and avoiding heat treatment. It is frequently preferred, especially for the integration of additives such as essential oils, anthocyanins, or silver nanoparticles (AgNP) into the film (Sultana, 2025).

In a study where silver nanoparticles were added to films prepared by casting using ginger starch, it was reported that the films both acquired antimicrobial properties and showed sensor properties by changing color in the presence of hydrogen peroxide (Teixeira et al., 2025). Similarly, cast films containing cassava starch and betalain pigments have been developed as smart labels that detect fish spoilage (Otálora González et al., 2022). However, the long drying times and low production speed of this method limit its industrial scalability (Garavito et al., 2024).

6.3.3.6 Extrusion

Extrusion, the industrial standard in thermoplastic starch production, allows for the continuous processing of starch under heat and mechanical shear forces. This method is the most suitable technique for commercial applications due to its elimination of solvent use and high production speed. However, high temperatures can cause degradation of thermolabile smart components such as anthocyanins. Nevertheless, it has been shown that smart packs containing cassava starch and anthocyanins can be produced on a large scale under appropriate processing conditions (e.g., twin-screw extrusion) (Vedove et al., 2021).

Extrusion is also critical for blending starch with other biopolymers such as PBAT or PLA and improving their properties. Cross-linking or modifying starch chains with hydrophobic groups using reactive extrusion techniques increases the material's resistance to moisture (Yudhistira et al., 2024; García-Guzmán et al., 2022).

6.3.3.7 Coating and lamination techniques

To overcome the moisture sensitivity, which is the biggest disadvantage of starch-based materials, multilayer structures (lamination) and coating techniques have been developed. In this approach, the starch film is used as the inner layer, usually carrying the active or smart component, while the outer layers (e.g., PLA or waxes) provide a barrier against moisture (Sultana, 2025).

In a multilayer approach, placing active components (e.g., berry extracts) in the inner layer in a "sandwich" model ensures the protection of these components from the external environment and their controlled release into the food (Sivapragasam et al., 2024). In addition, coating the surface of paper packaging with starch-based biocomposites improves the oil and gas barrier properties of the paper while giving the packaging antimicrobial functions. For example, papers coated with a mixture of chitosan and starch contribute to food safety as well as improved barrier properties (Adak et al., 2025).

6.3.4 Challenges and limitations

Due to growing environmental concerns, consumers seek products with less natural impact and a lower carbon footprint based on natural or renewable resources, which are "eco-friendly" or "green" and are in a sustainability trend. Starch-based films follow this course. Starch-based films are edible, biodegradable, and, at the same time, raw material for production is cheap. Unfortunately, despite all the benefits, there are some drawbacks and limitations to introducing starch-based films into the wide market.

6.4 Conclusion

The transition from traditional passive packaging to active and intelligent packaging systems marks a significant technological leap in contemporary food packaging. These advanced systems go beyond merely shielding food from external environmental factors; they actively interact with the packaged product and monitor its condition across the supply chain. Active packaging technologies contribute to extending shelf life and preserving food quality through mechanisms such as antimicrobial action, oxidation control, gas regulation, and moisture management. Meanwhile, intelligent packaging systems enhance these benefits by delivering real-time insights on product freshness, temperature variations, gas levels, and possible microbial contamination. This enables better quality control and more informed decision-making for producers, retailers, and consumers alike. With growing attention on environmental sustainability, the exploration of biodegradable biopolymers, especially starch-based materials, as alternatives to petroleum-based plastics has gained momentum. Starch is appealing due to its renewable nature, biodegradability, affordability, and effective film-forming properties. Beyond material and functional advancements, the adoption of cutting-edge analytical tools particularly Raman spectroscopy and its advanced variants such as Spatially Offset Raman Spectroscopy and Raman chemical imaging, has introduced non-invasive methods for evaluating packaging performance. These tools are instrumental in detecting food adulteration through packaging materials, assessing microbial degradation, and identifying polymers in recycling processes. Such technologies are vital for verifying product authenticity, ensuring compliance with regulations, and fostering circular economy practices within the packaging industry. In conclusion, the integration of sustainable materials, active preservation methods, intelligent monitoring systems, and advanced analytical technologies is driving the evolution of smart food packaging. These advancements promise to boost food safety, minimize waste, and support environmental sustainability efforts.

Acknowledgments

This research was carried out within the framework of the DurInnPack project, part of the PRIMA program supported by the European Union. This project has received funding from the French National Research Agency (ANR) under grant number ANR-24-P013-0002 and from the Algerian Ministry of Higher Education and Scientific Research (MESRS) through the General Directorate for Scientific Research and Technological Development (DGRSDT).

References

Abdullahi, A., Korumilli, T., Rao, K.J., 2025. Revolutionizing food packaging with bio-based polymers, active and intelligent materials for enhanced food safety and sustainability: review. Food Bioprocess Technol. 18 (8), 6836–6868. https://doi.org/10.1007/s11947-025-03868-9.

Abekoon, T., Buthpitiya, B.L.S.K., Sajindra, H., Samarakoon, E.R.J., Jayakody, J.A.D.C. A., Kantamaneni, K., Rathnayake, U., 2024. A comprehensive review to evaluate the synergy of intelligent food packaging with modern food technology and artificial intelligence field. Discov. Sustain. 5 (1). https://doi.org/10.1007/s43621-024-00371-7.

Adak, B., Baidya, S., Teramoto, Y., 2025. Biopolymers and their nanocomposites coated paper-based high barrier and sustainable food packaging materials. Carbohydr. Polym. 367, 123966. https://doi.org/10.1016/j.carbpol.2025.123966.

Adeyemi, J.O., Fawole, O.A., 2023. Metal-based nanoparticles in food packaging and coating technologies: a review. Biomolecules 13 (7). https://doi.org/10.3390/biom13071092.

Ahari, H., Soufiani, S.P., 2021. Smart and active food packaging: insights in novel food packaging. Front. Microbiol. 12. https://doi.org/10.3389/fmicb.2021.657233.

Ahmed, I., Lin, H., Zou, L., Li, Z., Brody, A.L., Mabood Qazi, I., Lv, L., Ramesh Pavase, T., Khan, M.U., Khan, S., Sun, L., 2018. An overview of smart packaging technologies for monitoring safety and quality of meat and meat products. Packag. Technol. Sci. 31 (7), 449–471. https://doi.org/10.1002/pts.2380.

Ahmed, M.W., Haque, M.A., Mohibbullah, M., Khan, M.S.I., Islam, M.A., Mondal, M.H. T., Ahmmed, R., 2022. A review on active packaging for quality and safety of foods: current trends, applications, prospects and challenges. Food Packag. Shelf Life 33. https://doi.org/10.1016/j.fpsl.2022.100913.

Akkemik, Y., Güner, A., 2020. Gıda ambalaj sistemlerinde yeni yaklaşımlar: akıllı ambalaj sistemleri. Türk Bilimsel Derlemeler Dergisi 13 (1), 9–22.

Alamdari, N.E., Aksoy, B., Aksoy, M., Beck, B., Jiang, Z., 2019. Using soybean hull in production of paper-based gas sensor for food packaging applications. In: AIChE Annual Meeting, Conference Proceedings. American Institute of Chemical Engineers.

Alamdari, N.E., Aksoy, B., Aksoy, M., Beck, B.H., Jiang, Z., 2021. A novel paper-based and pH-sensitive intelligent detector in meat and seafood packaging. Talanta 224. https://doi.org/10.1016/j.talanta.2020.121913.

Altay, B.N., Bloembergen, S., Aydemir, C., Karademir, A., Fleming, P.D., 2017. Use of nanoparticle binders for paper coatings: a review. J. Graph. Eng. Des. 8 (1), 39–43. https://doi.org/10.24867/jged-2017-1-039.

Altay, B.N., Klass, C., Chen, T., Fleck, A., Aydemir, C., Karademir, A., Fleming, P.D., 2021. The effect of green biobased binder on structural, mechanical, liquid absorption and wetting properties of coated papers. Clean. Eng. Technol. 5, 100274. https://doi.org/10.1016/j.clet.2021.100274.

Altay, B.N., Kubik, Al C.-, Williams, S., Huq, A., Sugiyama, M., Dhote, Y., Zunjarrao, A., Aydemir, C., Karademir, A., 2024. Starch-based nanoparticles as a replacement for synthetic latex: a comprehensive assessment of printability and colorimetric characteristics. Color Res. Appl. 49 (6), 538–552. https://doi.org/10.1002/col.22941.

Amin, U., Khan, M.K.I., Maan, A.A., Nazir, A., Riaz, S., Khan, M.U., Sultan, M., Munekata, P.E.S., Lorenzo, J.M., 2022. Biodegradable active, intelligent, and smart packaging materials for food applications. Food Packag. Shelf Life 33. https://doi.org/10.1016/j.fpsl.2022.100903.

Arruda, T.R., Machado, G.d.O., Marques, C.S., Souza, A.L.d., Pelissari, F.M., Oliveira, T. V.d., Silva, R.R.A., 2025. An overview of starch-based materials for sustainable food packaging: recent advances, limitations, and perspectives. Macromolecules 5 (2). https://doi.org/10.3390/macromol5020019.

Asgher, M., Qamar, S.A., Bilal, M., Iqbal, H.M.N., 2020. Bio-based active food packaging materials: sustainable alternative to conventional petrochemical-based packaging materials. Food Res. Int. 137. https://doi.org/10.1016/j.foodres.2020.109625.

Baietto, M., Wilson, A., 2015. Electronic-nose applications for fruit identification, ripeness and quality grading. Sensors 15 (1), 899–931. https://doi.org/10.3390/s150100899.

Bangar, S.P., Purewal, S.S., Trif, M., Maqsood, S., Kumar, M., Manjunatha, V., Rusu, A. V., 2021. Functionality and applicability of starch-based films: an eco-friendly approach. Foods 10 (9). https://doi.org/10.3390/foods10092181.

Bao, Y., Wang, M., Li, J., Yang, X., Wen, B., Wang, L., He, Y., Si, X., Li, B., 2024. A starch-based 3D printed intelligent colorimetric film co-loaded natural pigments for visualizing food freshness: effect of nozzle size on gel structure formation. Food Hydrocoll. 155, 110218. https://doi.org/10.1016/j.foodhyd.2024.110218.

Barnes, R., 2024. Plastics and the circular economy. In: Research Handbook on Plastics Regulation. Edward Elgar Publishing, pp. 375–397. https://doi.org/10.4337/9781802201529.00029.

Bedi, N., Srivastava, D.K., Srivastava, A., Mahapatra, S., Dkhar, D.S., Chandra, P., Srivastava, A., 2022. Marine biological macromolecules as matrix material for biosensor fabrication. Biotechnol. Bioeng. 119 (8), 2046–2063. https://doi.org/10.1002/bit.28122.

Bhargava, N., Sharanagat, V.S., Mor, R.S., Kumar, K., 2020. Active and intelligent biodegradable packaging films using food and food waste-derived bioactive compounds: a review. Trends Food Sci. Technol. 105, 385–401. https://doi.org/10.1016/j.tifs.2020.09.015.

Biji, K.B., Ravishankar, C.N., Mohan, C.O., Srinivasa Gopal, T.K., 2015. Smart packaging systems for food applications: a review. J. Food Sci. Technol. 52 (10), 6125–6135. https://doi.org/10.1007/s13197-015-1766-7.

Brizio, A.P.D.R., Prentice, C., 2015. Development of an intelligent enzyme indicator for dynamic monitoring of the shelf-life of food products. Innov. Food Sci. Emerg. Technol. 30, 208–217. https://doi.org/10.1016/j.ifset.2015.04.001.

Cao, P., Song, Y., Jin, D., Zhang, Y., Zhang, P., Tang, K., Li, Y., Wang, R., Zhang, K., 2025. Preparation and characterization of a starch-based active packaging with pH-sensitive release via tea tree essential oil loaded in a covalent organic framework. Food Chem. 486, 144643. https://doi.org/10.1016/j.foodchem.2025.144643.

Caon, T., Martelli, S.M., Fakhouri, F.M., 2016. , New trends in the food industry: application of nanosensors in food packaging. In: Nanobiosensors. Elsevier, Brazil, pp. 773–804. https://doi.org/10.1016/B978-0-12-804301-1.00018-7.

Chen, H.z., Zhang, M., Bhandari, B., Yang, C.h., 2020. Novel pH-sensitive films containing curcumin and anthocyanins to monitor fish freshness. Food Hydrocoll. 100. https://doi.org/10.1016/j.foodhyd.2019.105438.

Chen, Y., Song, Y., Zhang, Z., Chen, Y., Deng, Q., Wang, S., 2021. Thiol-functionalized covalent organic frameworks as thermal history indictor for temperature and time history monitoring. Adv. Funct. Mater. 31 (41). https://doi.org/10.1002/adfm.202104885.

Clemente, I., Aznar, M., Nerín, C., 2016. Raman imaging spectroscopy as a tool to investigate the cell damage on aspergillus ochraceus caused by an antimicrobial packaging containing benzyl isothiocyanate. Anal. Chem. 88 (9), 4772–4779. https://doi.org/10.1021/acs.analchem.6b00116.

Cui, C., Ji, N., Wang, Y., Xiong, L., Sun, Q., 2021. Bioactive and intelligent starch-based films: a review. Trends Food Sci. Technol. 116, 854–869. https://doi.org/10.1016/j.tifs.2021.08.024.

Cui, F., Zheng, S., Wang, D., Ren, L., Meng, Y., Ma, R., Wang, S., Li, X., Li, T., Li, J., 2024. Development of machine learning-based shelf-life prediction models for multiple marine fish species and construction of a real-time prediction platform. Food Chem. 450, 139230. https://doi.org/10.1016/j.foodchem.2024.139230.

Dang, X., Han, S., Wang, X., 2025. Versatile corn starch-based sustainable food packaging with enhanced antimicrobial activity and preservative properties. J. Colloid Interface Sci. 694, 137698. https://doi.org/10.1016/j.jcis.2025.137698.

D'Almeida, A.P., de Albuquerque, T.L., 2024. Innovations in food packaging: from bio-based materials to smart packaging systems. Processes 12 (10), 2085. https://doi.org/10.3390/pr12102085.

Eom, K.H., Kim, M.C., Lee, S., Lee, C.W., 2012. The vegetable freshness monitoring system using RFID with oxygen and carbon dioxide sensor. Int. J. Distributed Sens. Netw. 2012. https://doi.org/10.1155/2012/472986.

Evans, D.M., Parsons, R., Jackson, P., Greenwood, S., Ryan, A., 2020. Understanding plastic packaging: the co-evolution of materials and society. Glob. Environ. Change 65, 102166. https://doi.org/10.1016/j.gloenvcha.2020.102166.

Fadiji, A.E., Mthiyane, D.M.N., Onwudiwe, D.C., Babalola, O.O., 2022. Harnessing the known and unknown impact of nanotechnology on enhancing food security and reducing postharvest losses: constraints and future prospects. Agronomy 12 (7). https://doi.org/10.3390/agronomy12071657.

Fang, Z., Zhao, Y., Warner, R.D., Johnson, S.K., 2017. Active and intelligent packaging in meat industry. Trends Food Sci. Technol. 61, 60–71. https://doi.org/10.1016/j.tifs.2017.01.002.

Farrell, R., Cortese, Y.J., Devine, D.M., Gately, N., Rueda, M., Rodriguez, L., Pezzoli, R., 2024. The function and properties of common food packaging materials and their suitability for reusable packaging: the transition from a linear to circular economy. Curr. Res. Green Sustain. Chem. 9. https://doi.org/10.1016/j.crgsc.2024.100429.

Gao, T., Tian, Y., Zhu, Z., Sun, D.W., 2020. Modelling, responses and applications of time-temperature indicators (TTIs) in monitoring fresh food quality. Trends Food Sci. Technol. 99, 311–322. https://doi.org/10.1016/j.tifs.2020.02.019.

Garavito, J., Peña-Venegas, C.P., Castellanos, D.A., 2024. Production of starch-based flexible food packaging in developing countries: analysis of the processes, challenges, and requirements. Foods 13 (24). https://doi.org/10.3390/foods13244096.

García-Guzmán, L., Cabrera-Barjas, G., Soria-Hernández, C.G., Castaño, J., Guadarrama-Lezama, A.Y., Rodríguez Llamazares, S., 2022. Progress in starch-based materials for food packaging applications. Polysaccharides 3 (1), 136–177. https://doi.org/10.3390/polysaccharides3010007.

George, J., Kumar, R., Aaliya, B., Sunooj, K.V., 2023. Packaging solutions for monitoring food quality and safety. In: Engineering Aspects of Food Quality and Safety. Springer International Publishing, Cham, pp. 411–442. https://doi.org/10.1007/978-3-031-30683-9_14.

Ghaani, M., Cozzolino, C.A., Castelli, G., Farris, S., 2016. An overview of the intelligent packaging technologies in the food sector. Trends Food Sci. Technol. 51, 1–11. https://doi.org/10.1016/j.tifs.2016.02.008.

Girardeau, A., Biscola, V., Keravec, S., Corrieu, G., Fonseca, F., 2020. Application of lactic acid bacteria in time-temperature integrators: a tool to monitor quality and safety of perishable foods. In: Lactic Acid Bacteria. CRC Press, pp. 241–260.

Gregor-Svetec, D., 2018. Intelligent packaging. In: Nanomaterials for Food Packaging. Elsevier, pp. 203–247. https://doi.org/10.1016/B978-0-323-51271-8.00008-5.

Hailu, F.W., Fanta, S.W., Tsige, A.A., Delele, M.A., 2025. Current status and challenges of colorimetric intelligent packaging for fruit and vegetables: a review. J. Food Qual. 2025 (1). https://doi.org/10.1155/jfq/6669828.

Halonen, N., Pálvölgyi, P.S., Bassani, A., Fiorentini, C., Nair, R., Spigno, G., Kordas, K., 2020. Bio-Based smart materials for food packaging and sensors – a review. Front. Mater. 7. https://doi.org/10.3389/fmats.2020.00082.

Han, J.H., Ho, C.H.L., Rodrigues, E.T., 2005. Intelligent Packaging Innovations in Food Packaging. Elsevier, Canada, pp. 138–155. https://doi.org/10.1016/B978-012311632-1/50041-3.

Hashem, A.H., El-Naggar, M.E., Abdelaziz, A.M., Abdelbary, S., Hassan, Y.R., Hasanin, M.S., 2023. Bio-based antimicrobial food packaging films based on hydroxypropyl starch/polyvinyl alcohol loaded with the biosynthesized zinc oxide nanoparticles. Int. J. Biol. Macromol. 249, 126011. https://doi.org/10.1016/j.ijbiomac.2023.126011.

Hassan, S.A., Abbas, M., Mujahid, W., Ahmed, W., Ahmad, S., Maan, A.A., Shehzad, A., Bhat, Z.F., Aadil, R.M., 2023. Utilization of cereal-based husks to achieve sustainable development goals: treatment of wastewater, biofuels, and biodegradable packaging. Trends Food Sci. Technol. 140. https://doi.org/10.1016/j.tifs.2023.104166.

He, X., Pu, Y., Chen, L., Jiang, H., Xu, Y., Cao, J., Jiang, W., 2023. A comprehensive review of intelligent packaging for fruits and vegetables: target responders, classification, applications, and future challenges. Compr. Rev. Food Sci. Food Saf. 22 (2), 842–881. https://doi.org/10.1111/1541-4337.13093.

Heo, W., Lim, S., 2024. A review on gas indicators and sensors for smart food packaging. Foods 13 (19), 3047. https://doi.org/10.3390/foods13193047.

Holman, B.W.B., Hopkins, D.L., Fowler, S.M., 2025. Raman spectroscopy is a non-invasive method to predict the total viable microbial count through the packaging material of vacuum packaged lamb meat. Meat Sci. 230. https://doi.org/10.1016/j.meatsci.2025.109943.

Janjarasskul, T., Suppakul, P., 2018. Active and intelligent packaging: the indication of quality and safety. Crit. Rev. Food Sci. Nutr. 58 (5), 808–831. https://doi.org/10.1080/10408398.2016.1225278.

Jeantet, R., Croguennec, T., Schuck, P., Brule, G., 2016. In: Handbook of Food Science and Technology 2: Food Process Engineering and Packaging, 2. John Wiley & Sons.

Jiang, G., Hou, X., Zeng, X., Zhang, C., Wu, H., Shen, G., Li, S., Luo, Q., Li, M., Liu, X., Chen, A., Wang, Z., Zhang, Z., 2020. Preparation and characterization of indicator films from carboxymethyl-cellulose/starch and purple sweet potato (Ipomoea batatas (L.) lam) anthocyanins for monitoring fish freshness. Int. J. Biol. Macromol. 143, 359–372. https://doi.org/10.1016/j.ijbiomac.2019.12.024.

Jiang, C., Liu, G., Zhang, Q., Wang, S., Zou, Y., 2024. Preparation and characterization of intelligent packaging labels based on pea starch, κ-carrageenan and black raspberry extract for monitoring freshness of pork. Food Chem. X 24, 101850. https://doi.org/10.1016/j.fochx.2024.101850.

Kalpana, S., Priyadarshini, S.R., Maria Leena, M., Moses, J.A., Anandharamakrishnan, C., 2019. Intelligent packaging: trends and applications in food systems. Trends Food Sci. Technol. 93, 145–157. https://doi.org/10.1016/j.tifs.2019.09.008.

Karnwal, A., Rauf, A., Jassim, A.Y., Selvaraj, M., Al-Tawaha, A.R.M.S., Kashyap, P., Kumar, D., Malik, T., 2025. Advanced starch-based films for food packaging: innovations in sustainability and functional properties. Food Chem. X 29. https://doi.org/10.1016/j.fochx.2025.102662.

Katsara, K., Kenanakis, G., Viskadourakis, Z., Papadakis, V.M., 2021. Polyethylene migration from food packaging on cheese detected by Raman and infrared (Atr/ft-ir) spectroscopy. Materials 14 (14). https://doi.org/10.3390/ma14143872.

Khan, B., Bilal Khan Niazi, M., Samin, G., Jahan, Z., 2017. Thermoplastic starch: a possible biodegradable food packaging material—a review. J. Food Process. Eng. 40 (3). https://doi.org/10.1111/jfpe.12447.

Kılıç, N.N., Kahyaoglu, L.N., 2025. Smart thermoplastic starch films with dual functionality: quality preservation of tomatoes and spoilage detection in chicken. Int. J. Biol. Macromol. 330. https://doi.org/10.1016/j.ijbiomac.2025.147898.

Kim, K., Kim, E., Lee, S.J., 2012. New enzymatic time-temperature integrator (TTI) that uses laccase. J. Food Eng. 113 (1), 118–123. https://doi.org/10.1016/j.jfoodeng.2012.05.009.

Kumari, S., Yadav, B.S., Yadav, R., 2022. Morphological and thermo-mechanical characterization of sweet potato starch based nanocomposites reinforced with barley starch nanoparticles. J. Food Sci. Technol. 59 (12), 4924–4934. https://doi.org/10.1007/s13197-022-05581-3.

Kuswandi, B., Jumina, 2020. Active and Intelligent Packaging, Safety, and Quality Controls. Elsevier BV, pp. 243–294. https://doi.org/10.1016/b978-0-12-816184-5.00012-4.

Kuswandi, B., Jayus, Restyana, A., Abdullah, A., Heng, L.Y., Ahmad, M., 2012. A novel colorimetric food package label for fish spoilage based on polyaniline film. Food Control 25 (1), 184–189. https://doi.org/10.1016/j.foodcont.2011.10.008.

Kwok, K., Taylor, L.S., 2012. Analysis of the packaging enclosing a counterfeit pharmaceutical tablet using Raman microscopy and two-dimensional correlation spectroscopy. Vib. Spectrosc. 61, 176–182. https://doi.org/10.1016/j.vibspec.2012.02.018.

Lee, K., Baek, S., Kim, D., Seo, J., 2019. A freshness indicator for monitoring chicken-breast spoilage using a Tyvek® sheet and RGB color analysis. Food Packag. Shelf Life 19, 40–46. https://doi.org/10.1016/j.fpsl.2018.11.016.

Li, L., Zuo, J., Duan, X., Wang, S., Hu, K., Chang, R., 2021. Impacts and mitigation measures of plastic waste: a critical review. Environ. Impact Assess. Rev. 90, 106642. https://doi.org/10.1016/j.eiar.2021.106642.

Li, B., Bao, Y., Li, J., Bi, J., Chen, Q., Cui, H., Wang, Y., Tian, J., Shu, C., Wang, Y., Lang, Y., Zhang, W., Tan, H., Huang, Q., Si, X., 2022. A sub-freshness monitoring chitosan/starch-based colorimetric film for improving color recognition accuracy via controlling the pH value of the film-forming solution. Food Chem. 388, 132975. https://doi.org/10.1016/j.foodchem.2022.132975.

Liang, T., Sun, G., Cao, L., Li, J., Wang, L., 2019. A pH and NH3 sensing intelligent film based on Artemisia sphaerocephala Krasch. gum and red cabbage anthocyanins anchored by carboxymethyl cellulose sodium added as a host complex. Food Hydrocoll. 87, 858–868. https://doi.org/10.1016/j.foodhyd.2018.08.028.

Listyarini, A., Handayani, W., Fauzia, V., Imawan, C., 2020. Active starch/PVA composite films containing *Syzygium oleana* for ammonia vapor colorimetric indicators. Mater. Sci. Forum 990, 318–324. https://doi.org/10.4028/www.scientific.net/msf.990.318.

Liu, Y., Qin, Y., Bai, R., Zhang, X., Yuan, L., Liu, J., 2019. Preparation of pH-sensitive and antioxidant packaging films based on κ-carrageenan and mulberry polyphenolic extract. Int. J. Biol. Macromol. 134, 993–1001. https://doi.org/10.1016/j.ijbiomac.2019.05.175.

Liu, Z., Huang, M., Zhu, Q., Qin, J., Kim, M.S., 2021. Detection of adulterated sugar with plastic packaging based on spatially offset Raman imaging. J. Sci. Food Agric. 101 (15), 6281–6288. https://doi.org/10.1002/jsfa.11297.

Liu, D., Zhao, P., Chen, J., Yan, Y., Wu, Z., 2022a. Recent advances and applications in starch for intelligent active food packaging: a review. Foods 11 (18), 2879. https://doi.org/10.3390/foods11182879.

Liu, L., Wu, W., Zheng, L., Yu, J., Sun, P., Shao, P., 2022b. Intelligent packaging films incorporated with anthocyanins-loaded ovalbumin-carboxymethyl cellulose nanocomplexes for food freshness monitoring. Food Chem. 387, 132908. https://doi.org/10.1016/j.foodchem.2022.132908.

Liu, X., Xu, F., Yong, H., Chen, D., Tang, C., Kan, J., Liu, J., 2025. Recent advances in chitosan-based active and intelligent packaging films incorporated with flavonoids. Food Chem. X 25, 102200. https://doi.org/10.1016/j.fochx.2025.102200.

Lohumi, S., Lee, H., Kim, M.S., Qin, J., Cho, B.K., 2018. Through-packaging analysis of butter adulteration using line-scan spatially offset Raman spectroscopy. Anal. Bioanal. Chem. 410 (22), 5663–5673. https://doi.org/10.1007/s00216-018-1189-1.

Loucanova, E., Kalamarova, M., Parobek, J., 2017. The innovative approaches to packaging – comparison analysis of intelligent and active packaging perceptions in Slovakia. Studia Universitatis „Vasile Goldis" Arad – Economics Series 27 (2), 33–44. https://doi.org/10.1515/sues-2017-0007.

Lv, H., He, D., Xu, E., Jin, Z., Zhao, H., Yuan, C., Zhao, M., Gao, W., Li, J., Liu, Y., Wu, Z., Liu, P., Cui, B., 2025. Starch-based biodegradable active intelligent packaging with color superimposition via emulsion electrospinning nanofiber for meat freshness monitoring and shelf-life extension. Carbohydr. Polym. 367, 123940. https://doi.org/10.1016/j.carbpol.2025.123940.

Ma, Y., Yang, W., Xia, Y., Xue, W., Wu, H., Li, Z., Zhang, F., Qiu, B., Fu, C., 2022. Properties and applications of intelligent packaging indicators for food spoilage. Membranes 12 (5), 477. https://doi.org/10.3390/membranes12050477.

Mali, S.N., Pandey, A., 2024. Development of curcumin integrated smart pH indicator, antibacterial, and antioxidant waste derived Artocarpus lakoocha starch-based packaging film. Int. J. Biol. Macromol. 275 (Pt 2), 133827. https://doi.org/10.1016/j.ijbiomac.2024.133827.

Mallick, N., Pal, D., Soni, A.B., Jhariya, D.C., Singh, D., 2020. Starch based antimicrobial food packaging film towards a sustainable environment. IOP Conf. Ser. Earth Environ. Sci. 597 (1), 012019. https://doi.org/10.1088/1755-1315/597/1/012019.

Mary, S.K., Koshy, R.R., Daniel, J., Koshy, J.T., Pothen, L.A., Thomas, S., 2020. Development of starch based intelligent films by incorporating anthocyanins of butterfly pea flower and TiO2 and their applicability as freshness sensors for prawns during storage. RSC Adv. 10 (65), 39822–39830. https://doi.org/10.1039/d0ra05986b.

Mirza Alizadeh, A., Masoomian, M., Shakooie, M., Zabihzadeh Khajavi, M., Farhoodi, M., 2021. Trends and applications of intelligent packaging in dairy products: a review. Crit. Rev. Food Sci. Nutr. 62 (2), 383–397. https://doi.org/10.1080/10408398.2020.1817847.

Mkhari, T., Adeyemi, J.O., Fawole, O.A., 2025. Recent advances in the fabrication of intelligent packaging for food preservation: a review. Processes 13 (2). https://doi.org/10.3390/pr13020539.

Mohammadalinejhad, S., Almasi, H., Moradi, M., 2020. Immobilization of Echium amoenum anthocyanins into bacterial cellulose film: a novel colorimetric pH indicator for freshness/spoilage monitoring of shrimp. Food Control 113, 107169. https://doi.org/10.1016/j.foodcont.2020.107169.

Mohammadian, E., Alizadeh-Sani, M., Jafari, S.M., 2020. Smart monitoring of gas/temperature changes within food packaging based on natural colorants. Compr. Rev. Food Sci. Food Saf. 19 (6), 2885–2931. https://doi.org/10.1111/1541-4337.12635.

Moment, E.S.I., 2014. In: Communication from the Commission to the European Parliament, the Council, the European Economic and Social Committee and the Committee of the Regions.

Montoya, J., Medina, J., Molina, A., Gutiérrez, J., Rodríguez, B., Marín, R., 2021. Impact of viscoelastic and structural properties from starch-mango and starch-arabinoxylans hydrocolloids in 3D food printing. Addit. Manuf. 39, 101891. https://doi.org/10.1016/j.addma.2021.101891.

Montoya-Yepes, D.F., Jiménez-Rodríguez, A.A., Aldana-Porras, A.E., Velásquez-Holguin, L.F., Méndez-Arteaga, J.J., Murillo-Arango, W., 2024. Starches in the encapsulation of plant active ingredients: state of the art and research trends. Polym. Bull. 81 (1), 135–163. https://doi.org/10.1007/s00289-023-04724-6.

Moreira, L.P., Selinger, A., Sadauskas-Henrique, H., Choueri, R.B., Gusso-Choueri, P.K., Pacheco, M.T.T., Silveira, L., 2025. Raman spectroscopy for identifying postconsumer packaging plastics for recycling. J. Air Waste Manag. Assoc. 75 (9), 719–734. https://doi.org/10.1080/10962247.2025.2513935.

Müller, P., Schmid, M., 2019. Intelligent packaging in the food sector: a brief overview. Foods 8 (1), 16. https://doi.org/10.3390/foods8010016.

Muñoz-Gimena, P.F., Oliver-Cuenca, V., Peponi, L., López, D., 2023. A review on reinforcements and additives in starch-based composites for food packaging. Polymers 15 (13), 2972. https://doi.org/10.3390/polym15132972.

Mustafa, F., Andreescu, S., 2018. Chemical and biological sensors for food-quality monitoring and smart packaging. Foods 7 (10), 168. https://doi.org/10.3390/foods7100168.

Naghdi, S., Rezaei, M., Abdollahi, M., 2021. A starch-based pH-sensing and ammonia detector film containing betacyanin of paperflower for application in intelligent packaging of fish. Int. J. Biol. Macromol. 191, 161–170. https://doi.org/10.1016/j.ijbiomac.2021.09.045.

Niranjana Prabhu, T., Prashantha, K., 2018. A review on present status and future challenges of starch based polymer films and their composites in food packaging applications. Polym. Compos. 39 (7), 2499–2522. https://doi.org/10.1002/pc.24236.

Nopwinyuwong, A., Trevanich, S., Suppakul, P., 2010. Development of a novel colorimetric indicator label for monitoring freshness of intermediate-moisture dessert spoilage. Talanta 81 (3), 1126–1132. https://doi.org/10.1016/j.talanta.2010.02.008.

Onyeaka, H., Obileke, K.C., Makaka, G., Nwokolo, N., 2022. Current research and applications of starch-based biodegradable films for food packaging. Polymers 14 (6), 1126. https://doi.org/10.3390/polym14061126.

Otálora González, C.M., Schelegueda, L.I., Pizones Ruiz-Henestrosa, V.M., Campos, C. A., Basanta, M.F., Gerschenson, L.N., 2022. Cassava starch films with anthocyanins and betalains from agroindustrial by-products: their use for intelligent label development. Foods 11 (21). https://doi.org/10.3390/foods11213361.

Pacquit, A., Frisby, J., Diamond, D., Lau, K., Farrell, A., Quilty, B., Diamond, D., 2007. Development of a smart packaging for the monitoring of fish spoilage. Food Chem. 102 (2), 466–470. https://doi.org/10.1016/j.foodchem.2006.05.052.

Pal, M., Devrani, M., Hadush, A., 2019. Recent developments in food packaging technologies. Beverage Food World 46 (1), 21–25.

Pandian, A.T., Chaturvedi, S., Chakraborty, S., 2021. Applications of enzymatic time–temperature indicator (TTI) devices in quality monitoring and shelf-life estimation of food products during storage. J. Food Meas. Char. 15 (2), 1523–1540. https://doi.org/10.1007/s11694-020-00730-8.

Park, Y.W., Kim, S.M., Lee, J.Y., Jang, W., 2015. Application of biosensors in smart packaging. Mol. Cell. Toxicol. 11 (3), 277–285. https://doi.org/10.1007/s13273-015-0027-1.

Pereira, V.A., de Arruda, I.N.Q., Stefani, R., 2015. Active chitosan/PVA films with anthocyanins from Brassica oleraceae (Red Cabbage) as time-temperature indicators for application in intelligent food packaging. Food Hydrocoll. 43, 180–188. https://doi.org/10.1016/j.foodhyd.2014.05.014.

Poyatos-Racionero, E., Ros-Lis, J.V., Vivancos, J.L., Martínez-Máñez, R., 2018. Recent advances on intelligent packaging as tools to reduce food waste. J. Clean. Prod. 172, 3398–3409. https://doi.org/10.1016/j.jclepro.2017.11.075.

Rehman, A., Jafari, S.M., Aadil, R.M., Assadpour, E., Randhawa, M.A., Mahmood, S., 2020. Development of active food packaging via incorporation of biopolymeric

nanocarriers containing essential oils. Trends Food Sci. Technol. 101, 106–121. https://doi.org/10.1016/j.tifs.2020.05.001.
Romainor, A.N., Chin, S.F., Lihan, S., 2022. Antimicrobial starch-based film for food packaging application. Starch - Stärke 74 (3–4). https://doi.org/10.1002/star.202100207.
Roveda, A.C., Dias, B.C., Passini, L.N., Manzani, D., Petruci, J.F.d.S., 2024. Transparent, flexible, and eco-friendly starch-based films for reversible optoelectronic noses for food spoilage monitoring in smart packaging. Microchim. Acta 191 (6). https://doi.org/10.1007/s00604-024-06426-w.
Roy, S., Rhim, J.-W., 2021. Fabrication of bioactive binary composite film based on gelatin/chitosan incorporated with cinnamon essential oil and rutin. Colloids Surf. B Biointerfaces 204, 111830. https://doi.org/10.1016/j.colsurfb.2021.111830.
Rukchon, C., Nopwinyuwong, A., Trevanich, S., Jinkarn, T., Suppakul, P., 2014. Development of a food spoilage indicator for monitoring freshness of skinless chicken breast. Talanta 130, 547–554. https://doi.org/10.1016/j.talanta.2014.07.048.
Salgado, P.R., Di Giorgio, L., Musso, Y.S., Mauri, A.N., 2021. Recent developments in smart food packaging focused on biobased and biodegradable polymers. Front. Sustain. Food Syst. 5. https://doi.org/10.3389/fsufs.2021.630393.
Sani, M.A., Azizi-Lalabadi, M., Tavassoli, M., Mohammadi, K., McClements, D.J., 2021. Recent advances in the development of smart and active biodegradable packaging materials. Nanomaterials 11 (5). https://doi.org/10.3390/nano11051331.
Schaefer, D., Cheung, W.M., 2018. Smart packaging: opportunities and challenges. Proced. CIRP 72, 1022–1027. https://doi.org/10.1016/j.procir.2018.03.240.
Shafi, Z., Singh, R., Sidiqi, U.S., Bashir, B., Rasool, S., Dash, K.K., Zahoor, I., Ahmed, I., Nagaraja, S.K., Dar, A.H., 2025. Quercetin infused starch matrix as a sustainable approach to smart packaging: a comprehensive review. Int. J. Biol. Macromol. 320. https://doi.org/10.1016/j.ijbiomac.2025.145746.
Siddiqui, J., Taheri, M., Alam, A.U., Deen, M.J., 2022. Nanomaterials in smart packaging applications: a review. Small 18 (1). https://doi.org/10.1002/smll.202101171.
Sivapragasam, N., Maqsood, S., Rupasinghe, H.P.V., 2024. Berry bioactive compounds immobilized in starch matrix for active and intelligent packaging: a review. Future Foods 10, 100397. https://doi.org/10.1016/j.fufo.2024.100397.
Sohail, M., Sun, D.W., Zhu, Z., 2018. Recent developments in intelligent packaging for enhancing food quality and safety. Crit. Rev. Food Sci. Nutr. 58 (15), 2650–2662. https://doi.org/10.1080/10408398.2018.1449731.
Soltani Firouz, M., Mohi-Alden, K., Omid, M., 2021. A critical review on intelligent and active packaging in the food industry: research and development. Food Res. Int. 141, 110113. https://doi.org/10.1016/j.foodres.2021.110113.
Song, S.W., Jeong, Y.C., Park, C.R., Kim, H.M., 2023. Quantitative fat analysis of milk using a line-illumination spatially offset Raman probe through carton packaging. Analyst 148 (14), 3321–3329. https://doi.org/10.1039/d3an00416c.
Stoma, M., Dudziak, A., 2022. Eastern Poland consumer awareness of innovative active and intelligent packaging in the food industry: exploratory studies. Sustainability 14 (20), 13691. https://doi.org/10.3390/su142013691.
Sultana, A., 2025. Starch-based biocomposite films and coatings in paper-based food packaging: a review. Environ. Chem. Lett. 23 (6), 1689–1712. https://doi.org/10.1007/s10311-025-01862-3.
Sun, K., Liang, X., Zhou, S., Wang, J., Lu, F., Deng, Q., 2025. Thermal history indicator hydrogel based on starch with multiple signal channels for visualized monitoring thermal accumulation of perishable products. J. Food Sci. Technol. 62 (8), 1569–1579. https://doi.org/10.1007/s13197-024-06130-w.

Sundqvist-Andberg, H., Åkerman, M., 2021. Sustainability governance and contested plastic food packaging–an integrative review. J. Clean. Prod. 306, 127111. https://doi.org/10.1016/j.jclepro.2021.127111.

Teixeira, E.d.M., Stefani, R., Corrêa, A.C., de Campos, A., Luchesi, B.R., Bondancia, T.J., Marconcini, J.M., Corradini, E., Neto, W.P.F., Teodoro, K.B.R., Correa, D.S., 2025. Assessment of the properties of ginger starch-based films and their potential application as intelligent packaging. Int. J. Biol. Macromol. 330. https://doi.org/10.1016/j.ijbiomac.2025.148120.

Torche, A., Chouana, T., Sanah, I., Djeghim, F., Khodja, E.A.T., Mezreb, K., Elboutachfaiti, R., Delattre, C., D'Elia, M., Rastrelli, L., 2025. pH-sensitive cassava starch/onion peel powder films as colorimetric indicators for minced beef freshness monitoring. Foods 14 (17). https://doi.org/10.3390/foods14172974.

Union, I., 2014. In: Communication from the Commission to the European Parliament, the Council, the European Economic and Social Committee and the Committee of the Regions. Brussels.

Vedove, T.M.A.R.D., Maniglia, B.C., Tadini, C.C., 2021. Production of sustainable smart packaging based on cassava starch and anthocyanin by an extrusion process. J. Food Eng. 289. https://doi.org/10.1016/j.jfoodeng.2020.110274.

Versino, F., Ortega, F., Monroy, Y., Rivero, S., López, O.V., García, M.A., 2023. Sustainable and bio-based food packaging: a review on past and current design innovations. Foods 12 (5). https://doi.org/10.3390/foods12051057.

Wang, S., Liu, X., Yang, M., Zhang, Y., Xiang, K., Tang, R., 2015. Review of time temperature indicators as quality monitors in food packaging. Packag. Technol. Sci. 28 (10), 839–867. https://doi.org/10.1002/pts.2148.

Wang, M., Li, Y., Srinivasan, P., Hu, Z., Wang, R., Saragih, A., Repka, M.A., Murthy, S. N., 2018. Interactions between biological products and product packaging and potential approaches to overcome them. AAPS PharmSciTech 19 (8), 3681–3686. https://doi.org/10.1208/s12249-018-1184-z.

Wang, P., Qin, H., He, D., Zou, Z., Lin, X., Tang, Q., 2024. Developing colorimetric ammonia-sensing nanocomposite films based on potato starch/PVA and ZnCu-BTC nanorods for real-time monitoring food freshness. Int. J. Biol. Macromol. 277 (Pt 3), 134376. https://doi.org/10.1016/j.ijbiomac.2024.134376.

Wardejn, S., Wacławek, S., Dudek, G., 2024. Improving antimicrobial properties of biopolymer-based films in food packaging: key factors and their impact. Int. J. Mol. Sci. 25 (23), 12580. https://doi.org/10.3390/ijms252312580.

Widjaja, E., Garland, M., 2012. Detailed spectroscopic analysis of complex multi-component materials using a combination of Raman mapping with BTEM. J. Raman Spectrosc. 43 (7), 828–833. https://doi.org/10.1002/jrs.3110.

Wu, W., Zheng, L., Yu, J., Liu, L., Goksen, G., Shao, P., 2024a. High-sensitivity intelligent packaging films harnessing rose anthocyanins and hydrophilic silica aerogel for visual food freshness monitoring. Food Qual. Saf. 8. https://doi.org/10.1093/fqsafe/fyad051.

Wu, W., Liu, L., Zhou, Y., Shao, P., 2024b. Highly ammonia-responsive starch/PVA film with gas absorption system as the 'bridge' for visually spoilage monitoring of animal-derived food. Food Chem. 430, 137032. https://doi.org/10.1016/j.foodchem.2023.137032.

Xie, F., Qin, Z., Luo, Y., He, Z., Chen, Q., Cai, J., 2025. Synergistically engineered starch-based composite films: multifunctional platforms integrating quaternary ammonium chitosan and anthocyanins for intelligent food monitoring and sustainable packaging. Food Chem. 478, 143560. https://doi.org/10.1016/j.foodchem.2025.143560.

Yildirim, S., Röcker, B., Pettersen, M.K., Nilsen-Nygaard, J., Ayhan, Z., Rutkaite, R., Radusin, T., Suminska, P., Marcos, B., Coma, V., 2018. Active packaging applications

for food. Compr. Rev. Food Sci. Food Saf. 17 (1), 165–199. https://doi.org/10.1111/1541-4337.12322.
Young, E., Mirosa, M., Bremer, P., 2020. A systematic review of consumer perceptions of smart packaging technologies for food. Front. Sustain. Food Syst. 4. https://doi.org/10.3389/fsufs.2020.00063.
Yudhistira, B., Husnayain, N., Punthi, F., Gavahian, M., Chang, C.K., Hsieh, C.W., 2024. Progress in the application of emerging technology for the improvement of starch-based active packaging properties: a review. ACS Food Sci. Technol. 4 (9), 1997–2012. https://doi.org/10.1021/acsfoodscitech.4c00260.
Yun, D., Cai, H., Liu, Y., Xiao, L., Song, J., Liu, J., 2019. Development of active and intelligent films based on cassava starch and Chinese bayberry (Myrica rubra Sieb. et Zucc.) anthocyanins. RSC Adv. 9 (53), 30905–30916. https://doi.org/10.1039/c9ra06628d.
Zhai, X., Xue, Y., Sun, Y., Ma, X., Ban, W., Marappan, G., Tahir, H.E., Huang, X., Wu, K., Chen, Z., Zou, W., Liu, B., Zhang, L., Yang, Z., Katona, J., 2025. Colorimetric food freshness indicators for intelligent packaging: progress, shortcomings, and promising solutions. Foods 14 (16), 2813. https://doi.org/10.3390/foods14162813.
Zhang, J., Zou, X., Zhai, X., Huang, X.W., Jiang, C., Holmes, M., 2019. Preparation of an intelligent pH film based on biodegradable polymers and Roselle anthocyanins for monitoring pork freshness. Food Chem. 272, 306–312. https://doi.org/10.1016/j.foodchem.2018.08.041.
Zhang, Z., Sun, Y., Sang, S., Jia, L., Ou, C., 2022. Emerging approach for fish freshness evaluation: principle, application and challenges. Foods 11 (13), 1897. https://doi.org/10.3390/foods11131897.
Zhang, S., Guo, H., Li, H., Qiao, X., Wu, P., Li, Y., Ai, S., Shan, Y., Ding, S., 2025. A starch-based composite film with dual-modality essential oil controlled-release for perishable fruit preservation. Food Packag. Shelf Life 52, 101606. https://doi.org/10.1016/j.fpsl.2025.101606.
Zhao, L., Liu, Y., He, M., Zhang, D., Liu, H., Sun, B., 2025. From industry 4.0 to 5.0: exploring the opportunity of biodegradable freshness indicator packaging. Compr. Rev. Food Sci. Food Saf. 24 (4). https://doi.org/10.1111/1541-4337.70242.
Zuo, J., Feng, J., Gameiro, M.G., Tian, Y., Liang, J., Wang, Y., Ding, J., He, Q., 2022. RFID-based sensing in smart packaging for food applications: a review. Future Foods 6, 100198. https://doi.org/10.1016/j.fufo.2022.100198.

CHAPTER SEVEN

Biodegradability and environmental impacts of starch-based packaging

Emine Arman Kandirmaz[1], Magdolna Pal[2]
[1]Faculty of Applied Sciences, Marmara University, Istanbul, Turkiye
[2]Faculty of Technical Sciences, Department of Graphic Engineering and Design, University of Novi Sad, Novi Sad, Serbia

7.1 Introduction

With advancing technology and modern life, the rapid demands of consumption, the rapid growth of the packaging industry, and the excessive use of petroleum-based synthetic polymers are leaving traces on the global ecosystem that are very difficult to reverse. Traditional plastics such as polyethylene (PE), polypropylene (PP), and polyethylene terephthalate (PET), which we use in every aspect of our daily lives, have increased in use and determined the needs of the industry due to their superior mechanical resistance, chemical stability, and economic accessibility. However, alongside the high functional potential of these materials, their high resistance to natural conditions transforms into a persistent burden on nature that can last for hundreds of years when their lifespan ends. The difficulties that current waste management faces in eliminating the large amount of these waste materials lead to the uncontrolled accumulation of synthetic polymers in terrestrial and marine habitats, triggering an environmental pollution crisis that directly threatens biodiversity.

One of the biggest problems in waste management is the transformation of plastics into microplastics over time through physical, chemical, and photocatalytic degradation. As frequently emphasized in the literature (Yousafzai et al., 2025), the accumulation of these microplastic particles in the environment and their transport along the food chain disrupts the environmental balance and directly endangers human health through toxicological risks. Furthermore, the reliance on fossil fuels throughout the entire process from production to disposal of these polymers increases the carbon footprint and jeopardizes climate change mitigation goals and international

Starch-Based Sustainable Food Packaging
ISBN: 978-0-443-36346-7
https://doi.org/10.1016/B978-0-443-36346-7.00011-4

environmental protocols. The microplastic problem is a local-level issue affecting plastic pollution globally.

The ecological damage caused by microplastics has made strict international regulations inevitable and has created a societal paradigm shift by increasing the demand of conscious consumers for sustainable, ethical, and biodegradable materials. This pressure has compelled the packaging industry to develop alternatives that are biodegradable and made from renewable resources, without burdening the environment. Among the solutions to overcome these environmental challenges, biopolymers stand out as a strategic solution. Starch, a biopolymer, is a focal point of scientific research due to its numerous advantageous properties, including its high yield from agricultural sources, low raw material cost, and complete mineralization without leaving any toxic residue under industrial composting conditions. However, for starch to fully compete with synthetic polymers, technical limitations such as its hydrophilic character (moisture sensitivity) and limited mechanical barrier properties stemming from its natural structure need to be improved. Current R&D efforts focus on blending starch with biopolyesters such as polylactic acid (PLA), reinforcing it with nanocellulose or clay nanocomposites, and enhancing barrier performance through surface modification techniques. Consequently, these technological advancements are not only a technical achievement but also the most concrete academic and humanitarian reflection of humanity's responsibility to reconcile with nature and leave a cleaner, more breathable world for future generations.

Based on these explanations, biopolymeric packaging materials are considered the most important alternative to synthetic petroleum-based polymers. Starch demonstrates superiority among biopolymeric packaging materials.

Reasons why starch is a good candidate for biopolymeric material include (Hernández-Centeno et al., 2025):

- Biodegradability: Starch can be completely broken down in nature by amylase enzymes secreted by microorganisms. Starch, with its polysaccharide structure, is an environmentally friendly biopolymer. Through amylase and related hydrolytic enzymes secreted by microorganisms, starch chains are broken down into smaller sugar units, ultimately producing environmentally harmless products such as carbon dioxide, water, and biomass. This allows starch to be completely decomposed without creating persistent environmental waste, thus offering a significant advantage against the long-term environmental accumulation problems associated with fossil-based polymers.

- Modifiable Structure: Its functional properties can be improved through chemical and physical processes. Due to the reactive hydroxyl groups in its structure, starch is highly susceptible to chemical and physical modification. In addition to chemical processes such as esterification, etherification, and cross-linking, physical methods like plasticization, heat treatment, and mechanical mixing can improve the mechanical strength, flexibility, water resistance, and processability of starch. This susceptibility to structural modification increases its suitability for various packaging types, applications, and performance requirements, making it a versatile biopolymer.
- Low Cost: It can be obtained in large quantities and at low cost from agricultural products. Starch can be produced in large quantities from widely grown agricultural products such as corn, wheat, potatoes, and rice. Its ease and widespread availability, along with a supply chain based on renewable resources, reduces the cost of starch. These characteristics make starch a significant advantage for high-volume and high-turnover sectors like the packaging industry. Furthermore, producing starch from local agricultural sources helps reduce environmental impacts associated with logistics.

The key element that makes starch a strong candidate material for packaging applications is its unique chemical and molecular structure. Starch, a natural, renewable, and biodegradable polysaccharide, is a macromolecule synthesized in plants for energy storage and composed entirely of glucose units. This structural feature gives starch both environmental compatibility and functional diversity. The performance of starch is largely determined by the ratio and molecular organization of its two main components—amylose and amylopectin. Amylose is a polysaccharide with a mostly linear chain structure linked by α-(1 $\rightarrow$ 4) glycosidic bonds. This linear structure allows for the formation of strong hydrogen bonds between the chains, giving starch the ability to form films, gel, and maintain mechanical integrity. Especially in aqueous systems, the rearrangement of amylose chains and the formation of a network structure make it possible to obtain a continuous and homogeneous matrix that forms the basis of starch-based films. Therefore, starches with high amylose content generally exhibit better film-forming properties for packaging films. In contrast, amylopectin has a high molecular weight and complex structure with branched α-(1 $\rightarrow$ 6) bonds in addition to the α-(1 $\rightarrow$ 4) main chain. This densely branched structure plays a decisive role in the thermal behavior of starch and makes thermoplastic starch (TPS) production possible. In the presence of suitable plasticizers, amylopectin-dominant structures allow for partial dissolution of the crystalline regions of starch and increase chain mobility, thus improving its melt processability. This property allows starch-based materials to be shaped using conventional polymer processing techniques such as extrusion, injection molding, and film blowing. Therefore, the performance of starch as a packaging material depends on the delicate balance between the amylose and amylopectin components. Amylose provides structural integrity and film-forming ability, while amylopectin supports processability and thermoplastic behavior. The synergistic interaction of these two components makes it possible to use starch in both solvent-based film production and melt-processing applications, offering significant flexibility in the design of starch-based packaging. Consequently, the complementary roles of amylose and amylopectin, components in the chemical structure of starch, make this biopolymer not only an environmentally friendly alternative for sustainable packaging technologies but also a functionally adaptable and improvable material. This structural diversity constitutes one of the fundamental scientific grounds for starch-based packaging to find wider application areas in the future (Glenn et al., 2025).

In light of this information, this section investigates the recycling, sustainability, and environmental impacts of starch-based packaging films.

7.2 Biodegradation mechanisms of starch-based materials

Although starch is a renewable and biodegradable biopolymer, its naturally occurring, highly brittle structure and pronounced hydrophilic character make it insufficient to meet the performance requirements of the packaging and plastics industry. The presence of a high density of hydroxyl groups in its molecular structure leads to strong interaction of starch with environmental moisture; this results in rapid degradation of the material's mechanical integrity and significant fluctuations in its properties throughout its service life. In particular, starch's water absorption increases in humid environments, leading to plasticization, swelling, and loss of mechanical strength. This hydrophilic structure negatively impacts the mechanical properties of starch-based materials, which are critical for packaging applications, such as flexibility, impact resistance, and tear resistance. Consequently, starch is not a suitable material for use alone in applications requiring high deformation capability and long-term mechanical stability, especially in plastic bags and flexible food packaging. Furthermore, its inadequate water vapor and gas barrier performance significantly limits its use in packaging systems requiring moisture barriers. From the perspective of the packaging industry, it is not enough for a material to simply be environmentally friendly; it is also expected to possess properties that can preserve the product's shelf life, maintain its integrity during transportation and storage, and resist varying environmental conditions. The fact that starch, in its natural state, cannot meet these requirements is one of the main factors limiting its direct use in the packaging and plastics industry. These limitations do not mean that starch should be completely excluded from the packaging sector; on the contrary, they highlight the need for advanced material design strategies such as chemical modifications, the use of plasticizers, polymer blends, and nanocomposite approaches. Through such approaches, the brittleness of starch can be reduced, its moisture sensitivity can be controlled, and the mechanical and barrier properties required for packaging applications can be improved.

The most significant limitation of starch in film-extrusion methods in the packaging industry is its high melting point and the thermal decomposition that occurs in its chemical structure when the starch reaches this temperature (Arruda et al., 2025). The industrial use of starch is affected by this narrow thermal range. Overcoming these negative effects of starch and making it a potential raw material for the packaging industry can be achieved by converting it into thermoplastic starch (TPS) using polymer chemistry. This type of starch (TPS), produced using a plasticizer under high temperature and high shear strength, exhibits superior performance. This transformation in the relevant molecular arrangement involves many physical and chemical reactions, as outlined in Table 7.1.

TPS, with its desired properties, is a sustainable and valid candidate molecule that can be used as a replacement for other synthetic polymers due to its low production cost and abundant availability in nature. Thermoplastic starch (TPS), which can be designed to meet the required mechanical, thermal, and processability properties, offers a good alternative to petroleum-derived polymers due to its low production cost and availability from renewable resources. Thus, it emerges as a sustainable material. Its abundance and ease of production from natural agricultural products increase the continuity and economic availability of TPS as a candidate raw material, reducing the need for fossil fuels. The low energy requirements for TPS production and its easy adaptability to conventional polymeric

Table 7.1 Physical and chemical reactions observed in starch degradation.

Type of reaction	Impact	References
Water diffusion	Water diffusion into the starch matrix allows the plasticizer to penetrate the film, thus weakening the interchain hydrogen bonds.	Yang et al. (2024)
Granular expansion	Water absorption into starch and expansion of starch granules due to high temperature.	Rao et al. (1997)
Gelatinization	The process involves the initiation of heat-induced dissolution of the semi-crystalline segments in the starch structure, leading to swelling of the starch granules and the formation of a viscous fluid.	Pandey et al. (2025)
Melting	Dissolving in the crystalline segments of the starch chain causes the polymer chains to become completely fluid and form an amorphous structure.	Bertoft (2017)
Crystallization/ retrogradation	The rearrangement of starch segments during cooling or storage of films causes TPS to become brittle.	Khairy et al. (1963)

material manufacturing processes are factors that influence and facilitate its industrial applicability. In addition, the ease of biodegradability of starch and its environmentally friendly recycling at the end of its life cycle make TPS a strong candidate molecule both technically and in terms of sustainability. Due to the aforementioned properties, TPS has the potential to replace petroleum-derived synthetic polymers in high-volume applications such as plastic bags, single-use packaging, and short-lived consumer goods.

However, for TPS to be used as a packaging material throughout its shelf life, its mechanical weakness and high moisture permeability must be eliminated. This can only be achieved by blending or modifying TPS with other polymers that possess the desired properties. This blending, especially with biodegradable polyesters (polylactic acid—PLA, polyhydroxyalkanoates—-PHA and poly (butylene adipate-co-terephthalate—PBAT), can improve the desired mechanical and barrier properties of the resulting composite material (Jayarathna et al., 2022).

The degradation of biopolymeric packaging films made from starch is a multi-component process involving hydrolysis, microbial degradation, and the influence of environmental factors. Hydrophilic and naturally occurring starch is highly susceptible to enzymes. Thus, starch degrades rapidly in its natural environment. The degradation of naturally occurring starch-based packaging films is a complex process that cannot be explained by a single mechanism. This process involves a multi-layered sequence of simultaneous and interacting hydrolytic decomposition, microbial activity, and environmental factors. It begins with hydrolysis initiated by environmental moisture, weakening the glycosidic bonds in the starch polymer structure and causing a stepwise decrease in molecular structure and weight over time. This stage creates a suitable environment for microbial degradation, thus accelerating biological processes. Starch, a naturally occurring polysaccharide containing hydrophilic groups, readily absorbs water and moisture due to the hydroxyl groups on its chain. This makes it more susceptible to hydrolytic enzymes secreted by microorganisms. Among these enzymes, amylase and others break down the starch polymer into smaller units such as oligomers and monomers. These smaller units are more easily metabolized by microorganisms. This property allows the starch polymer to be quickly incorporated into the environment. Temperature, pH, humidity level, and microbial density are factors that directly affect the rate and type of degradation of starch films. Under suitable conditions, starch packaging films quickly lose their integrity, and their chemical transformation processes accelerate. While this may seem like a positive feature for the environment, it can lead to the elimination of the requirement for maintaining structural integrity throughout the product's lifespan; therefore, the process must be managed very carefully.

The starch film degradation mechanism occurs in several stages,

- *Water absorption and swelling*

Due to the hydrophilic segments in its structure, starch easily absorbs moisture from its environment. At this stage: the starch film absorbs water or moisture and swells. Along with this, the effect of hydrogen bonds between the starch chains decreases. Thus, the starch matrix becomes more permeable and suitable for enzyme activity. This stage is critically important in initiating the degradation of starch films (Nantambi et al., 2025).

This stage is one of the first and most important steps in the degradation process of starch films. It refers to the water absorption and subsequent swelling behavior that occurs when the starch film interacts with environmental moisture or water. The starch chain contains a high amount of hydrophilic hydroxyl groups in its structure. This property allows it to easily absorb water from the environment. The water absorption weakens the interaction between the starch chains, particularly causing a significant decrease in the effectiveness of hydrogen bonds.

The effect of reduced hydrogen bonds increases the mobility of the starch chains, loosens the main matrix structure, and causes a noticeable swelling in the volume of the packaging film. With swelling, the dense and compact structure of the starch film may be lost. Thus, water, oxygen, and biological agents can diffuse more easily into the material. As a result, the starch film becomes more vulnerable to hydrolytic enzymes secreted by microorganisms. This stage plays a critical role as the triggering step in the biological degradation process. Without water absorption and swelling behavior in starch films, enzymes cannot effectively reach the polymer, and the initiation of hydrolytic reactions becomes difficult. Consequently, this stage, considered one of the most important phases of starch film degradation, occurs according to the physical and molecular-level changes that take place. This demonstrates that water absorption and swelling properties are among the fundamental parameters that must be considered in the design of both the performance life and environmental degradation behavior of starch-based packaging films.

- *Chemical decomposition stage: Hydrolysis*

The most important degradation stage of starch films is hydrolysis. Water molecules break the α-1,4 and α-1,6 glycosidic bonds that make up starch, separating it into smaller glucose units. Hydrolysis is highly affected by external factors such as temperature, humidity, pH, and UV radiation. As a result, starch breaks down into its constituent smaller molecules (maltose, dextrin, glucose, etc.) (Bhati et al., 2025).

Another crucial chemical step in the degradation process of starch films is hydrolysis. During hydrolysis, water molecules penetrating the starch film via moisture or water cause the α-1,4 and α-1,6 glycosidic bonds in the starch's chemical structure to break down over time. The breaking of these α-1,4 and α-1,6 glycosidic bonds transforms the long starch chain structure into shorter oligomers and monomers. This weakens the molecular integrity of the starch polymeric film. Environmental conditions are crucial in hydrolysis. External factors such as temperature, humidity, pH, and ultraviolet (UV) radiation significantly influence hydrolysis. High humidity and temperature increase water mobility and reactivity, thus accelerating hydrolytic reactions. Additionally, acidic or basic environments can catalyze the breaking of glycosidic bonds. UV radiation can accelerate or initiate chain breaks, indirectly speeding up the hydrolysis process. As the hydrolysis step progresses, starch can be broken down into smaller, water-soluble components such as maltose, dextrins, and free glucose. These released small molecules are metabolized by microorganisms. Hydrolysis ensures the bioavailability necessary for subsequent stages of biological degradation. Based on this information, it can be concluded that the hydrolysis step is a fundamental degradation mechanism that bridges the gap between the loss of physical integrity of starch films and microbial mineralization.

- *Biological stage: Microbial decomposition*

The small molecules formed by the hydrolysis of starch are used as nutrients by microorganisms found in soil or compost. At this stage, bacteria (e.g., Bacillus, Pseudomonas), fungi (e.g., Aspergillus, Penicillium), and yeasts metabolize the starch film, producing CO_2, H_2O, and biomass (Leja and Lewandowicz, 2010).

The biological degradation phase of films made from starch occurs when the small, mobile components produced in the previous step are metabolized by microorganisms present in the environment. Water-soluble components such as glucose, maltose, and dextrin, produced in the previous step, act as readily available carbon and energy sources for microorganisms in this phase. Thus, films made from starch and its derivatives can easily participate in biochemical cycles in the natural environment. In the natural environment, various types of bacteria, fungi, and yeasts found in soil and compost media play more important roles at this stage. Bacteria, particularly Bacillus and Pseudomonas, with their high enzyme production capacity, enable the rapid metabolism of starch-based films, while filamentous fungi such as Aspergillus and Penicillium facilitate the breakdown of starch chains with their secreted amylolytic enzymes. Furthermore, yeasts, when given suitable conditions, ferment these simple sugars and accelerate the conversion process. During the microbial decomposition stage, small molecules produced in previous stages are used by microorganisms in cellular respiration and biosynthesis reactions, generating carbon dioxide (CO_2), water (H_2O), and microbial biomass. This carbon dioxide, water, and biomass, combined with the resulting microbial decomposition, mineralize the starch films and prevent the formation of persistent residues in nature. In light of this information, the microbial decomposition stage is considered the step that most significantly impacts the biodegradability of starch-based packaging from an environmental perspective.

The effectiveness of the microbial decomposition stage can vary depending on conditions such as microbial diversity, oxygen presence, humidity, and temperature. This results in different degradation profiles of starch-based films and materials in different disposal environments (soil, compost, controlled bioreactors, etc.).

- *Film structure degradation*

When the above stages occur, the mechanical strength of the starch films weakens, cracks form on the film surface, the film begins to acquire brittle properties, loses its barrier properties, and its integrity deteriorates. Finally, the starch film breaks into small pieces and becomes completely mineralized (Kramer, 2009).

The structural integrity of films produced from starch and its derivatives begins to deteriorate through the sequential processes of hydrolysis, microbial decomposition, and environmental interactions. Examined at the molecular level, these chain breaks and weakening of intra-matrix bonds lead to a decrease in the mechanical strength of the material. Ultimately, starch films lose their initial properties such as flexibility and tensile strength, becoming more brittle. As the structural weakening progresses, microcracks and irregularities form in the starch film structure and surface. These cracks allow water and microorganisms to easily penetrate the film, triggering side reactions and conditions that accelerate the degradation process. Furthermore, the breakdown of the starch film's integrity leads to a gradual decrease in its gas and water vapor barrier properties. This indicates that the packaging functions have irreversibly ceased. In the later stages of the degradation process, the starch film structure breaks down structurally into smaller particles. These particles are then consumed more rapidly in the environment by microbial activity. At the end of all processes, the film produced from starch completely mineralizes, transforming into environmentally harmless end products such as carbon dioxide, water, and biomass. This process supports the idea that starch films can be incorporated into environmental cycles without persistent residue and have a significant advantage as biodegradable packaging systems.

Biodegradation (see Fig. 7.1) is a naturally triggered process that explains the end of packaging material in nature. Here, the chemical structure of the biopolymer is broken down with the help of naturally occurring microorganisms such as fungi and bacteria. Thus, the biopolymer matrix acts as a source of nutrients, carbon, and energy for these organisms. These organisms use exoenzymes (hydrolases, esterases, etc.) to digest biopolymers. With the help of these enzymes, the covalent bonds in the biopolymer chain are hydrolyzed, creating smaller molecules (monomers and oligomers) that can pass through the cell membrane (Nasrollahzadeh et al., 2021).

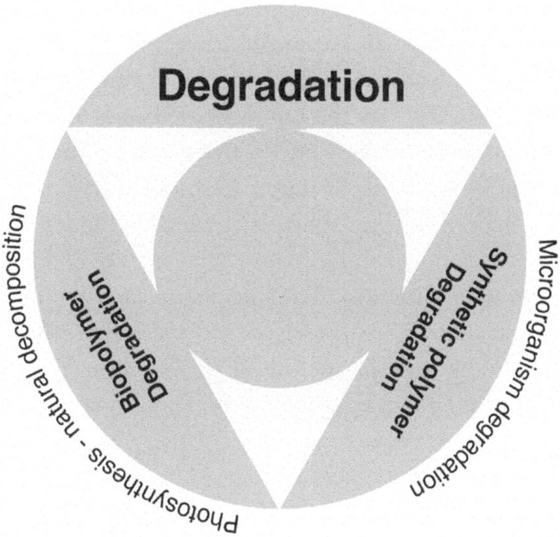

Figure 7.1 End products of the degradation of synthetic and natural polymers.

The oligomers and monomers formed because of the degradation described above and taken up by cells are mineralized (oxidized) through intracellular metabolic cycles (Krebs cycle). The oxidized compounds are broken down into carbon dioxide, water, and biomass under aerobic conditions (Fig. 7.2). Biodegradable polymers are thus recovered and returned to nature by participating in the biocycle without transforming into harmful chemicals. This facilitates waste management and helps reduce the accumulation of pollution (Stewart, 2018).

7.2.1 Enzymatic degradation mechanism of starch

Enzymes secreted by natural fungi and bacteria play a major role in the biodegradation of starch-based materials. Examples of these enzymes include α-amylase, β-amylase, glucoamylase, debranching enzymes (pullulanase, isoamylase), and various glycosidases. The α-1,4 and α-1,6 glycosidic bonds in starch undergo enzyme-catalyzed hydrolysis in an aqueous environment. Thus, starch is broken down into smaller oligosaccharides and then monosaccharides (Kumar and Satyanarayana, 2009).

The enzymatic biodegradation rate of starch films depends on factors such as molecular structure, morphology, and additives used in the biopolymeric matrix (Ruggero et al., 2020).

Degree of Crystallinity: Due to differences in its chemical structure, starch consists of crystalline and amorphous segments. In crystalline segments, the diffusion of water into the structure is restricted, consequently limiting the enzyme's effect on the bonds. A higher proportion of crystalline segments in starch generally leads to a decrease in the hydrolysis rate. However, amorphous segments in starch are more suitable for enzyme activity, therefore their biodegradability is generally higher (Dome et al., 2020).

Amylose–Amylopectin Ratio: The linear structure of amylose in starch facilitates the access of water and enzymes to the starch, thus accelerating biodegradation. The branched structure of amylopectin in starch, containing α-1,6 bonds, makes biodegradation more complex and slower (Liu et al., 2013).

Effect of Plasticizers: Plasticizers such as glycerol, sorbitol, and polyethylene glycol used in starch film production increase the free volume of the film in the polymeric matrix and enhance chain mobility, thereby increasing the film's interaction with water. The hydrophilic nature of the starch film facilitates water penetration and accelerates enzymatic hydrolysis (Mohammed et al., 2023)

Water Activity and Diffusion: The biodegradation of starch is directly dependent on the dispersibility of water within the starch film. Sufficient water penetration into the film is necessary for enzyme diffusion and hydrolysis to proceed thermodynamically. Therefore, the humidity of the environment affects the rate of biodegradation (Basiak et al., 2018).

7.2.2 Changes in microbial activity in response to environmental factors

The biodegradation performance of starch-based packaging materials is directly affected by the physical, chemical, and biological properties of their environment. Although the biodegradation mechanism fundamentally consists of water absorption, enzymatic hydrolysis, and mineralization of organic compounds by microorganisms, the rates of these processes vary significantly depending on environmental conditions. Environmental factors, particularly temperature, humidity, pH, oxygen levels, and the diversity and density of microbial populations, are the main variables determining the kinetics of degradation. Therefore, the performance of starch-based biopolymers should be evaluated not only through laboratory-scale controlled tests but also under real environmental conditions.

- **Microbial activity in compost medium**
  Composting media are systems with high biological activity where organic molecules undergo microbial decomposition. They typically consist of conditions that maximize biodegradation, such as a temperature range of 50–60°C, a humidity level of 50%–60%, optimal oxygen conditions, and high microbial diversity (bacteria, fungi, actinomycetes). These conditions increase and maximize the activity of enzymes used in starch degradation, accelerating hydrolysis. It is known that in industrial composting conditions, the mineralization of starch films takes several weeks (Mayekar et al., 2023). Composting media is the system in which starch films decompose most rapidly.
- **Microbial activity in soil medium**
  Soil media, which have a more heterogeneous composition and less controlled nature compared to compost media, are suitable but less suitable for biodegradation. Temperature in the soil can vary between 10 and 30°C depending on location, and moisture content also varies depending on rainfall, irrigation, and soil structure. In addition, the number and type of microorganisms in the soil may not be as suitable for biodegradation as compost. Furthermore, oxygen availability depends on soil permeability and varies depending on the layers. Considering all these conditions, the decomposition of starch films generally takes months (Sen and Das, 2018). Also, differences in soil structure (clay content, amount of organic matter, pH, degree of compaction) directly affect microbial activity, causing variability in the decomposition rate. In natural soil, biodegradation depends on parameters such as film thickness, film surface area, and the type and amount of additives used within the film.
- **Microbial activity in the marine environment**
  Marine starch films have lower biodegradation efficiency compared to soil and compost systems. Reasons for this low bioactivity include low microbial density, low temperature (typically 10–20°C), high salinity, decreasing oxygen levels with depth, and limited hydrolytic enzyme activity. These conditions in the marine environment cause slow and limited biodegradation of starch films. Laboratory experiments show that starch films can take up to hundreds of days to decompose in marine environments (Janik et al., 2018). However, the high hydrophilic structure of starch films increases water absorption, reducing their mechanical strength; but this does not mean that biodegradation is complete. The process occurring in the sea is generally dispersion-like, and mineralization is less common.

- **Microbial activity in wastewater and freshwater environments**

Wastewater treatment systems are biological reactors where microorganisms break down organic matter. In these systems: the microbial density is quite high, the temperature is controlled (usually 20–35°C), and the pH is close to neutral, within the suitable range for enzymatic hydrolysis. Therefore, starch-based materials decompose much faster in these systems compared to freshwater. In freshwater environments, the microbial load is lower and temperature variability is greater; consequently, decomposition occurs more slowly than in compost and wastewater.

- **The effect of environmental factors on microbial activity**

Key environmental parameters influence microbial activity and, consequently, degradation kinetics (Meng et al., 2014). Some of these environmental influences are examined below:

Temperature: This is one of the most important factors for enzyme activity and microbial growth. Temperatures below the optimum range slow down degradation. Temperature, a parameter affecting microbial activity, is one of the most influential parameters in terms of microbial metabolism and the enzymatic reactions that occur. Microorganisms require an optimum temperature range for their growth rate and enzyme activity, and microbial activity reaches its maximum level only within this range. At temperatures below this optimum range, enzymatic reactions slow down, microbial multiplication is limited, and the biological degradation process is disrupted. At excessively high temperatures above the optimum range, enzymes denature, and the degradation efficiency can be negatively affected by cell damage.

Humidity: This is a requirement for hydrolysis reactions and is also critically important for enzymatic reactions. Moisture, a parameter affecting microbial activity, plays a critical role in the degradation of films produced from starch and its derivatives. The presence of water is essential for hydrolysis reactions to occur. Microorganisms, bacteria, and yeasts also require optimum humidity to sustain their lives and carry out their enzymatic activities. Insufficient humidity limits both chemical hydrolysis and microbial activity. In environments with excessive humidity, oxygen diffusion decreases, and aerobic degradation is particularly negatively affected.

Oxygen: Aerobic degradation occurs in the presence of oxygen. Oxygen, a parameter affecting microbial activity, is a crucial factor determining the type and progression of biological decomposition mechanisms. Aerobic decomposition is a reaction that occurs in the presence of oxygen, in which organic matter is rapidly and efficiently converted into end products—carbon dioxide and water—by microorganisms. In environments with little or no oxygen, decomposition occurs via anaerobic mechanisms; this reduces the rate of decomposition and can lead to the formation of different intermediate products (e.g., methane).

pH: Many microorganisms are activated in the pH range of 5.5–8.0. pH, a parameter affecting microbial activity, directly influences the vital functions and enzymatic stability of microorganisms. Many bacterial and fungal species function more effectively in a pH range of approximately 5.5–8.0. In acidic or basic conditions below or above this range, enzyme functions are suppressed, microbial metabolism is restricted, and the rate of biological degradation is reduced. Therefore, the pH of the environment is critically important for the progression of degradation kinetics.

Microbial diversity and density: A larger microbial community indicates that more diverse enzymes participate in biodegradation. Microbial diversity and density, parameters affecting microbial activity, are among the parameters determining the effect of biological degradation. Rich and diverse microbial communities enable different enzymes to act simultaneously, thus ensuring faster degradation of starch-based films. High microbial density increases access to the starch film, while diversity allows complex structures to be broken down through different degradation pathways.

When all these parameters are considered together, it is clear that testing starch biofilms under real environmental conditions is essential. In particular, it can be concluded that different soil and water types encountered in different locations and climates alter the degradation rate.

7.2.3 Degradation in composite and multicomponent systems

To improve the mechanical and barrier properties of starch biofilms, composite and nanocomposite systems have been widely developed in the literature, incorporating thermoplastic starch (TPS) with polymers (e.g., Polylactic Acid—PLA, Polycaprolactone—PCL), cellulose derivatives, lignin, and especially nanofillers such as layered silicate clays (e.g., Montmorillonite—MMT) or cellulose nanocrystals (CNC/CNF) (Table 7.2) (Muñoz-Gimena et al., 2023). These additions aim to facilitate packaging production by reducing the hydrophilicity of the starch polymer or partially eliminating its mechanical weakness.

The addition of nanofillers (especially MMTs) to polymeric films imparts a "tortuous path" effect to the film, hindering the diffusion of water vapor and water molecules into the polymer; thus, reducing the water

Table 7.2 Mechanical, barrier and biodegradation properties of starch-based nanofilled composites.

Matrix	Nanofiller	Mechanical properties	Barrier properties	Biodegradation/ sustainability	References
Corn/potato starch	Montmorillonite (MMT)	Increased tensile strength (15%–92%), increased elastic modulus.	Water vapor permeability decreased (22%–67%).	Moderately degradable, biodegradable.	Tang et al. (2008)
Cornstarch	Nanocellulose (CNC/ CNF)	Increased tensile strength, increased elastic modulus.	Water vapor and O_2 permeability decreases.	It decomposes quickly but is edible.	Patil et al. (2022)
Thermoplastic starch	Nanocellulose (CNC/ CNF) from Agave Bagasse	Increased tensile strength, increased Young's modulus.	Water vapor permeability decrease.	Moderately degradable.	Lomelí-Ramírez et al. (2023)
Starch (derivative of potato peels)	Cellulose nanofiber	Increased elastic modulus leads to increased tensile strength.	Water vapor permeability decrease.	It degrades quickly, it is biodegradable.	Medina-Jaramillo et al. (2025)

permeability of the film and improving its water/moisture barrier performance. However, this is achieved only if the added filler is well distributed within the film and forms a good interface. It has been known that the added filler can improve both the mechanical and thermal stability of the nanocomposite and delay hydrolysis and degradation due to microorganisms.

However, in some cases, it has been observed that filler materials do not always slow down degradation; It is sometimes thought that incompatibility at the filler-matrix interface, agglomeration, or high surface area can increase water retention, leading to micromorphological irregularities and microcrack formation, thus facilitating enzyme access and accelerating degradation. Cellulose nanofiber (CNC/CNF), especially when used as an additive, is known to be easily degraded by enzymatically due to its increased surface area and hydrophilic properties. These negative effects in starch films can be overcome by surface modification of the added filler.

In light of all these studies, the degradation behavior of nanocomposites added to starch biofilms depends on many factors. The type and quantity of nanofillers, dispersion quality, filler-matrix interface, surface modifications, processing parameters (extrusion, mixing, crosslinking, etc.), and most importantly, life cycle tests conducted under real environmental conditions such as compost, soil, and water should be evaluated together. In this way, it becomes possible to optimize both the suitability for packaging and the biodegradability.

7.3 Environmental effects

7.3.1 Environmental impact assessment

Starch-based packaging materials, which fall under the category of biodegradable polymers and are derived from renewable biomass, should be examined using Life Cycle Analysis (LCA) to assess their environmental performance. LCA is a standardized method that systematically evaluates environmental impacts by quantifying energy and raw material usage and waste/emission loads throughout the entire life cycle of a product or process, from raw material to waste. Bio-based polymers derived from agricultural sources such as starch can be integrated into the renewable carbon cycle more effectively than fossil-based polymers (Hu et al., 2025).

Plants store carbon by binding atmospheric CO_2 through photosynthesis, potentially reducing reliance on fossil fuel-based raw materials. Therefore, bio-based packaging films may have a lower carbon footprint in terms of production and waste management compared to films produced from fossil fuels (Yadav and Nikalje, 2024).

To give an example of the generalization described above, in an LCA study of polyamide-starch biocomposite material, the carbon footprint of the composite's production process was analyzed along with other environmental impacts such as acidification, eutrophication, and land use; sensitive components in biocomposite production were identified, and ideas were put forward on how to improve them (Äkräs et al., 2024).

In general, meta-analyses and reviews on bioplastics emphasize that bio-based packaging solutions are potentially more sustainable than fossil-based plastics; however, this potential largely depends on life cycle stages such as raw material production, agricultural inputs, water and energy consumption, and waste management (composting, landfilling, incineration, etc.) (Sobeih et al., 2025). However, it is also observed that biodegradable polymers are not always considered environmentally friendly in LCA. For example, while bio-based packaging has advantages in terms of greenhouse gas emissions compared to fossil-based plastics, it is reported that compostable bioplastics may have environmental disadvantages due to their weight increase, copolymer additives, and energy-intensive additives (Folino et al., 2020)

When calculating LCA, the water, fertilizer/pesticides, land-use changes, and energy inputs used during the cultivation of the biopolymeric source must be considered. In addition, energy, water, and chemical inputs in the production processes of polymeric raw materials and packaging materials after the garden are considered in LCA (Senila et al., 2024)

Furthermore, waste management of biodegradable polymeric films (household composting, industrial composting, landfilling, incineration, recycling) is another building block determining the environmental impact in LCA. Incorrect waste management can turn the environmental advantage of biodegradable films into a disadvantage.

In the LCA study investigating the production of thermoplastic starch (TPS)/corn starch-based biopolymer, the environmental impacts of the plasticization process were analyzed, highlighting that water and energy consumption, as well as various indicators of toxicity and ecotoxicity, may be significant (Surendren et al., 2024).

In this context, the most sustainable version of starch-based packaging materials is possible with a holistic strategy such as "use of starch from agricultural waste/by-products + energy-efficient production + appropriate biodegradation/composting infrastructure." In conclusion, the following strategies can be suggested to optimize the environmental impact of packaging using starch films:

- Use of biomass from byproducts or agricultural waste in agricultural raw material production that does not create food-feed competition
- Technologies that increase energy and water efficiency in the production process, use of renewable energy
- At the end of the product life cycle — organic waste management, industrial/household composting, effective use of biodegradability
- Full cradle-to-grave (LCA) in life cycle assessment: not only raw materials and production but also use, waste management and potential recycling/composting should be included
- Environmental and technical optimization in polymer formulations (e.g., balancing biobased additives, filler-matrix ratios, in terms of heritability-sustainability performance)

7.3.2 Disposal scenarios: Compostability, biodegradability, and recycling

Packaging films produced using starch biopolymer are a good candidate material for sustainable packaging due to their renewable resources and high biodegradability. Starch is a hydrophilic polysaccharide consisting of amylose and amylopectin chains linked by α-1,4 and α-1,6 glycosidic bonds, which makes it susceptible to enzymatic hydrolysis. Starch films are rapidly broken down in biological environments by suitable enzymes (amylase and glucosidase).

Under industrial composting conditions, starch-based packaging films should be biodegradable according to EN 13432 and ASTM D6400 standards (Greene, 2014; Wojtowicz 2009). Literature reports indicate that starch biofilms reach over 90% mineralization within 60–90 days; however, in some low-thickness systems with optimized plasticizer content, this period extends to 2–6 weeks (Burelo et al., 2023). The degradation products of starch biofilms are generally water (H_2O), carbon dioxide (CO_2), and biomass produced by microorganisms; these products demonstrate that starch participates in the natural carbon cycle.

In home composting conditions, the temperature generally ranges between 20 and 30°C, which limits microbial activity and thus slows down decomposition kinetics. Furthermore, the fact that no toxic byproducts or microplastic residues are formed during home composting of starch supports the environmental reliability of these materials.

From a recycling perspective, starch films have limited mechanical recycling potential. This low recycling potential is due to starch's low thermal stability (degradation above ≈180–200°C), its extreme sensitivity to water, and the degradation of starch chains during remelting and reshaping processes. Therefore, packaging films made from pure starch are generally considered for biodegradation and composting and are not included in conventional plastic recycling.

In addition, blending starch with polylactic acid (PLA), polycaprolactone (PCL), or other biodegradable thermoplastics at ratios of 10%–40% can improve the process stability and re-processability of the material. Literature reports that starch-thermoplastic blends can be reprocessed several times in the presence of suitable plasticizers; however, mechanical strength and barrier properties gradually decrease with each reprocessing cycle (Pilla, 2011). Therefore, in these systems, recycling is addressed through a downcycling approach involving a limited number of reuses, rather than a closed-loop system.

In conclusion, while starch packaging films have environmental advantages in terms of biodegradability, compostability, and renewability, they have disadvantages in terms of recycling performance and long-term stability. Therefore, when conducting an environmental assessment of starch packaging films, prioritizing composting and integration into the biological cycle rather than recycling should be emphasized in waste management.

7.3.3 Ecotoxicological effects

The degradation products released during the biodegradation of starch packaging films are glucose and similar low molecular weight oligosaccharides, which are natural biochemical metabolic intermediates. These products are utilized by microorganisms and converted into water (H_2O), carbon dioxide (CO_2), and biomass; therefore, starch biofilms are considered to have a low ecotoxicological risk (Wojtowicz, 2009). This is one of the key arguments demonstrating the environmental compatibility of starch biofilms, exhibiting a contrast to the persistent or toxic products resulting from the degradation of synthetic petroleum-based polymers.

Starch packaging films are generally not used in their pure form in applications; they are modified with various additives to improve performance characteristics such as mechanical strength, moisture resistance, and UV stability. Among the additives used are hydrophobic additives, UV stabilizers, antioxidants, and functional nanoparticles (e.g., nanoclay, metal oxide nanoparticles, or nanocellulose). While these additives improve material properties, their environmental impact and potential ecotoxicological effects must be carefully considered. In particular, the release of additives within the film during degradation can lead to increased bioavailability in aquatic and terrestrial ecosystems (Lambert et al., 2014). Therefore, the use of heavy metal-free, biocompatible, and preferably biodegradable additives in the production of starch packaging films is considered to benefit environmental safety. Naturally derived UV absorbers, plant-based antioxidants, and organic-based stabilizers stand out in this context.

Furthermore, the behavior of low molecular weight intermediates released during the biodegradation process (e.g., short-chain organic acids, plasticizer derivatives, or additive residues) in the surrounding environment should not be overlooked. The potential effects of these compounds on aquatic organisms (algae, crustaceans, and fish larvae) and soil microorganisms should be evaluated using standardized ecotoxicity tests. Algal growth inhibition, Daphnia magna immobilization, and earthworm toxicity tests, as defined by the OECD and ISO, are among the commonly used methods for quantitatively demonstrating the environmental safety of starch-based packaging (Boros and Ostafe, 2020).

In conclusion, while starch-based packaging films offer ecotoxicological advantages due to the nature of starch polymers, the environmental safety of the final product should not be assessed solely based on starch. The ecotoxicological impact should be evaluated considering the components of the entire packaging film formulation and its degradation products. Therefore, additive selection, degradation scenarios, and ecotoxicological testing should be addressed within a holistic framework in sustainable packaging design.

7.3.4 Comparison with traditional polymers

Synthetic polymers produced from fossil fuels, such as polyethylene (PE), polypropylene (PP), and polyethylene terephthalate (PET), decompose slowly in nature due to their high molecular weight, hydrophobic structure, and carbon-carbon bonds that are resistant to degradation by microorganisms. However, current literature reports that these polymers can maintain their structural integrity for hundreds of years depending on the conditions, ultimately forming microplastics and nanoplastics (de Jong et al., 2025). These persistent particles accumulate in aquatic and terrestrial ecosystems and can enter the food chain, posing an ecotoxic risk.

In contrast, starch packaging films can decompose within months due to their biodegradable polysaccharide chemical structure. Under industrial and domestic composting conditions, it has been observed in the literature that starch-based films completely biodegrade within a few months to approximately 1 year without forming persistent microplastics (Tyagi et al., 2022). Considering the review above, starch packaging materials have become environmentally attractive alternatives for single-use and short-lived packaging.

One advantage of degrading starch packaging films is the integration of their degradation products into the soil ecosystem. The organic carbon released during starch degradation is thought to support microbial biomass formation and have positive effects on soil structure, water-holding capacity, and microbial activity.

Synthetic polymers produced from fossil fuels are generally considered to have significant advantages in terms of packaging performance. Polymers such as PE, PP, and PET are widely preferred in food and industrial packaging applications due to their properties such as high mechanical strength, low water vapor and gas permeability, dimensional stability over a wide temperature range, and long shelf life (Mtibe et al., 2021). Biofilms produced from starch, due to their hydrophilic structure, can exhibit moisture sensitivity, limited water vapor barrier performance, and fluctuations in mechanical properties.

The choice between starch-based and fossil-derived packaging films should be made not based on environmental superiority, but rather on application-specific performance requirements, service life, product characteristics, and waste management methods. Starch packaging films offer environmental advantages for single-use, compostable packaging with a short shelf life, while synthetic polymers provide advantages for packaging requiring a long shelf life, high barrier performance, and mechanical strength.

7.4 Challenges and limitations

Although biodegradation in starch biofilms is environmentally important, it presents some challenges in production and application. Due to the water-loving groups present in the starch structure, microorganisms form a suitable substrate in humid environments, making the biodegradation process dependent on environmental conditions. High humidity and active microbial environments can cause biodegradation to begin before the film's shelf life is complete.

Biodegradation kinetics vary widely depending on temperature, humidity, and microbial density. This makes predicting and standardizing the degradation behavior of starch-based films challenging (Su et al., 2023). Furthermore, the heterogeneous progression of weathering, mostly starting from the surface, is another factor that complicates the interpretation of mass loss and mineralization.

In multi-component starch packaging films, differences in the biodegradation rates of the various components within the film structure create significant environmental differences. Specifically, the starch component, due to its natural hydrophilic structure and high sensitivity to enzymes, degrades rapidly; however, other polymers or additives in the film exhibit slower or more limited degradation behavior, potentially disrupting the complete decomposition process. This results in the loss of physical integrity in the initial stages of biodegradation, but incomplete chemical and biological transformation, thus delaying the full mineralization process. This type of degradation behavior leads to the accumulation of residues and microplastics in the environment. Although these microplastics pose a lower environmental risk compared to synthetic plastics, the delayed full mineralization calls into question the claim of biodegradability. In this context, evaluating biodegradation solely through observable physical degradation or mass loss is insufficient to reveal the true environmental impact of multi-component starch-based systems. In evaluating the biodegradation performance of starch packaging films, approaches that assume the mineralization process, where all components are converted into inorganic products, should be considered correct. Mineralization-focused test methods, based on characteristics such as carbon dioxide formation, oxygen consumption, and biomass conversion, more reliably reveal the compatibility of the packaging material. Such analyses can explain not only whether the material has degraded, but also to what extent the degradation is complete and whether it leaves permanent or temporary residues in the environment. Consequently, for a more accurate explanation of the sustainability of multi-component starch packaging films, it is crucial that biodegradation analyses are not limited to physical decomposition-based indicators; they should be supported by time-dependent and mineralization-based evaluation approaches encompassing all components. This holistic approach will both increase the scientific validity of academic studies and contribute to the development of environmentally reliable packaging designs in industrial applications.

7.5 Future perspective

Controlled biodegradation of starch-containing packaging films is crucial, both as an environmentally friendly alternative and as a viable and reliable packaging solution on an industrial scale. Due to its naturally occurring polysaccharide structure, starch is biodegradable. However, this advantage limits its use because the material cannot maintain its integrity throughout its lifespan. In this context, biodegradation should be considered not as a passive outcome, but as a conscious and manageable parameter of material design (Sharma et al., 2025).

Nanocomposite starch films overcome traditional approaches that focus solely on the rate of biodegradation, allowing for control over the temporal and spatial dimensions of the degradation process. Thus, degradation ceases to be a random process dependent on specific environmental triggers; instead, it becomes a predictable and designable function where the onset time, concentration areas, and rate of progression can be foreseen. This type of control plays a crucial role in ensuring consistent performance, particularly in real-world use scenarios where packaging materials are exposed to varying humidity, temperature, and microbial activity conditions (Tian et al., 2024).

The integration of nano clays, cellulose nanocrystals, and other bio-based nano-additives into the starch matrix forms the fundamental building blocks of this controlled degradation approach. These dispersed additives directly influence the kinetics of biodegradation, by creating complex diffusion pathways that slow down and direct the movement of water and enzymes in the starch matrix. This ensures that, the degradation process becomes more resistant to environmental fluctuations, more homogeneous, and more repeatable. On the other hand; this structural arrangement provided by nano-additives positively affects not only the degradation behavior but also key parameters determining packaging performance, such as mechanical strength, barrier properties, and thermal stability. This multi dimensional improvement enables starch-based packaging to maintain its functionality throughout its lifespan and to degrade in a way that is compatible with environmental systems in the post use phase. Thus, biodegradation ceases to be a weakness of the material and becomes a conscious outcome of sustainable design.

In conclusion, starch-based packaging films with controlled biodegradation behavior achieved through nanocomposite approaches stand out as next-generation packaging materials with predictable and safe degradation profiles that respect the balance between human activities and the natural environment. Interdisciplinary academic studies and industrial applications in this field are laying a strong foundation for starch-based nanocomposites to become a fundamental component of sustainable packaging systems in the future.

7.6 Results and conclusion

Starch based packaging films are increasingly becoming important alternative materials within sustainable packaging technologies due to their derivation from renewable agricultural resources and their efficient biodegradability under suitable environmental conditions. Thanks to their natural polysaccharide based structure, these films decompose in a way that is compatible with biological cycles, transforming into environmentally harmless products such as carbon dioxide, water, and biomass, thus; they offer a strong potential solution to environmental problems such as persistent waste accumulation and micro plastic formation caused by fossil based polymers. These properties make starch-based packaging environmentally friendly materials compatible with a circular economy approach.

However, the biodegradation behavior of starch based packaging is highly sensitive to environmental conditions encountered in application areas. Factors such as temperature, humidity, microbial activity, and the chemical composition of the environment directly affect the degradation mechanisms and rate, making it difficult to predict and control the performance life of these materials. Ensuring controlled degradation in the post-use phase while maintaining the structural integrity of the packaging throughout its lifespan remains a significant area of research and development. Furthermore, in multi component packaging systems, how plasticizers, fillers, biopolymer blends, or functional additives used in conjunction with the starch matrix affect the complete mineralization process of the final product has not yet been fully elucidated.

The presence of moisture is a key factor accelerating the biodegradation of starch based films due to their hydrophilic nature, while the type and quantity of additives used also play a decisive role in the degradation kinetics. Therefore, the evaluation of the biodegradability of starch based packaging should not be approached from a narrow perspective focusing solely on the chemical structure of the material, but rather from a holistic perspective encompassing environmental conditions, use scenarios, and waste management processes. Increasing experimental studies reflecting real world use and disposal conditions, in addition to standard test methods, will contribute to deepening the body of knowledge in this field.

In recent years; advancements in production technologies, nanocomposite approaches, and life cycle-oriented design principles have enabled more controlled and predictable biodegradation behavior of starch-based packaging. Thanks to nanofiller materials, surface modifications, and multiphase system designs, packaging films can be developed that exhibit both improved mechanical and barrier properties and environmentally sensitive degradation profiles. Biodegradation is no longer a random process but a conscious function of material design. As a result, starch-based packaging films that are environmentally safe, compatible with human health, and have controlled biodegradation properties are expected to find a more widespread use in sustainable packaging systems in the future. With the integrated advancement of academic and industrial research, it is predicted that the competitiveness of these materials in terms of performance, cost, and environmental impact will increase, and they will constitute a strong alternative to fossil-based packaging.

References

Äkräs, L., Silvenius, F., Baniasadi, H., Vahvaselkä, M., Ilvesniemi, H., Seppälä, J., 2024. A cradle-to-gate life cycle assessment of polyamide-starch biocomposites: carbon footprint as an indicator of sustainability. Clean Technol. Environ. Policy 26 (10), 3297–3312. https://doi.org/10.1007/s10098-024-02884-1.

Arruda, T.R., Machado, G.d.O., Marques, C.S., Souza, A.L.d., Pelissari, F.M., Oliveira, T. V.d., Silva, R.R.A., 2025. An overview of starch-based materials for sustainable food packaging: recent advances, limitations, and perspectives. Macromol 5 (2). https://doi.org/10.3390/macromol5020019.

Basiak, E., Lenart, A., Debeaufort, F., 2018. How glycerol and water contents affect the structural and functional properties of starch-based edible films. Polymers 10 (4), 412. https://doi.org/10.3390/polym10040412.

Bertoft, E., 2017. Understanding starch structure: recent progress. Agronomy 7 (3), 56. https://doi.org/10.3390/agronomy7030056.

Bhati, N., Bhatt, P., Kumar, V., Tomar, S., Arora Sethi, V., Mittal, A., Maguluri, L.P., Goel, R., Tyagi, L.K., 2025. Eco-friendly enzymatic starch modification: precision strategies for tailored industrial and biomedical functionalities. Starch - Stärke 77 (9). https://doi.org/10.1002/star.70087.

Boros, B.V., Ostafe, V., 2020. Evaluation of ecotoxicology assessment methods of nanomaterials and their effects. Nanomaterials 10 (4). https://doi.org/10.3390/nano10040610.

Burelo, M., Hernández-Varela, J.D., Medina, D.I., Treviño-Quintanilla, C.D., 2023. Recent developments in bio-based polyethylene: degradation studies, waste management and recycling. Heliyon 9 (11). https://doi.org/10.1016/j.heliyon.2023.e21374.

Dome, K., Podgorbunskikh, E., Bychkov, A., Lomovsky, O., 2020. Changes in the crystallinity degree of starch having different types of crystal structure after mechanical pretreatment. Polymers 12 (3), 641. https://doi.org/10.3390/polym12030641.

Folino, A., Karageorgiou, A., Calabrò, P.S., Komilis, D., 2020. Biodegradation of wasted bioplastics in natural and industrial environments: a review. Sustainability 12 (15), 6030. https://doi.org/10.3390/su12156030.

Glenn, G.M., Shogren, R.L., Jin, X., 2025. Starch Plastification Handbook of Starch Science and Technology. CRC Press, United States, pp. 263–281. https://doi.org/10.1201/9781003464396-16.

Greene, J., 2014. Degradation and Biodegradation Standards for Starch-Based and Other Polymeric Materials. Elsevier BV, pp. 321–356. https://doi.org/10.1016/b978-0-444-53730-0.00025-7.

Hernández-Centeno, F., Hernández-González, M., López-De La Peña, H.Y., Rodríguez-González, C.A., Rios-Velasco, C., Zamudio-Flores, P.B., Tirado-Gallegos, J.M., 2025. Functionalized Starch Biomaterials as a Possible Alternative in the Future of Food Packaging: Advantages, Disadvantages, and Perspectives Food Security, Safety, and Sustainability: Validated Innovative Approaches and Technological Advancements. Apple Academic Press, Mexico, pp. 225–260.

Hu, Y., Huang, H., Chen, M., Pan, C., Alamerew, A.N., Zhang, J., He, M., 2025. Life cycle carbon footprint assessment of a typical bamboo-based fiber composite material. Fibers 13 (10). https://doi.org/10.3390/fib13100134.

Janik, H., Sienkiewicz, M., Przybytek, A., Guzman, A., Kucinska-Lipka, J., Kosakowska, A., 2018. Novel biodegradable potato starch-based compositions as candidates in packaging industry, safe for marine environment. Fibers Polym. 19 (6), 1166–1174. https://doi.org/10.1007/s12221-018-7872-1.

Jayarathna, S., Andersson, M., Andersson, R., 2022. Recent advances in starch-based blends and composites for bioplastics applications. Polymers 14 (21), 4557. https://doi.org/10.3390/polym14214557.

de Jong, E., Goumans, I., Visser, R.(H.A.), Puente, Á., Gruter, G.-J., 2025. The opportunities and challenges of biobased packaging solutions. Polymers 17 (16), 2217. https://doi.org/10.3390/polym17162217.

Khairy, M., Morsi, S., Sterling, C., 1963. Crystallization in starch: the role of ions in the lyotropic series. J. Polym. Sci. - Part A Gen. Pap. 1 (12), 3547–3559. https://doi.org/10.1002/pol.1963.100011203.

Kramer, M.E., 2009. Structure and function of starch-based edible films and coatings. In: Edible Films and Coatings for Food Applications. Springer New York, New York, NY, pp. 113–134. https://doi.org/10.1007/978-0-387-92824-1_4.

Kumar, P., Satyanarayana, T., 2009. Microbial glucoamylases: characteristics and applications. Crit. Rev. Biotechnol. 29 (3), 225–255. https://doi.org/10.1080/07388550903136076.

Lambert, S., Sinclair, C., Boxall, A., 2014. Occurrence, degradation, and effect of polymer-based materials in the environment. Rev. Environ. Contam. Toxicol. 227, 1–53. https://doi.org/10.1007/978-3-319-01327-5_1.

Leja, K., Lewandowicz, G., 2010. Polymer biodegradation and biodegradable polymers—a review. Pol. J. Environ. Stud. 19 (2), 255–266.

Liu, X., Wang, Y., Yu, L., Tong, Z., Chen, L., Liu, H., Li, X., 2013. Thermal degradation and stability of starch under different processing conditions. Starch - Stärke 65 (1–2), 48–60. https://doi.org/10.1002/star.201200198.

Lomelí-Ramírez, M.G., Reyes-Alfaro, B., Martínez-Salcedo, S.L., González-Pérez, M.M., Gallardo-Sánchez, M.A., Landázuri-Gómez, G., Vargas-Radillo, J.J., Diaz-Vidal, T., Torres-Rendón, J.G., Macias-Balleza, E.R., García-Enriquez, S., 2023. Thermoplastic starch biocomposite films reinforced with nanocellulose from Agave tequilana Weber var. Azul Bagasse. Polymers 15 (18). https://doi.org/10.3390/polym15183793.

Mayekar, P.C., Limsukon, W., Bher, A., Auras, R., 2023. Breaking it down: how thermoplastic starch enhances poly(lactic acid) biodegradation in compost—a comparative analysis of reactive blends. ACS Sustain. Chem. Eng. 11 (26), 9729–9737. https://doi.org/10.1021/acssuschemeng.3c01676.

Medina-Jaramillo, C., Coelho-Leandro, G., Ayala-Valencia, G., López-Córdoba, A., 2025. Starch-based biocomposites reinforced with cellulose nanofibers from potato peel byproducts. Sci. Rep. 15 (1). https://doi.org/10.1038/s41598-025-22303-9.

Meng, P., Pei, H., Hu, W., Shao, Y., Zheng, L., 2014. How to increase microbial degradation in constructed wetlands: influencing factors and improvement measures. Bioresour. Technol. 157, 316–326. https://doi.org/10.1016/j.biortech.2014.01.095.

Mohammed, A.A.B.A., Hasan, Z., Omran, A.A.B., Elfaghi, A.M., Khattak, M.A., Ilyas, R. A., Sapuan, S.M., 2023. Effect of various plasticizers in different concentrations on physical, thermal, mechanical, and structural properties of wheat starch-based films. Polymers 15 (1). https://doi.org/10.3390/polym15010063.

Mtibe, A., Motloung, M.P., Bandyopadhyay, J., Ray, S.S., 2021. Synthetic biopolymers and their composites: advantages and limitations—an overview. Macromol. Rapid Commun. 42 (15). https://doi.org/10.1002/marc.202100130.

Muñoz-Gimena, P.F., Oliver-Cuenca, V., Peponi, L., López, D., 2023. A review on reinforcements and additives in starch-based composites for food packaging. Polymers 15 (13), 2972. https://doi.org/10.3390/polym15132972.

Nantambi, H.R., Lubwama, M., Kigozi, J., 2025. Preparation and characterization of encapsulated biochar-blended compost pellets with biopolymer coating for controlled nutrient release and water retention. J. Mater. Cycles Waste Manag. 27 (5), 3762–3778. https://doi.org/10.1007/s10163-025-02338-w.

Nasrollahzadeh, M., Nezafat, Z., Shafiei, N., Soleimani, F., 2021. Biodegradability properties of biopolymers. Biopolymer-Based Metal Nanoparticle Chemistry for Sustainable Applications 1, 231. https://doi.org/10.1016/B978-0-12-822108-2.00010-7.

Pandey, S., Kumar, A., Kumar, Y., Shirsat, S., Rao, P.S., Mitra, J., 2025. Starch Structure and Functionality Advanced Starch Modification: Emerging Technologies and Industrial Applications. CRC Press, India, pp. 9–25. https://doi.org/10.1201/9781003564287-2.

Patil, S., Bharimalla, A.K., Nadanathangam, V., Dhakane-Lad, J., Mahapatra, A., Jagajanantha, P., Saxena, S., 2022. Nanocellulose reinforced corn starch-based biocomposite films: composite optimization, characterization and storage studies. Food Packag. Shelf Life 33, 100860. https://doi.org/10.1016/j.fpsl.2022.100860.

Pilla, S., 2011. Starch Based Composites for Packaging Applications Handbook of Bioplastics and Biocomposites Engineering Applications. Wiley, pp. 189–266. https://doi.org/10.1002/9781118203699.ch8.

Rao, M.A., Okechukwu, P.E., Da Silva, P.M.S., Oliveira, J.C., 1997. Rheological behavior of heated starch dispersions in excess water: role of starch granule. Carbohydr. Polym. 33 (4), 273–283. https://doi.org/10.1016/s0144-8617(97)00025-8.

Ruggero, F., Carretti, E., Gori, R., Lotti, T., Lubello, C., 2020. Monitoring of degradation of starch-based biopolymer film under different composting conditions, using TGA, FTIR and SEM analysis. Chemosphere 246, 125770. https://doi.org/10.1016/j.chemosphere.2019.125770.

Sen, C., Das, M., 2018. Biodegradability of starch based self-supporting antimicrobial film and its effect on soil quality. J. Polym. Environ. 26 (12), 4331–4337. https://doi.org/10.1007/s10924-018-1304-6.

Senila, L., Kovacs, E., Resz, M.-A., Senila, M., Becze, A., Roman, C., 2024. Life cycle assessment (LCA) of bioplastics production from lignocellulosic waste (study case: PLA and PHB). Polymers 16 (23), 3330. https://doi.org/10.3390/polym16233330.

Sharma, M., Kumar, P., Singh, J., 2025. Eco-friendly Nanofillers as Additives Handbook of Nanofillers. Springer Science+Business Media, India, pp. 1709–1728. https://doi.org/10.1007/978-981-96-2407-2_88.

Sobeih, M.O., Sawalha, S., Hamed, R., Ali, F., Kim, M.P., 2025. Starch-derived bioplastics: pioneering sustainable solutions for industrial use. Materials 18 (8). https://doi.org/10.3390/ma18081762.

Stewart, G.G., 2018. Energy metabolism by the yeast cell. In: Brewing and Distilling Yeasts. Springer International Publishing, Cham, pp. 77–107. https://doi.org/10.1007/978-3-319-69126-8_6.

Su, C.y., Li, D., Wang, L.j., Wang, Y., 2023. Biodegradation behavior and digestive properties of starch-based film for food packaging–a review. Crit. Rev. Food Sci. Nutr. 63 (24), 6923–6945. https://doi.org/10.1080/10408398.2022.2036097.

Surendren, A., Hasan, Y., Mohanty, A.K., Abbassi, B., Misra, M., 2024. Comparative environmental life cycle assessment on corn starch plasticization and co-plasticization processes. Sustainability 16 (17), 7406. https://doi.org/10.3390/su16177406.

Tang, X., Alavi, S., Herald, T.J., 2008. Barrier and mechanical properties of starch-clay nanocomposite films. Cereal Chem. 85 (3), 433–439. https://doi.org/10.1094/CCHEM-85-3-0433.

Tian, J., Kong, Y., Qian, S., Zhang, Z., Xia, Y., Li, Z., 2024. Mechanically robust multifunctional starch films reinforced by surface-tailored nanofibrillated cellulose. Compos. B Eng. 275, 111339. https://doi.org/10.1016/j.compositesb.2024.111339.

Tyagi, P., Agate, S., Velev, O.D., Lucia, L., Pal, L., 2022. A critical review of the performance and soil biodegradability profiles of biobased natural and chemically synthesized polymers in industrial applications. Environ. Sci. Technol. 56 (4), 2071–2095. https://doi.org/10.1021/acs.est.1c04710.

Wojtowicz, A., 2009. Biodegradability and Compostability of Biopolymers. Thermoplastic Starch: A Green Material for Various Industries, pp. 55–76. https://doi.org/10.1002/9783527628216.

Yadav, K., Nikalje, G.C., 2024. Comprehensive analysis of bioplastics: life cycle assessment, waste management, biodiversity impact, and sustainable mitigation strategies. PeerJ 12 (9), e18013. https://doi.org/10.7717/peerj.18013.

Yang, J., Wang, Z., Ren, J., Lin, X., Zhao, J., Jiang, X., Chen, Y., 2024. Molecular dynamics simulation and properties of thermoplastic starch – effect of water content on starch plasticization. Polymer 290, 126571. https://doi.org/10.1016/j.polymer.2023.126571.

Yousafzai, S., Farid, M., Zubair, M., Naeem, N., Zafar, W., Asam, Z.ul Z., Farid, S., Ali, S., 2025. Detection and degradation of microplastics in the environment: a review. Environ. Sci. Adv. 4 (8), 1142–1165. https://doi.org/10.1039/d5va00064e.

CHAPTER EIGHT

Life cycle assessment and carbon foot print of starch

Engin Kocatürk[1], M. Özgür Seydibeyoğlu[2]
[1]Department of Nanotechnology Engineering, Zonguldak Bülent Ecevit University, Zonguldak, Turkiye
[2]Department of Metallurgical and Materials Engineering, Izmir Katip Celebi University, Izmir, Turkiye

8.1 Introduction

The progress of studies toward a sustainable world is related to the inclusion of biopolymers into the ecosystem in addition to the recycling processes carried out to increase the efficient use and economic evaluation of polymers. Sustainable and renewable bio-based polymeric materials with extraordinary properties are the fundamental building blocks used in a promising way to create and sustain the circular plastic economies of today and the future. From this point on, less use of fossil resources can be achieved, and new recycling processes-methods can emerge. For this, it is important to use life cycle assessment (LCA) (Kabir et al., 2020; Karan et al., 2019; Zhu et al., 2016). Application of LCA and life cycle-based approaches has been extensively used to support policy formation, starting from defining policy agendas, assisting in defining policy targets and providing evidence for monitoring and impact assessment on products and services. Companies and industries have used LCA in product and process development; marketing, development and selection of indicators to monitor the environmental performance of products or plants; selection of suppliers or subcontractors; and strategic planning. Finally, LCA has also been applied for guiding and informing consumers through ecolabels and public reporting from industry and academia (Filippi et al., 2025; Sevigné-Itoiz et al., 2021). Due to the environmental impact caused by petroleum-based plastics, biopolymer-based plastics or bioplastics have been widely considered a sustainable alternative. The most commonly used bioplastics are starch blends, polylactic acid (PLA), polyhydroxyalkanoates (PHA), polybutylene succinate (PBS), poly (butylene adipate-co-terephthalate) (PBAT) and polytrimethylene terephthalate (PTT) (Amaraweera et al., 2021; Islam et al., 2025; Hong et al., 2025; Barletta et al., 2025; "Das et al.. (2025). Poly (butylene adipate-co-terephthalate)

Starch-Based Sustainable Food Packaging
ISBN: 978-0-443-36346-7
https://doi.org/10.1016/B978-0-443-36346-7.00008-4

(PBAT) in biomedical applications: a comprehensive review of material properties, fabrication methods, and biofunctional potential. Materials Research Express.," no date; Wang et al., 2025).

Starch-based plastics exhibit great potential to replace petroleum-based plastics. Since it is the second most abundant biopolymer on earth, approximately 50% of the commercial bioplastics are fabricated from starch (Sobeih et al., 2025; Zhang et al., 2024). The cost-efficient and scalable conventional starch-based plastics have been utilized in a variety of fields. However, they still face some major challenges, such as high hydrophilicity, poor thermal processability, and unsatisfactory mechanical behaviors. Various strategies have been investigated to overcome these limitations, including physical/chemical modifications of starch reinforcement with other polymers or nanoparticles and addition of plasticizers (Arruda et al., 2025; Zhang et al., 2024; Zamir et al., 2025). Although some of the approaches have resulted in improved properties, a great deal of efforts is still required to develop high-performance starch-based plastics (Zhang et al., 2024). In this book section, the sustainable aspect of starch-based materials, which is one of the most popular sustainable materials, is comprehensively discussed. Life Cycle Assessment (LCA) and Carbon Footprint (CFP) studies on starch-based materials are examined and evaluated in detail for a sustainable world.

8.2 The need for sustainability: Addressing critical problem

A major issue that has concerned mankind and the world post-COVID-19 is the problem of climate change which has alarmed the need for achieving sustainability in all aspects of life (Usman et al., 2021; Paudel et al., 2023; Verma et al., 2024). This is the only way to tackle the growing issue of climate change which poses a greater threat to mankind. In particular, the use of materials and the process by which they are manufactured needs to be sustainable as otherwise the reliance on oil and other fossil fuels to make plastic is not sustainable (Bharadwaj et al., 2023; Verma et al., 2024). To accomplish this goal, we need to accelerate the transition to use materials that are sustainable, degradable, renewable, and at the very least, capable of being recycled in a closed-loop system, thus enabling a system of bio based circular economy (Srivastava et al., 2025; Verma et al., 2024). Challenges such as increasing environmental concerns, sustainability issues, ineffective end-of-life (EoL) options and price of raw materials are impacting the plastics industry (Amulya et al., 2025; Heller et al., 2020). Currently, synthetic plastics are predominantly produced from fossil-based resources (90%) and 4%–8% of the world's oil reserves are used in their manufacture. If this usage continues, 20% of the total oil is expected to be consumed for plastics production by 2050 with a 3%–5% annual increment. This increment is much rapid than overall oil demand, which is projected to grow 0.5% annually. Moreover, their trademark properties such as durability and non-degradability have become a drawback in the EoL phase, resulting in their accumulation in the environment for a very long time. Their persistence in the environment is degrading the natural systems such as oceans, generating greenhouse gas (GHG) emissions during production and treatment (after use) and causing detrimental impacts on human health (Amulya et al., 2021; "Agenda (2016). The new plastics economy rethinking the future of plastics. In World Economic Forum (Vol. 36).," no date, 2025). The ubiquitous societal demand toward a sustainable economy belongs to some of present-day government agendas. To face this challenge, it is required to devise and set the ground rules that must be integrated in future business mindset. The most representative and accepted sustainability assessment benchmark corresponds the 12 principles of Green Chemistry introduced by Anastas and Eghbali in 2010. These principles serve as the basis to establish the main prerequisites to design sustainable products and operations (Cerdan et al., 2024; Anastas and Eghbali, 2010; Kurul et al., 2025). The ones currently most strongly mainstreamed as development and sustainability pathways at a global level are: Circular economy (CE), Green economy (GE) and Bioeconomy (BE). These three concepts were chosen because they all propose to adapt or transform the current economy to a more sustainable one (Radu et al., 2025; Ali and Russell, 2025; D'Amato et al., 2017).

(Fig. 8.1).

Generally, sustainability primarily involves cost and environmental assessments. Holistic sustainability assessment of a process includes green chemistry metrics, circularity assessment and life cycle assessment (LCA) (Amulya et al., 2021; Waniek et al., 2025). Cost assessments are typically given as techno-economic analysis (TEA) (Gangwar et al., 2025), life cycle costing (LCC) or cash flow analysis (Erokhin et al., 2025; Thebuwena et al., 2025). Environmental assessments are typically performed using life cycle assessment (LCA) (Lucas et al., 2025). More recently, circular economy (CE) (Wiedenhofer et al., 2025) metrics and environmental, social and governance (ESG) have completed the sustainability tool portfolio (Hessel

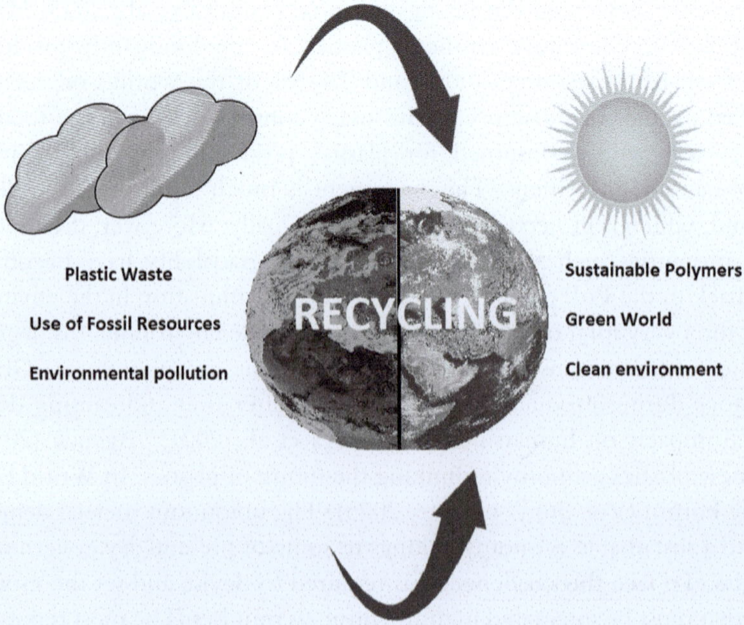

Figure 8.1 Green world: A sustainable future.

et al., 2024). From a sustainability perspective, encouraging new case studies to obtain data, validation of data, development and improvements in methodologies and evaluation in all dimensions using software tools should be encouraged. Therefore, to adapt to a low carbon economy, the entire life cycle of biopolymers needs to be taken into account to make concrete claims about advantages and disadvantages (Amulya et al., 2021; Kogje et al., 2025).

8.3 Sustainable polymers for green world

The polymer community is actively striving to create more sustainable plastics—materials that are valuable in use yet do not pose a threat to the planet and human health (Haider et al., 2019; Kanwal et al., 2025; Abel and Coates, 2025). To date, the focus has primarily been on discovering new materials that are biodegradable or chemically recyclable, and increasing attention is now being given to innovative methods for reusing and recycling existing commodity plastics through upcycling and mechanical recycling. These efforts aim to preserve the benefits of plastics while minimizing their environmental impact (Abel and Coates, 2025; Erickson et al., 2024). The use of polymeric materials from renewable resources has a long history, with naturally occurring polymers being among the first materials used by men. In the 19th century, natural materials, such as casein, natural rubber, and cellulose, were modified to obtain useful polymeric materials ("Olatunji (2016). Natural polymers (pp. 1–17). Cham, Switzerland: Springer International Publishing.," no date; Keya et al., 2019). Over the past few decades, the production and application of synthetic polymers have seen an almost exponential increase (Papageorgiou, 2018); ("Rashid et al. (2025). Microwave-Assisted Depolymerization of Natural and Synthetic Polymers. In Depolymerization: Concepts, Progress, and Challenges Volume 3: Emerging Trends (pp. 15–31). American Chemical Society.," no date). An impediment to the growth of green polymers is the daunting momentum of the existing fossil fuel-based polymer industry. The Chief Technology Officer of a Fortune 100 Company recently communicated that they are "highly interested in finding bio-based routes to [incumbent] commodity plastics which fit with current processing and disposal infrastructures," but are "less concerned with developing new-to-the-world biodegradable plastics for our packaging and products." (Miller, 2013; Coates and Getzler, 2020; Schirmeister and Mülhaupt, 2022). However, concerns regarding depletion of fossil resources, disposal and related issues, as well as government policies, have led to a continuously growing interest in the development of sustainable, safe, and environmentally friendly plastics from renewable resources (Trache et al., 2020; Ferreira-Filipe et al., 2021). Nowadays, green, premium and sustainable materials have become increasingly important for the production of various high-value products at low temperature. There are many types of products related to durable polymers in the literature. A common type of disease is divided into two basic classes, biobased polymers and biodegradable polymers, and many subclasses (Oliver-Cuenca et al., 2024; Cazan, 2022). Sustainable polymeric conditions are properties with great technological formations and are created to meet the lengths related to: mechanical resistance and efficiency; acceleration resistance; resistance to the effects of chemical components; low weight; dimensional stability; resistance to variable stress, shock and abrasion; climatic properties; and aesthetics (Kocaturk et al., 2023).

Biopolymers are natural polymers that are either produced directly by living organisms, such as plants and microorganisms, or derived from biological sources. Due to their intrinsic biodegradability and biocompatibility, biopolymers represent a promising alternative to conventional synthetic polymers (Balaji et al., 2017). Numerous strategies have been developed to enhance biopolymer properties, including adding plasticizers to increase flexibility, applying chemical cross-linking to improve strength and thermal stability, and incorporating fillers like nanoparticles, layered clays, natural fibers, and whiskers to reinforce the matrix and boost barrier performance. Collectively, these modifications enable the customization of biopolymer-based materials to meet specific industrial and environmental requirements (Bangar et al., 2025). Natural polymers derived from animal or plant sources possess unique structural properties and offer a range of benefits, including cost-effectiveness, renewability, non-toxicity, and biodegradability. These properties make them highly promising for the removal of pollutants from aquatic ecosystems. Starch, cellulose, chitosan, tannin, gum, sodium alginate, pectin, different types of proteins, and mucilage are common sources of natural polymers (Rabeie and Mahmoodi, 2024). Today, thanks to the remarkable advances in materials science, the continuous development of sustainable polymer materials enables their widespread use in critical areas such as composite design, material recycling, and next-generation material design, as well as in key industrial areas such as electronics, health, medicine, food, energy, purification, and textiles (Hamsan et al., 2025; Jiang et al., 2016; Zor et al., 2022; He et al., 2025b; Şen et al., 2022); ("Kocatürk et al. (2024b). Development and characterization of nanocellulose/carbonized waste rubber nanocomposites. BioResources, 19(2), 2670.," no date; Kocatürk et al., 2024a; Zeng et al., 2025; Zor et al., 2024). Fig. 8.2 represents examples of application areas of sustainable polymer materials.

In conclusion, sustainable polymers offer a feasible resolution to address key issues caused by unbalanced exercises in synthetic polymers (Edo et al., 2025; Kumar et al., 2024). Researchers are continuously working in developing new sustainable polymers and materials by considering their complexity in recycling and composting, on-demand fabrication scaling-up, reproducibility, and pricing for a circular economy strategy toward future sustainability. However, there are still various challenges or limitations in terms of energy cost, selectivity in recycling or depolymerization, recyclability, functional stability. Therefore, extensive research activities are required to focus on these challenges by designing new monomers, polymer constructs, and more green chemistry or sustainable protocols as well as energy-efficient methods for their production

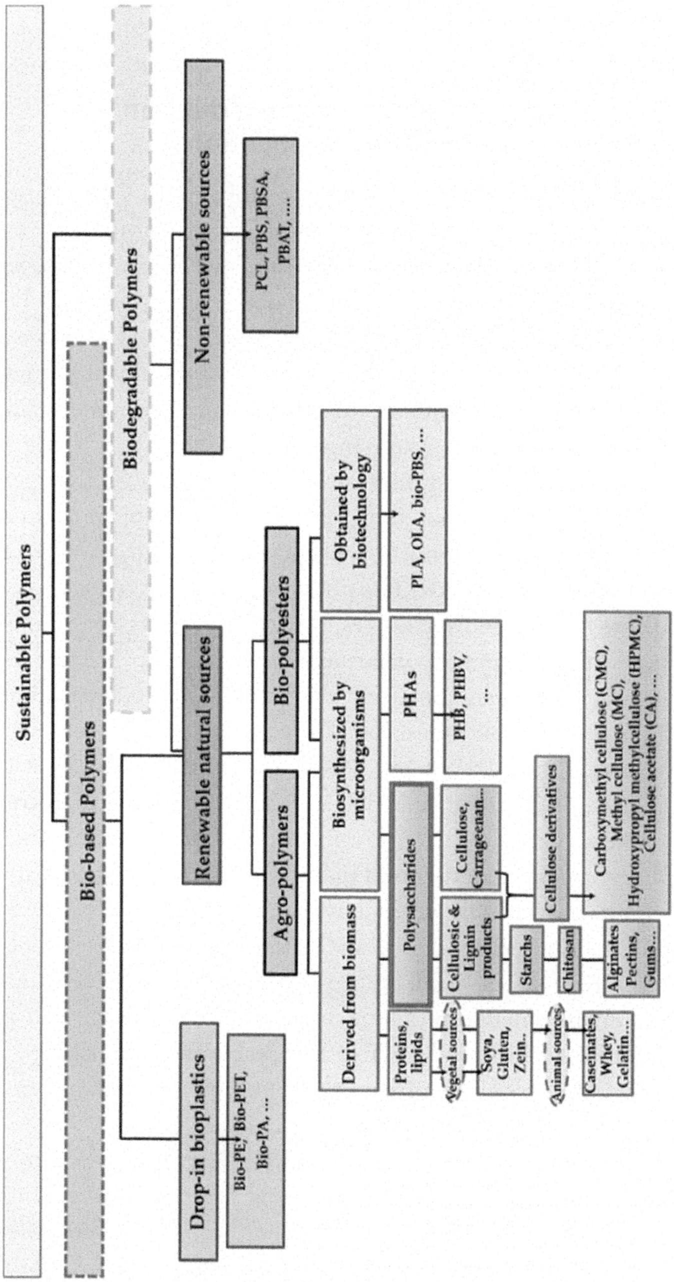

Figure 8.2 Applications of biodegradable materials (Samir et al., 2022).

and recycling of polymer wastes or composites (Kumar et al., 2024; Kristensen et al., 2024; Korley et al., 2022; Beach et al., 2009).

8.4 Starch: General properties, production, modification and common applications

Starch has a rich history as an adhesive globally. In China, historical evidence indicates the use of sticky rice as building mortar in the construction of The Great Wall. By the ninth century, starch was employed for sizing papers. In Egypt, it served as a binder for crafting, while in France, decorative wall hangings affixed with flour pastes emerged before 1630 (Chen et al., 2024). Starch is quantitatively the most dominant storage carbohydrate on Earth and is synthesized mostly in plants and some cyanobacteria. Starch is accumulated as water-insoluble particles, that is, the starch granules, whereas most other species produce water-soluble glycogen as a storage carbohydrate. Both polymers are similar in biological function and chemical composition, consisting of glucose units that are linked by α-1,4 and α-1,6 glycosidic bonds. The physical parameters of starch strongly differ from those of glycogen and are responsible for the high value of starch in various applications. Starch is a versatile biomaterial of special interest due to its abundance, cheapness, non-toxic properties, and biodegradability for various food and non-food industries. Thus, it is used in numerous dairy and bakery goods, soups, and sauces, as well as coatings and meat products. In addition, there is an increasing demand from non-food industries for starch as a renewable material (Apriyanto et al., 2022). Starch has been widely used as an additive in the food industry and other industrial applications, such as pharmaceuticals, drug delivery and composites, because of its low cost and easy availability (Figure x) (Marta et al., 2023). Starch plastics, which are blends of starch with other polymers, are being developed to contribute to environmental problems such as climate change and plastic pollution due to their biobased origin and possible biodegradability. They are among the oldest biobased plastics to be commercialized and are produced on an industrial scale today (Broeren et al., 2017).

Starch is the most widely used biodegradable materials for now. In Europe, 50% of the produced starch was used for non-food applications such as the composition in adhesives and paper binders, textiles, chemical production, the feedstock for fermentation and other industrial products (Mahfouzi et al., 2025; Xiang et al., 2022). Fig. 8.3 provides an overview of starch biopolymer. Although starch appears simple in terms of chemistry,

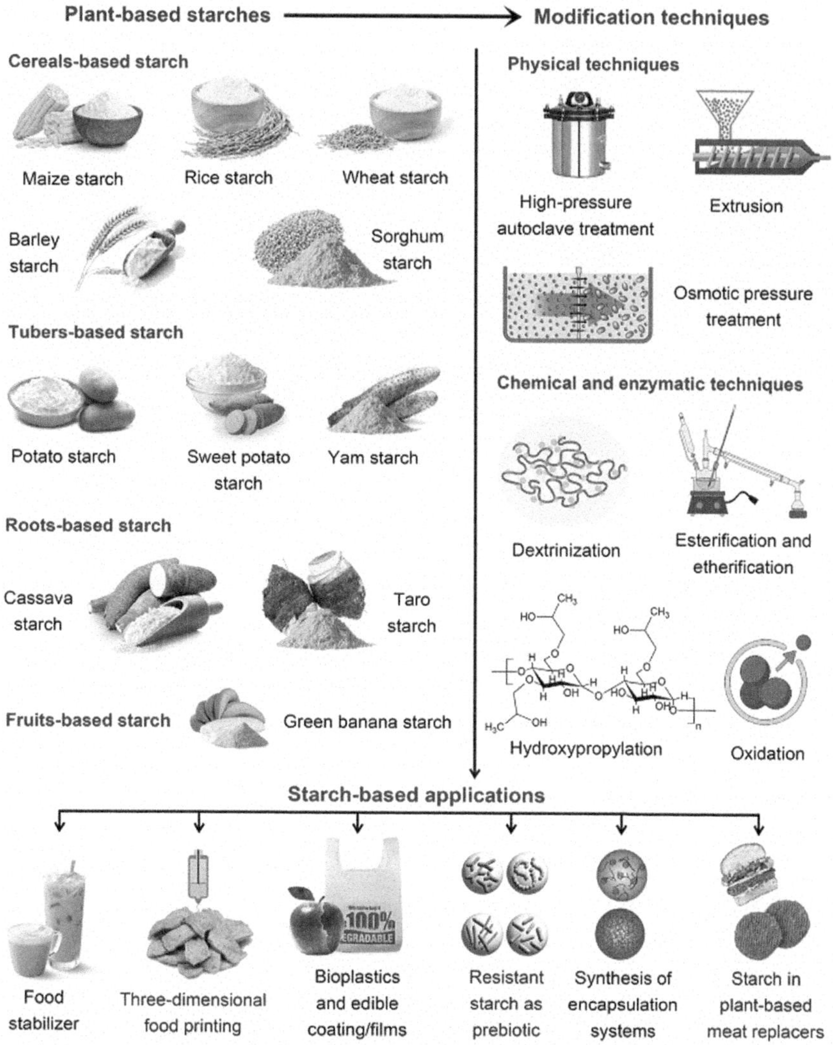

Figure 8.3 Starch overview (Rashwan et al., 2024).

it is a rather complex carbohydrate and various aspects such as its metabolism, internal organization or the connection of structure and physical properties are largely unclear (Apriyanto et al., 2022). The fine structure of starch is one of the major challenges that have plagued starch researchers in recent decades. Many researchers have been devoted to the study of starch multi-scale structures, including granule structure, growth ring structure, and molecular structure of starch, etc. In addition, many scholars, such as Eric, Apriyanto, Seung, Wang, and Copeland, etc. have summarized the progress of research on starch structure (Zhiguang et al., 2025).

Starch has excellent physical, chemical and functional properties including good water solubility, gelatinization, high temperature adhesion behavior and easy modification. Starch is one of the most abundant and widely used renewable/sustainable polymer materials worldwide with extremely low cost (It exhibits better solubility than cellulose and lower cost than proteins). It can be degraded in water and soil without leaving toxic residues, therefore it is considered as environmentally friendly materials. Starch is biocompatible and suitable for implantable electronics. Starch can be easily processed into flexible, foldable films and gels (Xiang et al., 2022; Marta et al., 2023). Starch modification can be carried out by genetic, enzymatic, physical or chemical techniques. There are numerous in-depth studies and reviews on starch modification in the literature (Amobonye et al., 2025; Téllez-Morales and Rodríguez-Miranda, 2025; Maniglia et al., 2021; Miao and BeMiller, 2025; Apriyanto et al., 2022; Punia Bangar et al., 2022; Zhao et al., 2024). Fig. 8.4 summarizes the main points of starch modification in a simple way. Although enzymatic and genetic methods are not used in polymer industries due to high production costs, they are mainly considered in the pharmaceutical and food industries. Physical methods are environmentally friendly and low-cost; however, they limit their industrial applications by disrupting the starch structure. Physically modified starch is used in food and non-food applications such as packaging and 3D printing. Chemical methods can change the structural properties of starch to obtain desired product qualities and are widely considered in the industry. Methods such as cross-linking, alkali treatment, etherification, esterification, acid hydrolysis, grafting and binary modifications can be used to prepare chemically modified starch and its derivatives. Chemical modification leads to changes in the crystallinity, morphology and molecular weight of starch, while changing the tensile strength, elongation at break and adhesive properties of starch. Moreover, chemical modifications can increase shelf life and improve quality, including excellent foaming, air and moisture barrier properties. Chemically modified starch shows potential in packaging, biobased adhesives, pharmaceutical, agricultural, superabsorbent and wastewater treatment applications. The main disadvantages of modified starch include high water sensitivity, low mechanical properties, formation of toxic substances and difficult processability (Amaraweera et al., 2021). Over the last decade, the modified starch market has been growing continuously. According to Markets data, the

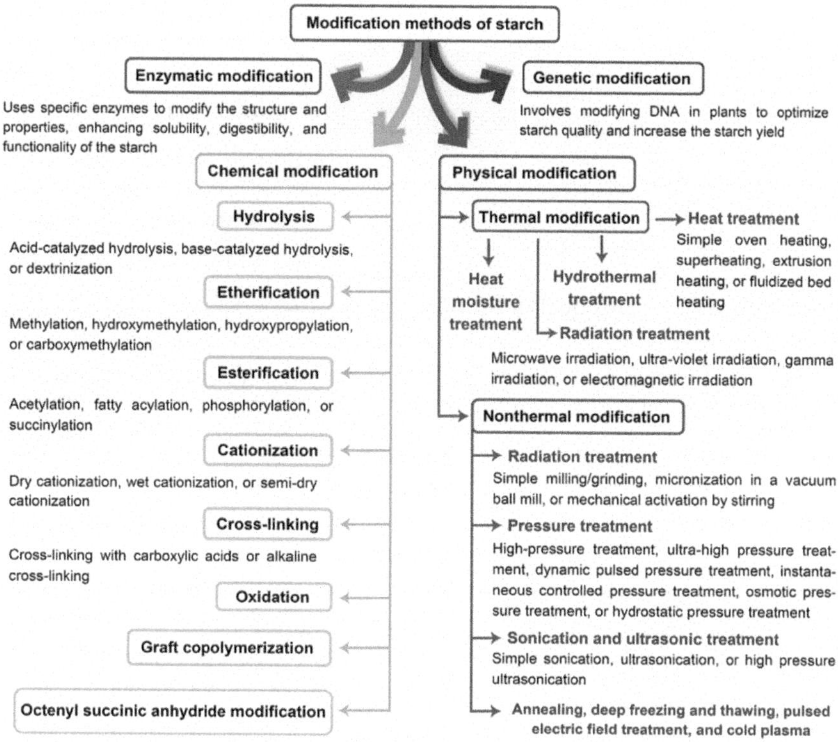

Figure 8.4 General classification modification of starch (Rashwan et al., 2024).

global modified starch trade is expected to be worth US$ 13.7 billion by 2022 and to earn US$ 15.9 billion by 2027, registering a compound annual growth rate (CAGR) of 3.1%. Furthermore, the demand for processed and prepared foods is driving the need for modified starch. Therefore, as innovative experiments and new technologies advance in many industrial sectors (food and non-food), the use of modified starch is expanding in numerous sectors (Compart et al., 2023).

Starch finds diverse applications in many sectors thanks to its functional properties (Seguí and Barrera, 2025; Yadav Prakruthi et al., 2025). In the cosmetics sector, its absorbency and smooth texture make it indispensable for powders and creams (Suhagia, T. A., Ayare, P., & Alle, M. (2025). *Biodegradable Beauty: Carbohydrate Polymers in Cosmetics. In Carbohydrate Polymer Nanotechnologies: Design, Development, Utilization and Perspectives (pp. 455–483). Singapore: Springer Nature Singapore.*, no date); (Olivato, J. B. (2024). *Starch: a natural, safe, and multifunctional ingredient for cosmetic formulations. In Starch Industries: Processes and Innovative Products in Food and Non-Food Uses* (pp. *255–269*). *Academic Press.*, no date). In paper production, it functions as a binder and coating agent that increases paper strength (Ren et al., 2021; Kang et al., 2025; Zieher et al., 2025). In the food industry, it is used as a thickener and stabilizer in products such as sauces and gluten-free foods and helps maintain moisture and texture in baked goods such as cakes and bread (Surendra Babu et al., 2018; Rahmadi Putri et al., 2023). The textile industry uses rice starch as a sizing agent to improve fabric finishes and yarn quality (Limeneh et al., 2025; Adewale et al., 2022). In the construction sector, rice starch is used in environmentally friendly adhesives and binders by utilizing its natural adhesive properties (Tourtelot et al., 2023; Gandhi et al., 2025; Seeponkai et al., 2024). In the pharmaceutical sector, rice starch acts as a binder and disintegrant in tablets, providing consistent drug release (Miculescu et al., 2017; Trisopon et al., 2025). In addition, rice starch has many diverse and high-tech applications, such as the development of biodegradable plastics, food packaging applications, biomedical applications, and electronics, offering a sustainable alternative to traditional materials (Dang et al., 2025; Si and Zhang, 2024; Kapoor et al., 2025; Fang et al., 2025; Compart et al., 2023; de Carvalho et al., 2025; Nazrin et al., 2025; Yan et al., 2024; Shan et al., 2025; Radhakrishnan and Panicker, 2025).

8.5 Fundamentals of starch life cycle assessment (LCA)

LCA can guide the possible co-location of biorefineries with existing petrochemical refineries by analyzing scenarios of preserving and using existing petrochemical infrastructures, process and technologies in relation to new biobased supply chains and emerging technologies. LCA can also be applied to promote sustainability at early stages in the development of a technology or suite of technologies in a value chain, influence design and ensure that sustainability goals of innovation are achieved and not compromised (Karka et al., 2021; Pathania et al., 2025; Blanco et al., 2020). Possible limitations to this include lack of data, incumbents against which to compare, and uncertainty with respect to both how the emerging technology will be deployed as well as the market conditions into which the technology will be deployed (Sevigné-Itoiz et al., 2021; Blanco et al., 2020). Life cycle assessment (LCA) is a quantitative methodology to evaluate the potential environmental impacts related to a product, a process, or a service.

It can be used to consider the impacts of something that took place in the past, or in a prospective manner to assess potential evolution pathways of a system. In both cases, practitioners tend to make the hypothesis that the system is in a deterministic state, even though this is often not the case (Falsafi et al., 2025; Singh et al., 2024; Cerchione et al., 2025). Today, life cycle assessment (LCA) is the most widely used approach to model and calculate the environmental impacts of products and processes. The results of LCAs are often said to be deterministic, even though the real-life applications are uncertain and vague (Cornago et al., 2022). The sustainability of a material is usually assumed as a function of its origin. However, only by using a life cycle assessment different materials can be compared and ensure that the chosen alternative is more sustainable than the substituted material (Barahmand and Eikeland, 2022; Zikulnig et al., 2025).

When the literature is examined, it is known that the environmental impact categories of polymers and different polymers are analyzed in terms of environmental impact. From the literature search, it is found that GW, Ozone Depletion, Acidification, Eutrophication, Respiratory Effects, Ecotoxicity and Fossil Fuel Depletion are commonly focused on in LCA (Norris, 2002; Luo et al., 2024; Aisyah et al., 2025). Many articles used SimaPro software for LCIA for polymers (Baltrocchi et al., 2025; "Das et al., 2021. Life cycle assessment of polymers and their recycling. In Circular Economy of Polymers: Topics in Recycling Technologies (pp. 143–170). American Chemical Society.," no date). Many articles used ReCiPe method for the analysis of the environmental impact of polymers (Ita-Nagy et al., 2020; Obaid et al., 2025). Cradle to gate and cradle to grave were the most tried system boundaries in the research studies. Most authors analyzed the environmental impact in the European context. Eco Invent Database was used to analyze the environmental impact for biopolymers and fossil-based polymers. Most authors analyzed product-based polymers for environmental impact assessment. The tried waste disposal scenarios are known as landfill, incineration, composting and recycling (Ramesh and Vinodh, 2020).

The sustainability of a material is often assumed to be a function of its origin. However, only by using a life cycle assessment can different materials be compared and one can be sure that the selected alternative is more sustainable than the substitute material. However, conducting full life cycle assessment studies requires a large amount of information that is not always available. Therefore, conducting preliminary life cycle analyses using literature data and examining the main impact factors can be used as a clear indicator of the sustainability of the proposed alternatives (Serra-Parareda et al., 2022). A preliminary LCA was conducted to assess the environmental impacts of the newly developed PLA/starch biocomposite granules in their early development stages. This analysis revealed significant reductions in the carbon footprint of these biocomposite granules, up to 18% compared to plain PLA granules and 3%, 38% and 63% compared to fossil-based ABS, HIPS and PA66, respectively (Baniasadi et al., 2024). Another study discussed alternatives to lime-hemp concrete that would achieve similar thermal properties with equivalent or lower environmental impact. Binders such as gypsum, geopolymers and starch were suggested as alternatives for biocomposite production due to their performance and low environmental impact. In addition, a preliminary life cycle assessment was conducted to assess the environmental impact of the studied biocomposites. Preliminary LCA results show that geopolymer and starch binders have a similar environmental impact to traditionally used binders, while gypsum has a significantly lower impact. According to previous studies, biocomposites with such binders have been proven to have lower CO2 emissions than standard building materials such as aerated concrete (Fig. 8.5) (Bumanis et al., 2020).

A study has been conducted that nicely demonstrates the impacts of pure, eco-friendly, or productive plastics produced from wastewater recovery facilities. Cradle-to-factory gate life studies have been conducted for six different plastic compositions that include representative amounts of additives such as compatibilizers. Starch plastics reduce greenhouse gas (GHG) emissions and non-renewable energy use (NREU) reductions but have higher eutrophication potential and allow for greater agricultural land use than common petrochemical plastics. GHG savings are largely due to the composition of the plastic, with 85% reductions in some grades and 80% increases (by weight) in others compared to petrochemical

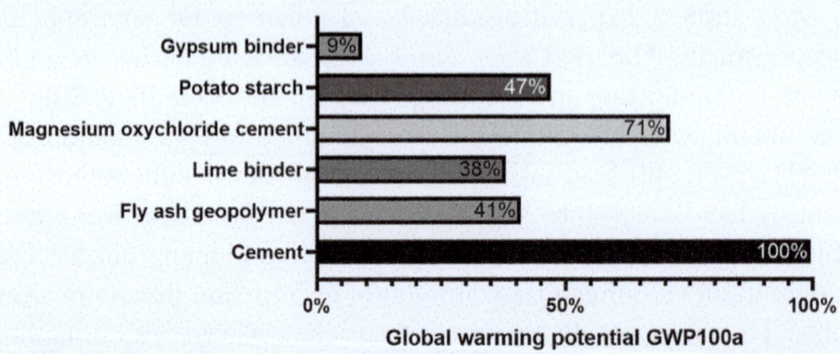

Figure 8.5 GWP100a impact of various bio-composite binders (Bumanis et al., 2020).

counterparts. Additives can account for up to 40% of the GHG emissions of polluting plastics. The greatest GHG savings are achieved when programs such as PBAT and PBS are minimized, while rich, natural fibers and mineral fillers are maximized. Using recovered nitrogen instead of pure nitrogen results in modest reductions in NREU and greenhouse gas emissions (<10% in most cases), while reducing eutrophication and managed land use by up to 60%, the study reported (Broeren et al., 2017).

(Fig. 8.6).

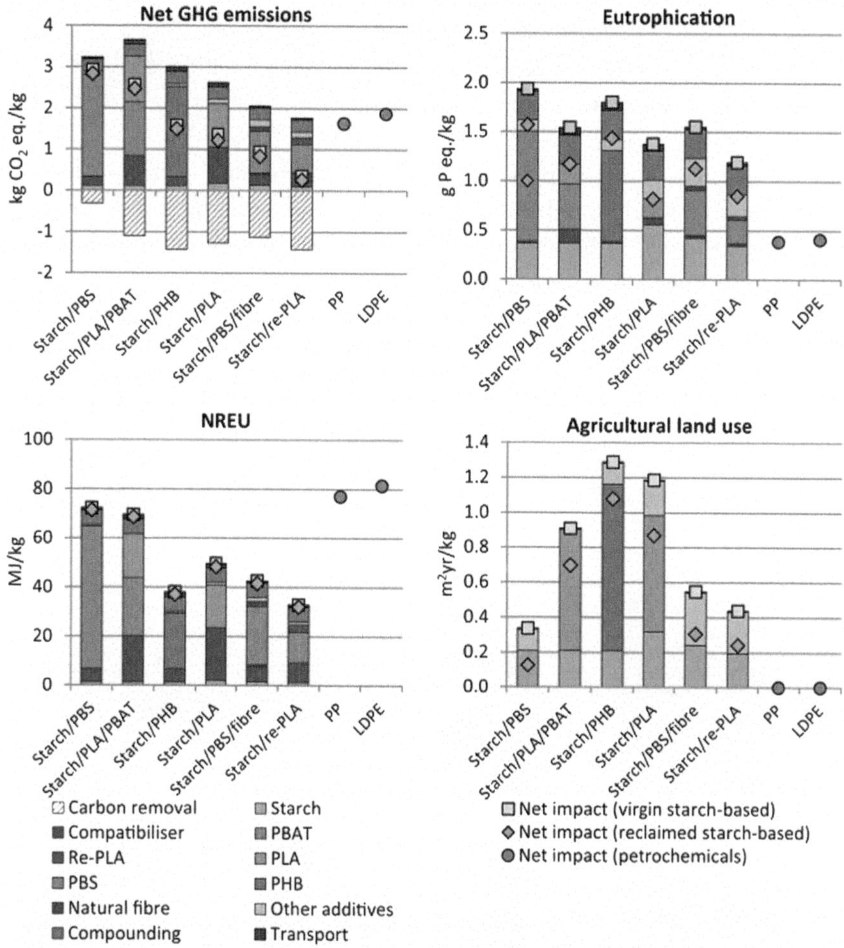

Figure 8.6 Impact assessment results cradle-to-factory gate production of starch plastics and petrochemical references (Broeren et al., 2017).

The sustainability of glucose production from two different feedstocks, corn starch (MS) and woody biomass residues (WBR), was assessed by life cycle assessment (LCA) methodology. The LCA results proved that there are benefits in all environmental impacts assessed when using WBR as a glucose source alternative. Environmental damages associated with the starch production process, which account for more than 60% of the calculated impacts for glucose production from corn starch, were identified as the main step where improvement efforts should be focused for this process. On the other hand, pretreatment of biomass residues was the most contributing stage in the WBR process, mainly due to the large heat and electricity requirements associated with this stage. Finally, it was concluded that the WBR process proposed here can be considered a valuable alternative for glucose production from a sustainability perspective within the biorefinery concept (Blanco et al., 2020).

The environmental impacts of particleboard production using renewable materials, namely starch and cardon processing waste, were determined by a "cradle to gate" LCA study. General summary of the work performed is presented in Fig. 8.7. A case study based on experimental/pilot scale data was used to obtain primary data and was complemented by information from the LCI database EcoInvent V.3.5 and literature. Calculations were performed with Simapro software V 8.5.2 using the RECIPE methodology. The results show that energy consumption is the dominant inventory item and the main contributor to most environmental impact categories. The use of renewable energy and more efficient equipment or equipment

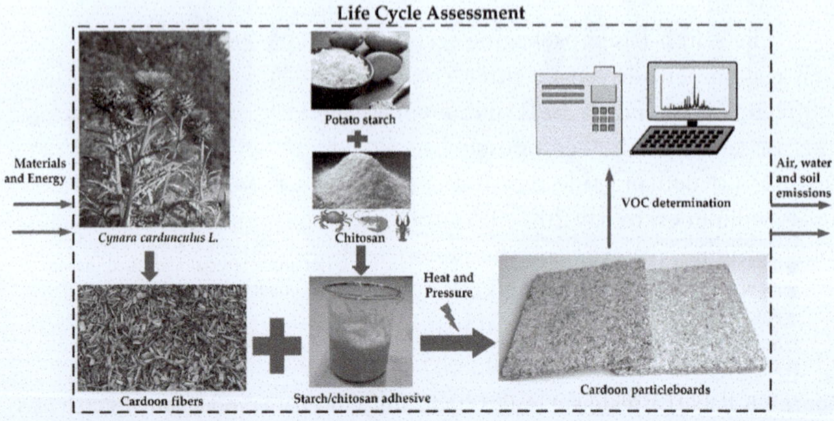

Figure 8.7 Way of assessing the environmental impacts of green particleboard (Mata et al., 2023).

better suited to the materials to be processed would be an interesting option to improve the environmental performance of the process. It significantly reduces the impacts in most environmental impact categories, for example, the carbon footprint is reduced by 34% (Mata et al., 2023).

One study showed the environmental impacts of co-plasticizer starch compared to glycerol-plasticized starch in developing starch-based biodegradable plastic. From the LCA analysis, it is clear that succinic anhydride and citric acid as co-plasticizers exhibited a similar environmental impact to glycerol as a plasticizer for plasticization processes. However, a relatively higher environmental impact was observed for the plasticization process with urea as a co-plasticizer. Therefore, starch glycerol-urea plasticization as co-plasticizer is not recommended from an environmental perspective. Energy was a significant factor affecting the environmental impacts among all input categories. In conclusion, the LCA study showed that co-plasticization of starch with citric acid and succinic anhydride was more environmentally sustainable than urea as a co-plasticizer. From the environmental impact perspective, citric acid and succinic anhydride could be the best alternative for glycerol-plasticized starch. In addition, one of the most important findings in the study was that by changing the energy form, during the co-plasticization process, a significant change in environmental impact can be achieved, especially when produced in large quantities in an industry (Surendren et al., 2024).

The depletion of released energy sources and increasing environmental concerns have increased the interest in biobased, environmentally friendly and high-tech materials. Starch nanocrystals (SNC), which are among these materials, consist of crystalline nanoplatelets obtained by the hydrolysis of starch and are generally used as nanofiller in polymer matrices. The most recent application areas have led to the need to scale up the SNC production process. However, for this new biobased nanomaterial to be sustainable, the environmental impacts of the production and processing processes must be limited. Therefore, while making analyzes and recommendations for scaling up the SNC production process, it is important to determine the environmentally sensitive stages with the life cycle assessment (LCA) method. For this purpose, various scenarios were created and comparisons were made in terms of different environmental impact indicators. In addition, a comparison was presented with organic modified nanoclay (OMMT), the main competitor of SNC. When evaluated in terms of life cycle assessment, SNC production requires less energy than OMMT extraction. However, SNC has higher values in terms of indicators such as global warming potential and acidification. However, while SNC offers significant advantages in terms of being renewable and biodegradable, OMMT contributes to the depletion of non-renewable energy and mineral resources. Therefore, when used as a filling material, SNC positively affects the environmental impact of the filling material at the end of its life (LeCorre et al., 2013).

The development of starch plastics has been driven by the potential environmental benefits resulting from the use of bio-based feedstocks or the desired biodegradability functionality. The environmental benefits and disadvantages of starch plastics compared to traditional petrochemical plastics can be quantified using life cycle assessment (LCA). Almost all LCA studies focus on the production of starch plastics from pure native starch (Amulya et al., 2021). Thermoplastic starch (TPS) is produced from shredded native starch in the presence of a certain amount of plasticizer and under certain extrusion conditions. The type of starch polymer varies from 100% pure starch to different types of blends with different petrochemical copolymer ratios. When the literature is reviewed, LCA studies for almost all TPS have proven that starch polymers require less non-renewable energy use (NREU) per kg than petrochemical polymers and in terms of greenhouse gas emissions, 1 kg of TPS results in lower GHG emissions than petrochemical plastics (Shen and Patel, 2008).

8.6 Carbon footprint (CFP) assessment for starch

Today's megatrends are two words: 'carbon footprint' and 'greenhouse gas emissions'. Carbon footprint is usually determined through a greenhouse gas assessment and expressed as tonnes of carbon dioxide equivalent (tCO_2e) (Serra-Parareda et al., 2022; Elwan et al., 2025). Total amount of greenhouse gas emissions and removals used as CO_2 alternatives, based on a life-scope assessment using the single impact category of climate change, constitutes a product system carbon footprint (Wright et al., 2011; Nabipour Afrouzi et al., 2023; Schmidt, 2009). In the 1960s, with increasing concerns about climate change, the concept of carbon footprint (CFP) gained popularity. The term quickly spread in business, media, and politics because it greatly affects the environment (Skrzypczak et al., 2025; Mathur et al., 2022; Agarwal and Sethi, 2026). While the concept of carbon footprint has been around for decades, its precise definition is still debated. Environmental footprints are often used to represent water, land, and carbon footprints. A carbon footprint is defined as the total amount of carbon dioxide emitted in units of mass (kg, tonnes, etc.) over the life cycle of a product or activity (Zhang et al., 2025; Pandey et al., 2011). A product's carbon footprint represents its contribution to greenhouse gas emissions during its supply chain. Carbon footprints do not include emissions from land-use change and industrial processes such as tractor manufacturing and diesel production (Mathur et al., 2022). This methodology can be widely applied to different industrial sectors such as the construction sector, water desalination, wastewater treatment, food sector and waste management (Pasqualino et al., 2011). CFP is directly linked to climate change, which has caused environmental problems and food security to become common concerns worldwide. The ever-increasing global population has led to an increased demand for resources, subsequently intensifying food and energy-related crises (Nabipour Afrouzi et al., 2023). It is critical to consider sustainability aspects such as global warming, energy consumption, user waste output and improvements in production and recycling procedures. The impact of the carbon footprint on global trade links for products and services should be determined and the various parameters governing it should be analyzed (Shabir et al., 2023; Gutowski et al., 2013). Excessive greenhouse gas emissions causing global climate change have posed serious challenges to humanity's survival and development. Reducing carbon emissions and controlling greenhouse gas concentrations has become a global consensus. Consequently, many countries are actively developing low-carbon strategies and committing to net-zero emissions (He et al., 2025a; 2025b; Soeder, D. J. (2025). *Greenhouse gas and climate change. In Energy futures: The story of fossil fuel, greenhouse gas, and climate change* (pp. *97–141). Cham: Springer Nature Switzerland.*, no date) (Kanna I et al., 2024). The United States and the European Union plan to achieve net-zero emissions by 2050, while China has announced its goal of peaking its carbon emissions by 2030 and achieving carbon neutrality by 2060 (He et al., 2025a; 2025b; Nieto, 2022; Zhao et al., 2024). By the end of 2023, 150 countries have made "carbon neutrality" commitments covering 88% of global greenhouse gas emissions. In this context, the Carbon Footprint (CFP) has become an important indicator for assessing the greenhouse gas contributions of products, services, or organizations. The European Commission's (EC) Carbon Border Adjustment Mechanism (CBAM) aims to reduce the carbon and environmental footprints of export products to support the reduction of global greenhouse gas emissions (Rus et al., 2025; He et al., 2025a; 2025b; Zheng et al., 2023; Erdogdu, 2025).

Starch-based materials generally have a lower carbon footprint than certain conventional plastics. The production of conventional plastics often involves energy-intensive processes and the use of fossil fuels, which contribute to greenhouse gas (GHG) emissions. Plant-derived starch-based materials may offer a more environmentally friendly alternative ("Saishree et al. (2025). From Plastics to Bioplastics: Technological Innovations and Nature based Solutions to achieve Global Sustainability. In Micro-Nano Plastics Exposure, Environmental Degradation and Public Health Crisis: Perspectives and Concerns for Sustainable Development (pp. 257–268). Singapore: Springer Nature Singapore.," no date; Chalannavar et al., 2025; Singh et al., 2024). Starch is typically produced from corn, cassava, wheat, and potatoes, which involves various conversion processes and their associated greenhouse gas emissions. Figure x summarizes the yields of starch from maize, cassava, wheat and potato, along with their carbon footprints. This trend shows the inverse relationship between carbon footprint and starch production yield. Lower starch production yields mean that a higher amount of biomass is required for the same functional unit (one tonne of starch). Therefore, even crops with a significantly lower carbon footprint, such as potatoes, have the highest environmental impact, requiring approximately 5.9 t of biomass. In contrast, maize starch has the highest production yield, requiring 1.3 t of biomass. Cassava starch presents a unique situation: despite requiring 2.7 t of biomass, its cultivation has a relatively lower environmental impact, resulting in the lowest carbon footprint (Forfora et al., 2024).

The carbon footprint of cassava starch product in Thailand was calculated. The aim of this study was to assess and compare the global warming impacts of cassava starch production in three factories located in Northeast Thailand. The system boundary was defined from cassava cultivation and harvesting to cassava starch production from cradle to factory gate using the life cycle assessment (LCA) methodology based on ISO 14,040 series. The selected functional unit was 1000 kg dry native starch excluding packaging. Input and output data were analyzed and the results were shown in terms of global warming potential (GWP) impact categories. The average greenhouse gas (GHG) emission value of the three factories based on the LCA methodology was 594 $kgCO_2e$ per ton of dry native starch. The results showed that if the factory used biogas from the wastewater treatment plant, the cultivation and harvesting procedure was the main source of GHG emission, which was around 40%–59% of the total. According to this result, the total greenhouse gas emissions of cassava starch production in Thailand in 2013 were estimated to be approximately 10.5 billion tons of CO2e (Usubharatana and Phungrassami, 2015).

The increasing demand for sustainable materials has intensified efforts to evaluate by-products such as broad bean starch (FBS), a by-product of seed fractionation that currently lacks high-value applications. A study investigates the development and environmental impact of ecofilms from cross-linked air-sorted (65% starch content) and isolated (95% starch content) FBS reinforced with tunicate cellulose nanocrystals (t-CNCs, 0–6 wt% based on starch dry weight). In terms of environmental impact, the production process resulted in a low carbon footprint (<5 kg CO_2e), further supporting the sustainability of the process. In addition, the use of air-sorted starch, which is less refined, could provide a reduction in greenhouse gas emissions of approximately 50% according to our findings (Falua et al., 2025).

(Fig. 8.8).

It aims to evaluate the environmental impacts caused by conventionally processed sago starch products using the product life cycle (LCA) approach. The environmental impacts are estimated by analyzing the carbon footprint of greenhouse gas emissions from 1 ton of dry sago starch. The scope of the

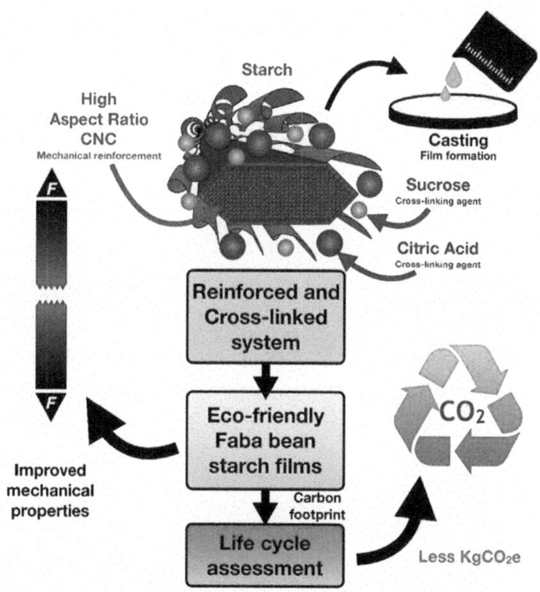

Figure 8.8 Characterization of starch-based ecofilms and determination of environmental footprint (Falua et al., 2025).

product life cycle is limited to the harvesting, transportation and extraction stages to produce dry sago starch. The results of the study show that the greenhouse gas emissions impact to produce 1 ton of sago starch is 17.9 kgCO_2eq. The largest contribution came from the drying process, especially the extraction process, which was 12.25 kgCO_2Eq. (68%). This impact value is much lower compared to corn starch (2700 kg CO_2eq), potato starch (2402 kg CO_2eq) and cassava starch (4310 kg CO_2eq). This is mainly due to the low energy input requirement of this conventional technology (Yusuf et al., 2019).

Circular economy and low carbon economy need to go hand in hand to be successful. A low carbon economy can help achieve the Paris agreement climate goals as well as the 2030 UN sustainable development goals. Since biobased polymers are here to stay and are important carriers for industrial development, their production should address low carbon development as a goal to increase sustainability and provide scope for innovation, competitiveness and job creation (Amulya et al., 2021). The present paper presents the cradle-to-gate life cycle assessment of a newly developed polyamide-starch biocomposite with varying potato starch content as biofiller (ranging from 0 to 70 wt%). The primary objective was to quantitatively measure the total carbon footprint of the selected biocomposite. The results showed that the gradual addition of potato starch as biofiller into the copolyamide matrix significantly reduced the total carbon footprint of the biocomposite, reaching a maximum reduction of 42%–43% with the highest starch content at 70 wt%. Furthermore, the newly developed polyamide-starch biocomposite showed excellent performance with 29%, 39%, 42%, 59% and 79% carbon footprint reduction compared to the reference fossil-based polyamides polyamide 6 (PA6), polyamide 12 (PA12) and polyamide 6.6 (PA6.6) and reactive extrusion-modified PA610/80 wt% polylactic acid (REX-PLA) and PA40/30 wt% glass fiber composites, respectively. Based on these findings, the polyamide-starch biocomposite, especially with the highest potato starch content (70% wt.%), shows significant potential as a new material solution to reduce the carbon footprint of several existing fossil and biobased polyamides together with polyamide-based composites. In doing so, it contributes to the development of a more climate-friendly future for plastics through reductions in their carbon footprint (Äkräs et al., 2024).

(Fig. 8.9).

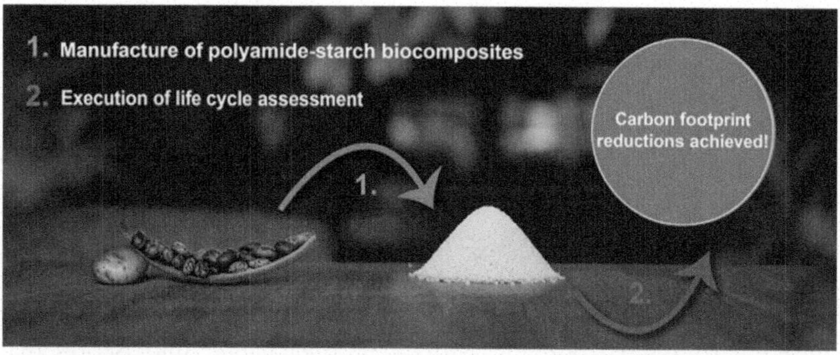

Figure 8.9 Cradle-to-gate life cycle assessment of polyamide-starch biocomposites (Äkräs et al., 2024).

8.7 Concluding remarks and future directions

Starch-based biopolymers have gained considerable attention as viable and sustainable alternatives to petroleum-derived polymers, largely due to their renewability, biodegradability, low toxicity, and broad availability across numerous agricultural systems. Their potential spans a wide spectrum of applications, including packaging, biomedical materials, coatings, and structural composites, positioning starch as a key contributor to the evolving bio-based materials market. However, despite the increasing global interest in environmentally responsible material design, qualitative statements on sustainability remain insufficient to fully demonstrate the environmental advantages of starch biopolymers. For this reason, the scientific community has increasingly emphasized the need for robust quantitative tools capable of objectively assessing environmental performance. Life Cycle Assessment (LCA) and carbon footprint analysis stand out as standardized, comprehensive, and scientifically validated methodologies for evaluating the environmental impacts of starch-based materials throughout their entire life cycle. These assessments encompass raw material cultivation, feedstock conversion, polymerization or modification processes, product use, and end-of-life options such as biodegradation, composting, or recycling. The literature examined in this chapter reveals that starch-based biopolymers frequently exhibit lower greenhouse gas emissions, reduced energy demand, and diminished ecological impacts compared with conventional fossil-based polymers. Nonetheless, the extent of these benefits is strongly influenced by multiple variables, including agricultural management practices, fertilizer and pesticide use, land and water requirements, processing efficiency, transportation logistics, and waste management strategies. By presenting a comprehensive review of life cycle data, this chapter provides a scientifically grounded understanding of the true environmental implications of employing starch as a bio-resource. Importantly, the discussion highlights the necessity of integrating LCA outcomes into material development strategies, regulatory frameworks, and industrial production systems. Such integration not only enhances the transparency and credibility of sustainability claims but also enables researchers and manufacturers to identify critical hotspots within supply chains, optimize resource use, and reduce environmental burdens. Moreover, the chapter underscores that the adoption of quantitative assessment methods is essential for guiding future innovation in biopolymer technology. As industries increasingly shift toward circular economy principles and low-carbon material solutions, LCA-guided design will play a central role in shaping next-generation starch biopolymers with superior environmental profiles. Ultimately, the widespread implementation of LCA and carbon footprint methodologies is expected to strengthen global efforts toward achieving sustainability targets, support evidence-based decision-making, and accelerate the transition toward more responsible, resilient, and environmentally compatible material systems.

References

Abel, B.A., Coates, G.W., 2025. Introduction: The future of plastics sustainability, 15206890 Chem. Rev. 125 (3), 1255–1256. https://doi.org/10.1021/acs.chemrev.4c00958.

Adewale, P., Yancheshmeh, M.S., Lam, E., 2022. Starch modification for non-food, industrial applications: Market intelligence and critical review, 01448617 Carbohydr. Polym. 291. https://doi.org/10.1016/j.carbpol.2022.119590.

Agarwal, S., Sethi, S., 2026. Carbon footprint analysis of traditional and innovative cellulosic textile fibers: A review. Renew. Sustain. Energy Rev. 226, 116208. https://doi.org/10.1016/j.rser.2025.116208, 13640321.

Agenda, I., January 2016. The new plastics economy rethinking the future of plastics. In: World Economic Forum, 36.

Aisyah, N., Khotimah, K., Madyaningarum, N., Wahyono, Y., Hakim, S.U.E., Sasongko, N.A., Nasruddin, N., 2025. Environmental and human health impact evaluation of nuclear power plant by using life cycle assessment (LCA) in Indonesia, 15567230 Energy Sources, Part A Recovery, Util. Environ. Eff. 47 (1), 11326–11350. https://doi.org/10.1080/15567036.2025.2501238.

Äkräs, L., Silvenius, F., Baniasadi, H., Vahvaselkä, M., Ilvesniemi, H., Seppälä, J., 2024. A cradle-to-gate life cycle assessment of polyamide-starch biocomposites: Carbon footprint as an indicator of sustainability. Clean Technol. Environ. Policy 26 (10), 3297–3312. https://doi.org/10.1007/s10098-024-02884-1, 1618-954X.

Ali, A., Russell, J.D., 2025. Accelerating the transition to wood-based circular bioeconomy: A literature review of current state, trends, opportunities, and priorities for future research, 21986436 Curr. For. Rep. 11 (1). https://doi.org/10.1007/s40725-025-00255-7.

Amaraweera, S.M., Gunathilake, C., Gunawardene, O.H.P., Fernando, N.M.L., Wanninayaka, D.B., Dassanayake, R.S., Rajapaksha, S.M., Manamperi, A., Fernando, C.A.N., Kulatunga, A.K., Manipura, A., 2021. Development of starch-based materials using current modification techniques and their applications: A review, 14203049 Molecules 26 (22). https://doi.org/10.3390/molecules26226880.

Amobonye, A., Bendoraitiene, J., Peciulyte, L., Rutkaite, R., 2025. Review of recent advancements in starch modification: Improving the functionality of starch-based films. Int. J. Biol. Macromol. 315 (Pt 2), 144354. https://doi.org/10.1016/j.ijbiomac.2025.144354, 01418130.

Amulya, K., Katakojwala, R., Ramakrishna, S., Venkata Mohan, S., 2021. Low carbon biodegradable polymer matrices for sustainable future, 26666820 Composites Part C: Open Access 4. https://doi.org/10.1016/j.jcomc.2021.100111.

Amulya, K., Verma, P., Sheikh, Z.U.D., Devi, A., Singh, A., Pathania, D., Mehta, P., Bhat, M.A., 2025. Comprehensive strategies for mitigating microplastic pollution: From detection to remediation, 13652486 Glob. Change Biol. 31 (10). https://doi.org/10.1111/gcb.70526.

Anastas, P., Eghbali, N., 2010. Green chemistry: principles and practice. Chem. Soc. Rev. 39 (1), 301–312. https://doi.org/10.1039/b918763b, 14604744.

Apriyanto, A., Compart, J., Fettke, J., 2022. A review of starch, a unique biopolymer – structure, metabolism and in planta modifications. Plant Sci. 318, 111223. https://doi.org/10.1016/j.plantsci.2022.111223, 01689452.

Arruda, T.R., Machado, G.d.O., Marques, C.S., Souza, A.L.d., Pelissari, F.M., Oliveira, T. V.d., Silva, R.R.A., 2025. An overview of starch-based materials for sustainable food packaging: Recent advances, limitations, and perspectives, 26736209 Macromol 5 (2). https://doi.org/10.3390/macromol5020019.

Balaji, A.B., Pakalapati, H., Khalid, M., Walvekar, R., Siddiqui, H., 2017. Natural and synthetic biocompatible and biodegradable polymers biodegradable and biocompatible polymer composites: Processing, properties and applications. Elsevier, Malaysia, pp. 3–32. https://doi.org/10.1016/B978-0-08-100970-3.00001-8.

Baltrocchi, A.P.D., Carnevale Miino, M., Maggi, L., Rada, E.C., Torretta, V., 2025. A comprehensive critical review of Life Cycle Assessment applied to thermoplastic polymers for mechanical and electronic engineering, 21622523 Environ. Technol. Rev. 14 (1), 458–470. https://doi.org/10.1080/21622515.2025.2491632.

Bangar, S.P., Chaudhary, V., Khubber, S., Whiteside, W.S., 2025. Kaolinite-based nanocomposites for enhancing starch and other biodegradable polymer applications in food packaging, 18790003 Int. J. Biol. Macromol. 320. https://doi.org/10.1016/j.ijbiomac.2025.145889.

Baniasadi, H., Äkräs, L., Madani, Z., Silvenius, F., Fazeli, M., Lipponen, S., Vapaavuori, J., Seppälä, J., 2024. Development and characterization of polylactic acid/starch biocomposites – From melt blending to preliminary life cycle assessment, 01418130 Int. J. Biol. Macromol. 279 (Pt 1), 135173. https://doi.org/10.1016/j.ijbiomac.2024.135173.

Barahmand, Z., Eikeland, M.S., 2022. Life cycle assessment under uncertainty: A scoping review. World 3 (3), 692–717. https://doi.org/10.3390/world3030039, 2673-4060.

Barletta, M., Genovesi, A., Desole, M.P., Gisario, A., 2025. Melt processing of biodegradable poly(butylene succinate) (PBS)—A critical review, 16189558 Clean Technol. Environ. Policy 27 (2), 683–725. https://doi.org/10.1007/s10098-024-03005-8.

Beach, E.S., Cui, Z., Anastas, P.T., 2009. Green Chemistry: A design framework for sustainability, 17545706 Energy Environ. Sci. 2 (10), 1038–1049. https://doi.org/10.1039/b904997p.

Bharadwaj, G., Chopde, S., Taluja, R., Lalitha, G., Chandrashekar, R., Ali Dhahi, H., Saxena, K.K., 2023. Closing the loop: Advances in materials, energy, and waste management. E3S Web Conf. 453, 01024. https://doi.org/10.1051/e3sconf/202345301024, 2267-1242.

Blanco, J., Iglesias, J., Morales, G., Melero, J.A., Moreno, J., 2020. Comparative life cycle assessment of glucose production from maize starch and woody biomass residues as a feedstock. Appl. Sci. 10 (8), 2946. https://doi.org/10.3390/app10082946, 2076-3417.

Broeren, M.L.M., Kuling, L., Worrell, E., Shen, L., 2017. Environmental impact assessment of six starch plastics focusing on wastewater-derived starch and additives. Resour. Conserv. Recycl. 127, 246–255. https://doi.org/10.1016/j.resconrec.2017.09.001, 09213449.

Bumanis, G., Vitola, L., Pundiene, I., Sinka, M., Bajare, D., 2020. Gypsum, geopolymers, and starch—alternative binders for bio-based building materials: A review and life-cycle assessment. Sustainability 12 (14), 5666. https://doi.org/10.3390/su12145666, 2071-1050.

Cazan, C., 2022. Advances in sustainable polymeric materials. Polymers 14 (22), 4972. https://doi.org/10.3390/polym14224972, 2073-4360.

Cerchione, R., Morelli, M., Passaro, R., Quinto, I., 2025. A critical analysis of the integration of life cycle methods and quantitative methods for sustainability assessment. Corp. Soc. Responsib. Environ. Manag. 32 (2), 1508–1544. https://doi.org/10.1002/csr.3010, 1535-3958.

Cerdan, K., Thys, M., Costa Cornellà, A., Demir, F., Norvez, S., Vendamme, R., Van den Brande, N., Van Puyvelde, P., Brancart, J., 2024. Sustainability of self-healing polymers: A holistic perspective towards circularity in polymer networks. Prog. Polym. Sci. 152, 101816. https://doi.org/10.1016/j.progpolymsci.2024.101816, 00796700.

Chalannavar, R.K., Divakar, M.S., Malabadi, R.B., Kamble, A.A., Swathi, Karamchand, K. S., Kolkar, K.P., Viviana Castaño Coronado, K., Neidilê Ribeiro Munhoz, A., Mammadova, S.S., 2025. Plant derived starch for the production of biodegradable plastic. Glob. J. Eng. Technol. Adv. 22 (3), 202–215. https://doi.org/10.30574/gjeta.2025.22.3.0058, 25825003.

Chen, Y., Rao, Y., Liu, P., Wu, L., Zhang, G., Zhang, J., Xie, F., 2024. High-amylose starch-based gel as green adhesive for plywood: Adhesive property, water-resistance, and flame-retardancy. Carbohydr. Polym. 339, 122247. https://doi.org/10.1016/j.carbpol.2024.122247, 01448617.

Coates, G.W., Getzler, Y.D.Y.L., 2020. Chemical recycling to monomer for an ideal, circular polymer economy, 20588437 Nat. Rev. Mater. 5 (7), 501–516. https://doi.org/10.1038/s41578-020-0190-4.

Compart, J., Singh, A., Fettke, J., Apriyanto, A., Starch Properties, C., 2023. A review of starch modifications and their applications. Polymers 15 (16), 3491. https://doi.org/10.3390/polym15163491, 2073-4360.

Cornago, S., Tan, Y.S., Brondi, C., Ramakrishna, S., Sze Choong Low, J., 2022. Systematic literature review on dynamic life cycle inventory: Towards industry 4.0 applications. Sustainability 14 (11), 6464. https://doi.org/10.3390/su14116464, 2071-1050.

D'Amato, D., Droste, N., Allen, B., Kettunen, M., Lähtinen, K., Korhonen, J., Leskinen, P., Matthies, B.D., Toppinen, A., 2017. Green, circular, bio economy: A comparative analysis of sustainability avenues. J. Clean. Prod. 168, 716–734. https://doi.org/10.1016/j.jclepro.2017.09.053, 09596526.

Dang, X., Han, S., Tang, J., Wang, X., 2025. Functional starch-based conductive hydrogel for flexible electronics: Design, construction, and applications. Aggregate 6 (10). https://doi.org/10.1002/agt2.70121, 2692-4560.

Das, S., Liang, C., Dunn, J.B., 2021. Life cycle assessment of polymers and their recycling. In: Circular economy of polymers: Topics in recycling technologies. American Chemical Society, pp. 143–170.

Das, A., Kn, C., Salins, S., Shetty, R., Shetty, S., 2025. Poly (butylene adipate-co-terephthalate)(PBAT) in biomedical applications: a comprehensive review of material properties, fabrication methods, and biofunctional potential. Mater. Res. Express 12, 062002.

de Carvalho, G.R., Marques, G.S., de Matos Jorge, L.M., Jorge, R.M.M., 2025. Application of cassava by-product as fillers in starch-based biocomposite films for packaging materials, 14362449 Polym. Bull. 82 (8), 3047–3062. https://doi.org/10.1007/s00289-025-05657-y.

Edo, G.I., Mafe, A.N., Ali, A.B.M., Akpoghelie, P.O., Yousif, E., Isoje, E.F., Igbuku, U.A., Zainulabdeen, K., Owheruo, J.O., Essaghah, A.E.A., Umar, H., Ahmed, D.S., Alamiery, A.A., 2025. Advancing sustainable food packaging: the role of green nanomaterials in enhancing barrier properties, 18667929 Food Eng. Rev. 17 (3), 573–607. https://doi.org/10.1007/s12393-025-09407-8.

Elwan, A.H., Tantawi, M.E., Fouda, A.M., 2025. Carbon footprint of private dental clinics in Egypt: A cross-sectional study, 14726831 BMC Oral Health 25 (1). https://doi.org/10.1186/s12903-024-05413-0.

Erdogdu, E., 2025. The carbon border adjustment mechanism: Opportunities and challenges for non-EU countries, 2041840X Wiley Interdiscip. Rev.: Energy Environ. 14 (1). https://doi.org/10.1002/wene.70000.

Erickson, E.D., Tominac, P.A., Ma, J., Aguirre-Villegas, H., Zavala, V.M., 2024. Evaluating the economic and environmental benefits of deploying a national-scale, thermochemical plastic waste upcycling infrastructure in the United States, 00981354 Comput. Chem. Eng. 189. https://doi.org/10.1016/j.compchemeng.2024.108800.

Erokhin, V., Bobryshev, A., Manzhosova, I., Frolov, A., Shamrina, S., Agafonova, N., 2025. Adaptation of life cycle costing practices to financial performance analysis in transitional economies: The experience of Russian manufacturing firms. J. Knowl. Econ. 16 (1), 1129–1152. https://doi.org/10.1007/s13132-024-02051-3, 1868-7873.

Falsafi, A., Togiani, A., Colley, A., Varis, J., Horttanainen, M., 2025. Life cycle assessment in circular design process: A systematic literature review. J. Clean. Prod. 521, 146188. https://doi.org/10.1016/j.jclepro.2025.146188, 09596526.

Falua, K.J., Babaei-Ghazvini, A., Acharya, B., 2025. Characterization and environmental footprint of crosslinked faba bean starch-based ecofilms reinforced with tunicate cellulose nanocrystals. Sustain. Mater. Technol. 43, e01236. https://doi.org/10.1016/j.susmat.2025.e01236, 22149937.

Fang, K., Pei, L., Huang, X., Wang, H., Li, Y., Zhu, D., Luo, B., 2025. Recent advancements in magnetic starch-based composites for biomedical applications: A review. Carbohydr. Polym. 362, 123689. https://doi.org/10.1016/j.carbpol.2025.123689, 01448617.

Ferreira-Filipe, D.A., Paço, A., Duarte, A.C., Rocha-Santos, T., Silva, A.L.P., 2021. Are biobased plastics green alternatives?—a critical review, 16604601 Int. J. Environ. Res. Publ. Health 18 (15). https://doi.org/10.3390/ijerph18157729.

Filippi, E., Niero, M., Corsini, F., Frey, M., 2025. Beyond the life cycle: assessing business models for the circular economy through life-cycle-based methodologies. Sustain. Prod. Consum. 60, 18–35. https://doi.org/10.1016/j.spc.2025.09.006, 23525509.

Forfora, N., Azuaje, I., Kanipe, T., Gonzalez, J.A., Lendewig, M., Urdaneta, I., Venditti, R., Gonzalez, R., Argyropoulos, D., 2024. Are starch-based materials more eco-friendly than fossil-based? A critical assessment. Clean. Environ. Syst. 13, 100177. https://doi.org/10.1016/j.cesys.2024.100177, 26667894.

Gandhi, V.C.S., Sreeraman, N., Surendiran, T., Kumaravelan, R., 2025. Enhancing sustainability in construction: Investigating starch/clay composites for green material development, 21906823 Biomass Convers. Biorefinery 15 (7), 9907–9914. https://doi.org/10.1007/s13399-024-05845-2.

Gangwar, A., Rawat, S., Rautela, A., Yadav, I., Singh, A., Kumar, S., 2025. Current advances in produced water treatment technologies: A perspective of techno-economic analysis and life cycle assessment. Environ. Dev. Sustain. 27 (7), 15077–15111. https://doi.org/10.1007/s10668-024-04558-w, 1573-2975.

Gutowski, T.G., Allwood, J.M., Herrmann, C., Sahni, S., 2013. A global assessment of manufacturing: Economic development, energy use, carbon emissions, and the potential for energy efficiency and materials recycling. Annu. Rev. Environ. Resour. 38, 81–106. https://doi.org/10.1146/annurev-environ-041112-110510, 15435938.

Haider, T.P., Völker, C., Kramm, J., Landfester, K., Wurm, F.R., 2019. Plastics of the future? The impact of biodegradable polymers on the environment and on society, 15213773 Angew. Chem. - Int. Ed. 58 (1), 50–62. https://doi.org/10.1002/anie.201805766.

Hamsan, M.H., Hama, P.O., Aziz, S.B., Hadi, J.M., Ahmed, M.B., Abdullah, R.M., Sadiq, N.M., Kadir, M.F.Z., 2025. Tailoring polymer electrolytes with PVA for improved performance in electrical double layer capacitors, 1226086X J. Ind. Eng. Chem. 146, 506–521. https://doi.org/10.1016/j.jiec.2024.11.035.

He, L., Zhang, T., Xie, W., Zhao, X., Wang, Z., He, Z., Yi, S., 2025a. Improving structural integrity and dynamics of lignocellulosic polymers in heat-treated wood with tung oil. Int. J. Biol. Macromol. 305 (Pt 1), 140949. https://doi.org/10.1016/j.ijbiomac.2025.140949, 01418130.

He, Q., Wu, X., Ding, X., 2025b. Uncertainty in the Carbon Footprint accounting and evaluation of textile and apparel products: A systematic review. J. Clean. Prod. 492, 144885. https://doi.org/10.1016/j.jclepro.2025.144885, 09596526.

Heller, M.C., Mazor, M.H., Keoleian, G.A., 2020. Plastics in the US: Toward a material flow characterization of production, markets and end of life, 17489326 Environ. Res. Lett. 15 (9). https://doi.org/10.1088/1748-9326/ab9e1e.

Hessel, V., Mukherjee, S., Mitra, S., Goswami, A., Tran, N.N., Ferlin, F., Vaccaro, L., Galogahi, F.M., Nguyen, N.T., Escribà-Gelonch, M., 2024. Sustainability of flow chemistry and microreaction technology, 14639270 Green Chem. 26 (18), 9503–9528. https://doi.org/10.1039/d4gc01882f.

Hong, S.J., Ahn, K., Kim, J.T., Kim, Y.T., 2025. Polyhydroxyalkanoates for sustainable food packaging: A comprehensive review on production, applications, and end-of-life scenarios for food packaging, 09242244 Trends Food Sci. Technol. 165. https://doi.org/10.1016/j.tifs.2025.105294.

Islam, M.S., Elahee, G.M.F., Fang, Y., Yu, X., Advincula, R.C., Cao, C., 2025. Polylactic acid (PLA)-based multifunctional and biodegradable nanocomposites and their applications, 13598368 Compos. B Eng. 306. https://doi.org/10.1016/j.compositesb.2025.112842.

Ita-Nagy, D., Vázquez-Rowe, I., Kahhat, R., Chinga-Carrasco, G., Quispe, I., 2020. Reviewing environmental life cycle impacts of biobased polymers: Current trends and methodological challenges. Int. J. Life Cycle Assess. 25 (11), 2169–2189. https://doi.org/10.1007/s11367-020-01829-2, 0948-3349.

Jiang, S., Liu, C., Wang, X., Xiong, L., Sun, Q., 2016. Physicochemical properties of starch nanocomposite films enhanced by self-assembled potato starch nanoparticles. LWT Food Sci. Technol. (Lebensm.-Wiss.-Technol.) 69, 251–257. https://doi.org/10.1016/j.lwt.2016.01.053, 00236438.

Kabir, E., Kaur, R., Lee, J., Kim, K.H., Kwon, E.E., 2020. Prospects of biopolymer technology as an alternative option for non-degradable plastics and sustainable management of plastic wastes, 09596526 J. Clean. Prod. 258. https://doi.org/10.1016/j.jclepro.2020.120536.

Kang, S., Guo, J., Makarov, I.S., Khan, M.R., Xiao, H., Song, J., 2025. Optimization of amylose to amylopectin ratio and degree of substitution in quaternized starch as a tool to improve paper strength. Starch - Stärke 77 (3). https://doi.org/10.1002/star.202400138, 0038-9056.

Kanna I, V., Roseline, S., Balamurugan, K., Jeeva, S., Augastin Santhiyagu, I., 2024. The effects of greenhouse gas emissions on global warming. Encycl. Renew. Energy, Sustain. Environ. 1–4 (1), 143–154. https://doi.org/10.1016/B978-0-323-93940-9.00216-4.

Kanwal, A., Sharaf, F., Zafar, T., Khurshaid, S.k., Farooq, S., 2025. Navigating solutions: a comprehensive review on plastic pollution in marine ecosystems and the vital role of biodegradable polymers, 14362449 Polym. Bull. 82 (15), 9623–9646. https://doi.org/10.1007/s00289-025-05938-6.

Kapoor, D.U., Pareek, A., Sharma, M., Prajapati, B.G., Suttiruengwong, S., Sriamornsak, P., 2025. Exploring starch-based excipients in pharmaceutical formulations: versatile applications and future perspectives, 18733441 Eur. J. Pharm. Biopharm. 212. https://doi.org/10.1016/j.ejpb.2025.114727.

Karan, H., Funk, C., Grabert, M., Oey, M., Hankamer, B., 2019. Green bioplastics as part of a circular bioeconomy. Trends Plant Sci. 24 (3), 237–249. https://doi.org/10.1016/j.tplants.2018.11.010, 13601385.

Karka, P., Johnsson, F., Papadokonstantakis, S., 2021. Perspectives for greening European fossil-fuel infrastructures through use of biomass: The case of liquid biofuels based on lignocellulosic resources. Front. Energy Res. 9. https://doi.org/10.3389/fenrg.2021.636782, 2296-598X.

Keya, K.N., Kona, N.A., Koly, F.A., Madina Maraz, K., Naimul Islam, Md, Khan, R.A., 2019. Natural fiber reinforced polymer composites: History, types, advantages, and applications. Mater. Eng. Res. 1 (2), 69–87. https://doi.org/10.25082/mer.2019.02.006, 26304503.

Kocaturk, E., Salan, T., Ozcelik, O., Alma, M.H., Candan, Z., 2023. Recent advances in lignin-based biofuel production, 19961073 Energies 16 (8). https://doi.org/10.3390/en16083382.

Kocatürk, E., Özgür Seydibeyoğlu, M., Zor, M., 2024a. Lignin-Containing Cellulose Nanofibrils (LCNF) for Polymer Nanocomposites. Elsevier BV. https://doi.org/10.1016/b978-0-323-95486-0.00068-5.

Kocatürk, E., Şen, F., Zor, M., Candan, Z., 2024b. Development and characterization of nanocellulose/carbonized waste rubber nanocomposites. Bioresources 19 (2), 2670.

Kogje, M., Satdive, A., Mestry, S., Mhaske, S.T., 2025. Biopolymers: A comprehensive review of sustainability, environmental impact, and lifecycle analysis. Iran. Polym. J. 34 (9), 1481–1524. https://doi.org/10.1007/s13726-024-01449-9, 1026-1265.

Korley, L.T.J., McNeil, A.J., Coates, G.W., 2022. Challenges and opportunities in sustainable polymers, 15204898 Acc. Chem. Res. 55 (18), 2543–2544. https://doi.org/10.1021/acs.accounts.2c00534.

Kristensen, S.K., Ahrens, A., Donslund, B.S., Skrydstrup, T., 2024. Perspective on the development of monomer recovery technologies from plastics designed to last. ACS Org. Inorg. Au 4 (4), 373–386. https://doi.org/10.1021/acsorginorgau.4c00009, 2694-247X.

Kumar, A., Thakur, V.K., Nezhad, H.Y., Lee, K.S., 2024. Prospects of sustainable polymers, 20452322 Sci. Rep. 14 (1). https://doi.org/10.1038/s41598-024-59439-z.

Kurul, F., Doruk, B., Nur Topkaya, S., 2025. Principles of green chemistry: Building a sustainable future. Discover Chemistry 2 (1). https://doi.org/10.1007/s44371-025-00152-9, 3005-1193.

LeCorre, D., Hohenthal, C., Dufresne, A., Bras, J., 2013. Comparative sustainability assessment of starch nanocrystals. J. Polym. Environ. 21 (1), 71–80. https://doi.org/10.1007/s10924-012-0447-0, 1566-2543.

Limeneh, D.Y., Tesema, A.F., Rajan, K., Yilma, K.T., 2025. Comparing the cost-effectiveness and yarn quality improvement with PVA versus conventional maize starch as the sizing agents, 17542340 J. Textil. Inst. 116 (6), 1002–1009. https://doi.org/10.1080/00405000.2024.2364126.

Lucas, K.R.G., Caldarelli, C.E., Ventura, M.U., Tang, L., Hayashi, K., Yoshikawa, N., 2025. Mapping life cycle assessment (LCA) scientific research in agriculture: what do we still have to do?, 21945411 Environ. Syst. Decis. 45 (1). https://doi.org/10.1007/s10669-024-09997-4.

Luo, Y., Ye, M., Zhou, Y., Su, R., Huang, S., Wang, H., Dai, X., 2024. Assessing the environmental impact of municipal waste on energy incineration technology for power generation using life cycle assessment methodology. Toxics 12 (11), 786. https://doi.org/10.3390/toxics12110786, 2305-6304.

Mahfouzi, M., Zhang, H., Haoran, L., McClements, D.J., Hadidi, M., 2025. Starch-based particles as stabilizers for Pickering emulsions: Modification, characteristics, stabilization, and applications, 15497852 Crit. Rev. Food Sci. Nutr. 65 (10), 1841–1856. https://doi.org/10.1080/10408398.2024.2312285.

Maniglia, B.C., Castanha, N., Le-Bail, P., Le-Bail, A., Augusto, P.E.D., 2021. Starch modification through environmentally friendly alternatives: a review, 15497852 Crit. Rev. Food Sci. Nutr. 61 (15), 2482–2505. https://doi.org/10.1080/10408398.2020.1778633.

Marta, H., Rizki, D.I., Mardawati, E., Djali, M., Mohammad, M., Cahyana, Y., Nanoparticles, S., 2023. Preparation, properties and applications. Polymers 15 (5), 1167. https://doi.org/10.3390/polym15051167, 2073-4360.

Mata, T.M., Freitas, C., Silva, G.V., Monteiro, S., Martins, J.M., de Carvalho, L.H., Silva, L.M., Martins, A.A., 2023. Life cycle analysis of a particleboard based on cardoon and starch/chitosan, 20711050 Sustainability 15 (23). https://doi.org/10.3390/su152316179.

Mathur, S., Waswani, H., Singh, D., Ranjan, R., 2022. Alternative fuels for agriculture sustainability: Carbon footprint and economic feasibility. AgriEngineering 4 (4), 993–1015. https://doi.org/10.3390/agriengineering4040063, 2624-7402.

Miao, M., BeMiller, J.N., 2025. An Overview of Starch Handbook of Starch Science and Technology. CRC Press, China, pp. 1–6. https://doi.org/10.1201/9781003464396-1.

Miculescu, F., Maidaniuc, A., Voicu, S.I., Thakur, V.K., Stan, G.E., Ciocan, L.T., 2017. Progress in hydroxyapatite-starch based sustainable biomaterials for biomedical bone substitution applications, 21680485 ACS Sustain. Chem. Eng. 5 (10), 8491–8512. https://doi.org/10.1021/acssuschemeng.7b02314.

Miller, S.A., 2013. Sustainable polymers: opportunities for the next decade, 21611653 ACS Macro Lett. 2 (6), 550–554. https://doi.org/10.1021/mz400207g.

Nabipour Afrouzi, H., Ahmed, J., Mobin Siddique, B., Khairuddin, N., Hassan, A., 2023. A comprehensive review on carbon footprint of regular diet and ways to improving lowered emissions. Results Eng. 18, 101054. https://doi.org/10.1016/j.rineng.2023.101054, 25901230.

Nazrin, A., Ilyas, R.A., Rajeshkumar, L., Hazrati, K.Z., Jamal, T., Mahardika, M., Aisyah, H.A., Atiqah, A., Radzi, A.M., 2025. Lignocellulosic fiber-reinforced starch thermoplastic composites for food packaging application: A review of properties and food packaging abetted with safety aspects. Food Packag. Shelf Life 47, 101431. https://doi.org/10.1016/j.fpsl.2025.101431, 22142894.

Nieto, M.J., 2022. Whatever it takes to reach net zero emissions around 2050 and limit global warming to 1.5c: the cases of United States, China, European union and Japan. SSRN Electron. J. https://doi.org/10.2139/ssrn.4007974, 1556-5068.

Norris, G.A., 2002. Impact characterization in the tool for the reduction and assessment of chemical and other environmental impacts: Methods for acidification, eutrophication, and ozone formation, 10881980 J. Ind. Ecol. 6 (3–4), 79–101. https://doi.org/10.1162/108819802766269548.

Obaid, A., Parvez Mahmud, M.A., Tabraiz, S., 2025. Life cycle impact assessment of hydrogen production from plastic waste polymers. Clean Technol. Environ. Policy 27 (11), 6809–6825. https://doi.org/10.1007/s10098-025-03253-2, 1618-954X.

Olatunji, O., 2016. Natural polymers. Springer International Publishing., Cham, Switzerland, pp. 1–17.
Olivato, J.B., 2024. Starch: a natural, safe, and multifunctional ingredient for cosmetic formulations. In: Starch Industries: Processes and Innovative Products in Food and Non-Food Uses. Academic Press, pp. 255–269.
Oliver-Cuenca, V., Salaris, V., Muñoz-Gimena, P.F., Agüero, Á., Peltzer, M.A., Montero, V.A., Arrieta, M.P., Sempere-Torregrosa, J., Pavon, C., Samper, M.D., Crespo, G.R., Kenny, J.M., López, D., Peponi, L., 2024. Bio-based and biodegradable polymeric materials for a circular economy, 20734360 Polymers 16 (21). https://doi.org/10.3390/polym16213015.
Pandey, D., Agrawal, M., Pandey, J.S., 2011. Carbon footprint: Current methods of estimation. Environ. Monit. Assess. 178 (1–4), 135–160. https://doi.org/10.1007/s10661-010-1678-y, 15732959.
Papageorgiou, G.Z., 2018. Thinking green: Sustainable polymers from renewable resources, 20734360 Polymers 10 (9). https://doi.org/10.3390/polym10090952.
Pasqualino, J., Meneses, M., Castells, F., 2011. The carbon footprint and energy consumption of beverage packaging selection and disposal. J. Food Eng. 103 (4), 357–365. https://doi.org/10.1016/j.jfoodeng.2010.11.005, 02608774.
Pathania, R., Pritam, P., Guleria, R., *Life Cycle Assessment of Microbial Biofuel: Environmental and Economic Impacts Biowaste to Biofuel: A Microbial Approach*. CRC Press, United States, 275–307, https://doi.org/10.1201/9781003633709-11.
Paudel, D., Chandra Neupane, R., Sigdel, S., Poudel, P., Khanal, A.R., 2023. COVID-19 pandemic, climate change, and conflicts on agriculture: A trio of challenges to global food security. Sustainability 15 (10), 8280. https://doi.org/10.3390/su15108280, 2071-1050.
Punia Bangar, S., Ashogbon, A.O., Singh, A., Chaudhary, V., Whiteside, W.S., 2022. Enzymatic modification of starch: A green approach for starch applications, 01448617 Carbohydr. Polym. 287. https://doi.org/10.1016/j.carbpol.2022.119265.
Rabeie, B., Mahmoodi, N.M., 2024. Green and environmentally friendly architecture of starch-based ternary magnetic biocomposite (Starch/MIL100/CoFe2O4): synthesis and photocatalytic degradation of tetracycline and dye. Int. J. Biol. Macromol. 274 (Pt 1), 133318. https://doi.org/10.1016/j.ijbiomac.2024.133318, 01418130.
Radhakrishnan, A., Panicker, U.G., 2025. Sustainable chitosan-aminated starch blends as innovative materials for advanced optical applications, 15728919 J. Polym. Environ. 33 (2), 1047–1068. https://doi.org/10.1007/s10924-024-03462-8.
Radu, I.C., Vadureanu, A.M., Cozorici, D.E., Blanzeanu, E., Zaharia, C., 2025. Advancing sustainability in modern polymer processing: strategies for waste resource recovery and circular economy integration, 20734360 Polymers 17 (4). https://doi.org/10.3390/polym17040522.
Rahmadi Putri, T., Adhitasari, A., Paramita, V., Endy Yulianto, M., Dwi Ariyanto, H., 2023 1. Effect of different starch on the characteristics of edible film as functional packaging in fresh meat or meat products: a review, 2023/01 Mater. Today Proc. 87. https://doi.org/10.1016/j.matpr.2023.02.396, 22147853 192 199.
Ramesh, P., Vinodh, S., 2020. State of art review on life cycle assessment of polymers. Int. J. Sustain. Eng. 13 (6), 411–422. https://doi.org/10.1080/19397038.2020.1802623, 1939-7038.
Rashid, M.A.M., Kim, K.H., Jeong, K., 2025. Microwave-assisted depolymerization of natural and synthetic polymers. In: Depolymerization: Concepts, Progress, and Challenges Volume 3: Emerging Trends. American Chemical Society, pp. 15–31.
Rashwan, A.K., Younis, H.A., Abdelshafy, A.M., Osman, A.I., Eletmany, M.R., Hafouda, M.A., Chen, W., 2024. Plant starch extraction, modification, and green

applications: a review, 16103661 Environ. Chem. Lett. 22 (5), 2483–2530. https://doi.org/10.1007/s10311-024-01753-z.

Ren, Y., Yuan, T.Z., Chigwedere, C.M., Ai, Y., 2021. A current review of structure, functional properties, and industrial applications of pulse starches for value-added utilization, 15414337 Compr. Rev. Food Sci. Food Saf. 20 (3), 3061–3092. https://doi.org/10.1111/1541-4337.12735.

Rus, T., Moldovan, R.-P., Beu, D., Pop, M., Ceclan, A., 2025. Towards a climate-neutral campus: carbon footprint assessment in higher education institutions. Appl. Sci. 15 (7), 3695. https://doi.org/10.3390/app15073695, 2076-3417.

Saishree, R., Das, S.K., Panda, N., Prusty, B.A.K., Mahata, C., 2025. From plastics to bioplastics: technological innovations and nature based solutions to achieve global sustainability. In: Micro-Nano Plastics Exposure, Environmental Degradation and Public Health Crisis: Perspectives and Concerns for Sustainable Development. Springer Nature Singapore, Singapore, pp. 257–268.

Samir, A., Ashour, F.H., Hakim, A.A.A., Bassyouni, M., 2022. Recent advances in biodegradable polymers for sustainable applications, 23972106 npj Mater. Degrad. 6 (1). https://doi.org/10.1038/s41529-022-00277-7.

Schirmeister, C.G., Mülhaupt, R., 2022. Closing the carbon loop in the circular plastics economy, 15213927 Macromol. Rapid Commun. 43 (13). https://doi.org/10.1002/marc.202200247.

Schmidt, H.-J., 2009. Carbon footprinting, labelling and life cycle assessment. Int. J. Life Cycle Assess. 14 (S1), 6–9. https://doi.org/10.1007/s11367-009-0071-y, 0948-3349.

Seeponkai, N., Pundi, S., Fuggate, P., Kanla, S., Khammuang, K., Seephonkai, P., 2024. StarchCrete: a starch-based biocomposite concrete for lightweight building material applications. Ind. Crop. Prod. 222, 119991. https://doi.org/10.1016/j.indcrop.2024.119991, 09266690.

Seguí, L., Barrera, C., 2025. Functional ingredients from food waste and by-products: Processing technologies, functional characteristics and value-added applications, 23048158 Foods 14 (5). https://doi.org/10.3390/foods14050847.

Şen, F., Kocatürk, E., Çakmakçı, E., Vezir Kahraman, M., 2022. Quaternary imidazolium-functionalized reactive silica nanoparticles-containing thiol-ene photocured antibacterial hybrid coatings. React. Funct. Polym. 170, 105149. https://doi.org/10.1016/j.reactfunctpolym.2021.105149, 13815148.

Serra-Parareda, F., Delgado-Aguilar, M., Espinach, F.X., Mutjé, P., Boufi, S., Tarrés, Q., 2022. Sustainable plastic composites by polylactic acid-starch blends and bleached kraft hardwood fibers. Compos. B Eng. 238, 109901. https://doi.org/10.1016/j.compositesb.2022.109901, 13598368.

Sevigné-Itoiz, E., Mwabonje, O., Panoutsou, C., Woods, J., 2021. Life cycle assessment (LCA): informing the development of a sustainable circular bioeconomy? Philos. Trans. R. Soc. A Math. Phys. Eng. Sci. 379 (2206). https://doi.org/10.1098/rsta.2020.0352, 1364-503X.

Shabir, I., Dash, K.K., Dar, A.H., Pandey, V.K., Fayaz, U., Srivastava, S., 2023. Carbon footprints evaluation for sustainable food processing system development: A comprehensive review, 26668335 Future Foods 7. https://doi.org/10.1016/j.fufo.2023.100215.

Shan, Y., Li, J., Nie, M., Li, D., Zhang, Y., Li, Y., Wang, L., Liu, L., Wang, F., Tong, L.-T., 2025. A comprehensive review of starch-based technology for encapsulation of flavor: From methods, materials, and release mechanism to applications. Carbohydr. Polym. 348 (Pt A), 122816. https://doi.org/10.1016/j.carbpol.2024.122816, 01448617.

Shen, L., Patel, M.K., 2008. Life cycle assessment of polysaccharide materials: A review. J. Polym. Environ. 16 (2), 154–167. https://doi.org/10.1007/s10924-008-0092-9, 15662543.

Si, W., Zhang, S., 2024. The green manufacturing of thermoplastic starch for low-carbon and sustainable energy applications: a review on its progress. Green Chem. 26 (3), 1194–1222. https://doi.org/10.1039/d3gc03671e, 1463-9262.

Singh, P., Pandey, V.K., Singh, R., Singh, K., Dash, K.K., Malik, S., 2024. Unveiling the potential of starch-blended biodegradable polymers for substantializing the eco-friendly innovations, 26661543 J. Agric. Food Res. 15. https://doi.org/10.1016/j.jafr.2024.101065.

Skrzypczak, D., Gorazda, K., Mikula, K., Mironiuk, M., Kominko, H., Sawska, K., Evrard, D., Trzaska, K., Moustakas, K., Chojnacka, K., 2025. Towards carbon neutrality: enhancing CO2 sequestration by plants to reduce carbon footprint. Sci. Total Environ. 966, 178763. https://doi.org/10.1016/j.scitotenv.2025.178763, 00489697.

Sobeih, M.O., Sawalha, S., Hamed, R., Ali, F., Kim, M.P., 2025. Starch-derived bioplastics: Pioneering sustainable solutions for industrial use, 19961944 Materials 18 (8). https://doi.org/10.3390/ma18081762.

Soeder, D.J., 2025. Greenhouse gas and climate change. In: Energy futures: The story of fossil fuel, greenhouse gas, and climate change. Springer Nature Switzerland, Cham, pp. 97–141.

Srivastava, R.K., Sarangi, P.K., Singh, A.K., Parveen, A., Sahoo, U.K., Padil, V.V.T., Abass, K.S., Jumaniyozov, K., Onn, C.W., Kurniawan, T.A., 2025. Innovative recycling strategies for non-recycled plastics: advancing the circular economy for a sustainable future, 27538125 RSC Sustain. 3 (12), 5433–5458. https://doi.org/10.1039/d5su00421g.

Suhagia, T.A., Ayare, P., Alle, M., 2025. Biodegradable beauty: Carbohydrate polymers in cosmetics. In: Carbohydrate Polymer Nanotechnologies: Design, Development, Utilization and Perspectives. Springer Nature Singapore, Singapore, pp. 455–483.

Surendra Babu, A., Parimalavalli, R., Mohan, R.J., 2018. Effect of modified starch from sweet potato as a fat replacer on the quality of reduced fat ice creams. J. Food Meas. Char. 12 (4), 2426–2434. https://doi.org/10.1007/s11694-018-9859-4, 2193-4126.

Surendren, A., Hasan, Y., Mohanty, A.K., Abbassi, B., Misra, M., 2024. Comparative environmental life cycle assessment on corn starch plasticization and Co-plasticization processes. Sustainability 16 (17), 7406. https://doi.org/10.3390/su16177406, 2071-1050.

Téllez-Morales, J.A., Rodríguez-Miranda, J., 2025. Improved extrusion cooking technology: A mini review of starch modification, 15428044 J. Culin. Sci. Technol. 23 (1), 1–10. https://doi.org/10.1080/15428052.2022.2163952.

Thebuwena, A.C.H.J., Samindi, S.M., Samarakoon, M.K., Chandima Ratnayake, R.M., 2025. Assessing the economic and environmental impact of glazed façades in tropical office buildings: A case study using life cycle costing. J. Build. Eng. 108, 112911. https://doi.org/10.1016/j.jobe.2025.112911, 23527102.

Tourtelot, J., d'Espinose de Lacaillerie, J.B., Duc, M., Mertz, J.D., Bourgès, A., Keita, E., 2023. Strengthening mechanisms of clay building materials by starch, 09500618 Constr. Build. Mater. 405. https://doi.org/10.1016/j.conbuildmat.2023.133215.

Trache, D., Tarchoun, A.F., Derradji, M., Hamidon, T.S., Masruchin, N., Brosse, N., Hussin, M.H., 2020. Nanocellulose: From Fundamentals to Advanced Applications, 22962646 Front. Chem. 8. https://doi.org/10.3389/fchem.2020.00392.

Trisopon, K., Kittipongpatana, N., Doungsaard, P., Chomchoei, N., Kittipongpatana, O.S., 2025. A novel directly compressible co-processed excipient, based-on rice starch for extended-release of tablets. Eur. J. Pharm. Biopharm. 208, 114623. https://doi.org/10.1016/j.ejpb.2024.114623, 09396411.

Usman, M., Husnain, M., Riaz, A., Riaz, A., Ali, Y., 2021. Climate change during the COVID-19 outbreak: Scoping future perspectives. Environ. Sci. Pollut. Control Ser. 28 (35), 49302–49313. https://doi.org/10.1007/s11356-021-14088-x, 0944-1344.

Usubharatana, P., Phungrassami, H., 2015. Carbon footprint of cassava starch production in north-eastern Thailand. Proced. CIRP 29, 462–467. https://doi.org/10.1016/j.procir.2015.02.031, 22128271.

Verma, J., Petru, M., Goel, S., 2024. Cellulose based materials to accelerate the transition towards sustainability. Ind. Crop. Prod. 210, 118078. https://doi.org/10.1016/j.indcrop.2024.118078, 09266690.

Wang, A., Duan, Z., Qin, S., Shen, X., Liu, Q., Li, D., Deng, B., 2025. Effect of processing conditions on the structure and properties of poly(trimethylene terephthalate) fibers and nonwovens produced in the spunbonding process. Fibers Polym. 26 (3), 1049–1060. https://doi.org/10.1007/s12221-025-00876-6, 1229-9197.

Waniek, K., Kappe, C.O., Baumgartner, R.J., 2025. From qualitative analysis to quantitative insights: A systematic review of early phase sustainability assessments of chemical processes, 14639270 Green Chem. 27 (36), 10944–10968. https://doi.org/10.1039/d5gc02565f.

Wiedenhofer, D., Wieland, H., Leipold, S., Aoki-Suzuki, C., Watari, T., Aguilar-Hernandez, G.A., Graf, S., Edelenbosch, O.Y., Zanon-Zotin, M., Kaufmann, L., Fortes, P., Haas, W., Streeck, J., 2025. The circular economy and climate change: The state of national and global evidence on mitigation potential, 15435938 Annu. Rev. Environ. Resour. 50 (1), 563–592. https://doi.org/10.1146/annurev-environ-111523-102441.

Wright, L.A., Kemp, S., Williams, I., 2011. Carbon footprinting': Towards a universally accepted definition. Carbon Manag. 2 (1), 61–72. https://doi.org/10.4155/cmt.10.39, 17583012.

Xiang, H., Li, Z., Liu, H., Chen, T., Zhou, H., Huang, W., 2022. Green flexible electronics based on starch. Npj Flexible Electronics 6 (1). https://doi.org/10.1038/s41528-022-00147-x, 2397-4621.

Yadav Prakruthi, L., Krishnan, H., Medha, T., Kumarakuru, K., Vasantha Kumari, P., Surekha, C., Alavilli, H., Kaushik, D., Hashem, A., Alotaibi, N.H., Avila-Quezada, G.D., Abd_Allah, E.F., Kumar, M., Reddy, C.K., 2025. Enhancing starch properties through dual modification: Ultrasonication and acetic acid treatment of non-conventional starches. Ultrason. Sonochem. 115, 107301. https://doi.org/10.1016/j.ultsonch.2025.107301, 13504177.

Yan, X., McClements, D.J., Luo, S., Liu, C., Ye, J., 2024. Recent advances in the impact of gelatinization degree on starch: Structure, properties and applications. Carbohydr. Polym. 340, 122273. https://doi.org/10.1016/j.carbpol.2024.122273, 01448617.

Yusuf, M.A., Romli, M., Suprihatin, E.I.W., 2019. Potential of traditional sago starch: Life cycle assessment (LCA) perspective. IOP Conf. Ser. Mater. Sci. Eng. 507 (1), 012014. https://doi.org/10.1088/1757-899x/507/1/012014, 1757-899X.

Zamir, S.S., Sadeghi, A., Razavi, S.M.A., 2025. Dual non-thermal physical and chemical modification methods of starch: Unlocking new potentials for high-performance polymer composites. Starch - Stärke 77 (7). https://doi.org/10.1002/star.70026, 0038-9056.

Zeng, Y., Wang, Y., Zhan, L., Shi, Z., Xiong, C., Yang, Q., 2025. High-performance chitin-based carbon aerogels with mesoporous structure for electromagnetic interference shielding. Diam. Relat. Mater. 154, 112198. https://doi.org/10.1016/j.diamond.2025.112198, 09259635.

Zhang, X., Zhang, H., Zhou, G., Su, Z., Wang, X., 2024. Flexible, thermal processable, self-healing, and fully bio-based starch plastics by constructing dynamic imine network. Green Energy Environ. 9 (10), 1610–1618. https://doi.org/10.1016/j.gee.2023.08.002, 24680257.

Zhang, X., Zhu, H., Chen, D., Miao, Y., Xiao, P., 2025. Full life cycle carbon emission accounting of the power system: A case study of a typical 500 kV substation. Carbon Neutraliz. 4 (4). https://doi.org/10.1002/cnl2.70025, 2769-3333.

Zhao, X., Mei, T., Cui, B., 2024. Lipids-modified starch: Advances in structural characteristic, physicochemical property, and application. Food Res. Int. 197 (Pt 1), 115146. https://doi.org/10.1016/j.foodres.2024.115146, 09639969.

Zheng, Y., Li, Z., Chai, J., 2023. Progress and prospects of international carbon peaking and carbon neutral research –based on bibliometric analysis (1991–2022). Front. Energy Res. 11. https://doi.org/10.3389/fenrg.2023.1121639, 2296-598X.

Zhiguang, C., Zhong, H., Chen, M., Fayong, G., Jing, L., 2025. The fine structure of starch: a review. Npj Science of Food 9 (1). https://doi.org/10.1038/s41538-025-00414-x, 2396-8370.

Zhu, Y., Romain, C., Williams, C.K., 2016. Sustainable polymers from renewable resources, 14764687 Nature 540 (7633), 354–362. https://doi.org/10.1038/nature21001.

Zieher, J., Stingl, R., Veigel, S., Gindl-Altmutter, W., Beaumont, M., Rosenau, T., Hettegger, H., 2025. Environmentally friendly approaches for the tuning of cellulosic paper properties through polysaccharide-reinforced silica coatings. Cellulose 32 (18), 10811–10830. https://doi.org/10.1007/s10570-025-06826-y, 0969-0239.

Zikulnig, J., Carrara, S., Kosel, J., 2025. A life cycle assessment approach to minimize environmental impact for sustainable printed sensors. Sci. Rep. 15 (1). https://doi.org/10.1038/s41598-025-95682-8, 2045-2322.

Zor, M., Mengeloglu, F., Aydemir, D., Sen, F., Kocatürk, E., Candan, Z., Ozcelik, O., 2022. Wood Plastic Composites (WPCs): Applications of Nanomaterials Emerging Nanomaterials: Opportunities and Challenges in Forestry Sectors. Springer International Publishing, Turkey, pp. 97–133. https://doi.org/10.1007/978-3-031-17378-3_4.

Zor, M., Kocatürk, E., Şen, F., Oran, B., Candan, Z., 2024. Characterization of nanocellulose/pyrolysis oil nanocomposite films, 0718221X Maderas Cienc. Tecnol. 26. https://doi.org/10.22320/s0718221x/2024.25.

CHAPTER NINE

Global initiatives and collaborations, regulation and policies, consumer awareness and industry uptake

Özlem Kübra Akdoğan[1, 2]
[1]Rutgers, The State University of New Jersey, Global Food Technology and Innovation, New Brunswick, NJ, United States
[2]Eastern Michigan University, College of Technology, Polymers and Coatings Technology, Ypsilanti, MI, United States

9.1 Global market outlook

Across the globe, the packaging industry is shifting as regulators tighten standards and everyday consumers increasingly favor sustainable options. Breakthroughs in polymers and alternative high-performance materials are accelerating that shift. As starch is the planet's second-most abundant polymer after cellulose, it is attracting interest as a natural, widely available option that can ease regulatory pressure tied to fossil-based materials while answering public demand for more environmentally responsible packaging.

The worldwide starch market is valued at USD 65.02 billion in 2025 and is projected to reach USD 95.85 billion by 2032, advancing at a compound annual growth rate of 5.7% (Maximize Market Research, 2026). Within this broader landscape, the global food starch market specifically is valued at USD 24.82 billion in 2024 and is forecast to reach USD 26.21 billion by 2025 to USD 40.53 billion by 2033, advancing at a 5.6% CAGR (Sharma, 2025). Simultaneously, starch-based packaging represents one of the fastest-growing segments within the sustainable materials, valued at USD 8.2 billion in 2025 and projected to reach USD 15.2 billion by 2035, growing at a steady compound annual growth rate of 6.4% (Sutaria, 2025). This convergence of market growth, regulatory pressure, consumer demand for clean ingredients, and rising environmental consciousness is reshaping how the industry operates across production, distribution, and end-of-life management systems.

Starch-Based Sustainable Food Packaging
ISBN: 978-0-443-36346-7
https://doi.org/10.1016/B978-0-443-36346-7.00004-7

This chapter argues that the future of starch-based packaging will not be determined by polymer performance alone. Instead, adoption and scale will be governed by the interaction of four independent forces: (1) cross-sector collaborations and global initiatives, (2) policy and regulatory framework, (3) industry readiness and market dynamics, and (4) consumer awareness and trust (Fig. 9.1).

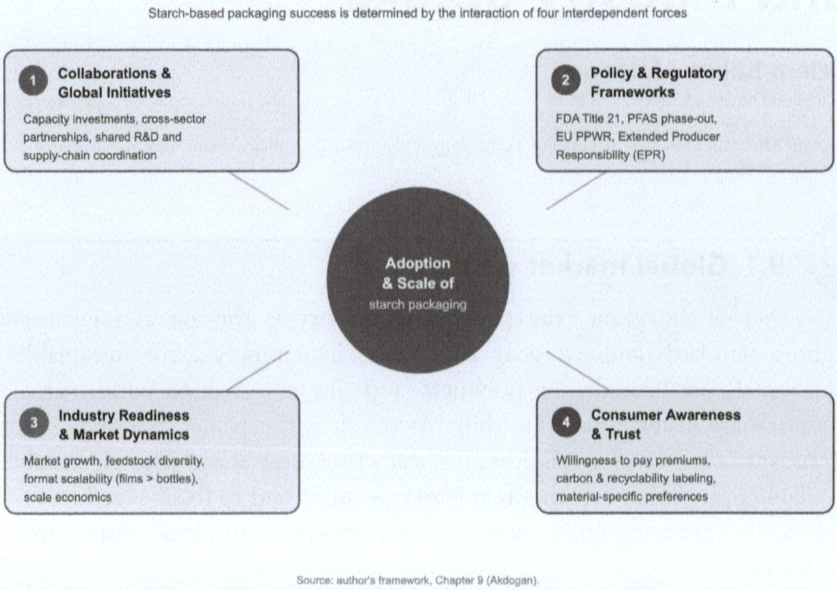

Figure 9.1 ***The four forces governing adoption and scale.*** Starch-based packaging success is determined by the interaction four interdependent forces. *Author's framework, Chapter 9 (Akdogan).*

9.2 Global market dynamics and strategic initiatives

9.2.1 Market growth and regional distribution

The global starch industry demonstrates strong geographic and segmental diversity. North America held 31.55% of global food starch revenue in 2025, while Asia–Pacific is forecast to record the strongest 5.60% compound annual growth rate between 2026 and 2031 (Mordor Intelligence, n.d). North America and Europe represent mature markets characterized by advanced technological infrastructure, stringent regulatory frameworks, and rising consumer awareness, while Asia–Pacific regions present explosive growth potential driven by urbanization, expanding middle-class demographics, and changing food preferences toward processed and convenience foods (Mordor Intelligence, n.d.).

This regional divergence has direct implications for starch-based packaging strategy: while North America and Europe prioritize regulatory compliance, certification, and performance parity with plastics, Asia–Pacific markets are shaping scale economics, feedstock diversification, and cost-driven innovation. Companies that serve these regions as interchangeable risk misaligning product design and market reality.

Recent regulatory changes across Asia–Pacific are pushing packaging rules forward from waste management, labeling to producer responsibility. While implementation timelines vary, trend is clear. Packaging systems are expected to demonstrate traceability, material safety, and end-of-life accountability. For starch-based packaging, this regulatory momentum creates both opportunity for scale and pressure to ensure system compatibility beyond material substitution.

The typical types of starch used in packaging applications include modified starch and blended starches (Liu et al., 2022). Modified starch family includes acetylated, hydroxypropyl, oxidized and enzymatically modified versions that belong the largest 67.9% share of the food starch market in 2024 and is also the fastest-growing segment, advancing around 5.5 to 7.5% compound annual growth rate through 2030 (Market.US, 2025). Importance of modified starch for packaging comes from the outcome of its superior resistance to degradation and chain dissociation under stress conditions included but not limited to thermal denaturation, hydrolytic environments and freeze-thaw cycling compared to natural corn starch (Cheng et al., 2025; Park et al., 2024). This superiority highlights modified starches having improved functional properties, especially greater stability under extreme conditions such as heat, acidity and freezing (Chen et al., 2025). As a conclusion, the dominance of modified starch in packaging applications does not only come from a technical outcome, it reflects industry's shift from material substitution to functional equivalence, where performance under regulatory, processing, and end-use stress conditions determines commercial viability.

The dominance of modified starch in packaging applications is not solely reflection of material performance requirements; it is increasingly shaped by regulatory expectations around chemical safety and functional transparency. The global acceleration of restrictions on intentionally added substances, most notably per- and polyfluoroalkyl substances (PFAS) in food contact and fiber-based packaging has shifted industry priorities toward inherently compliant material architectures. As a result, starch-based systems that achieve barrier performance without reliance on restricted chemistries are gaining strategic relevance across multiple regions.

Starch-based blend formulations frequently include polylactic acid (PLA), polybutylene adipate terephthalate (PBAT), polyhydroxyalkanoates (PHA), and polybutylene succinate (PBS). Those blends offer adjustable strength and flexibility for both rigid and flexible packaging. They can serve as substitutes for fossil-derived plastics, reducing reliance on nonrenewable feedstocks and lowering plastic waste. While scaling production remains a challenge, regulatory measures and circular-economy initiatives are encouraging researchers and industry to prioritize bioplastics in their R&D and development strategies (Grand View Research, 2025; Jayarathna et al., 2022; Ellen MacArthur Foundation, 2019; Dammak et al., 2020; Moreno-Bohorquez et al., 2026).

Starches can be used as films, bags and pouches, bottles, tableware and disposable foodservice items in packaging industry (GoneSheels n.d.; Garavito et al., 2024; Cruz et al., 2022; BioPackTech n.d.; Bioleader n.d.). Among them, starch-based films and foodservice tableware represent the most immediate pathways to scale, as they align simultaneously with regulatory pressure, consumer visibility and existing composting infrastructure. Bottles, while promising, remain constrained by performance and cost barriers that will likely limit near-term adaptation to niche applications (Royte, 2006).

Taken together, market growth alone does not guarantee environmental impact. Regions showing the fastest adaptation of starch-based packaging are also those where policy enforcement, waste collection systems, and composting infrastructure are evolving in parallel underscoring the system-level nature of material success (Royte, 2006). As indicated earlier, the most telling signal in current market data is not the absolute growth rate of starch-based packaging, but where that growth is structurally by policy, infrastructure and consumer behavior.

9.2.2 Major collaborations and capacity investments

The starch-based packaging market demonstrates competitive characteristics reflecting industrial partnerships and capacity improvements. Major multinational chemical and agricultural companies have established dominant market positions through substantial research and development investment, established manufacturing infrastructure, and distribution networks. Key industry players include Novamont (Italy), Agrana (Austria), BASF (Germany), Cargill (United States), NatureWorks (United States), Mitsubishi Chemical (Japan), Cardia Bioplastics (Australia), Biome Bioplastics (United Kingdom), Green Dot Bioplastics (United States), and Stora Enso (Finland).

Cargill opened a new corn milling plant in Gwalior, Madhya Pradesh, in March 2025 in partnership with Saatvik Agro to meet increasing demand from India's confectionery, infant formula, and dairy industries (Cargill, 2025).

Agrana, a leading Austrian starch producer, uses its access to raw materials and polymer technology to make starch-based bioplastic resins and compounds for various uses in European markets. The company developed AGENACOMP compounds combining thermoplastic starch with biodegradable polymers for flexible packaging applications, with biobased content reaching up to 60% fully based on Agrana's thermoplastic starch expertise (AGRANA 2026).

Simultaneously, Ingredion partnered with the Agrana in March 2025 for strategic collaboration to increase starch production in Romania, expanding its manufacturing footprint in Eastern Europe to address rising regional demand for specialty starches (Ernst, 2025). Apart from that, Ingredion announced a $50 million investment in a facility in Cedar Rapids, Iowa, to expand the company's specialty industrial starch capacity (Barton, 2025). These moves prove the industry expects lasting demand and requires production near emerging markets.

Roquette and Bonumose announced a new partnership agreement in July 2024 aimed at advancing the development of tagatose, a natural-origin sweetener with clinically proven health benefits (Roquette, 2024). This collaboration combines Roquette's extensive experience in starch and starch-based sweetener production with Bonumose's cutting-edge enzymatic technology, demonstrating how traditional starch manufacturers are pivoting toward innovative solutions (Roquette, 2024).

9.3 Regulatory landscape and policy frameworks

9.3.1 One planet, many rulebooks: a global tour of packaging regulation

The global regulatory landscape for packaging is rapidly evolving due to public awareness, climate commitments, and waste infrastructure limitations. Although end goal is the same, regions are fragmented and countries are moving at different speeds while trying to optimize political agenda with innovative solutions. Under this diverse challenge, it is important more than ever for businesses to understand international regulations that shape material and design decisions for their next generation packaging portfolio for remaining competitive and compliant.

The regulatory environment for starch-based packaging materials in the United States has undergone a fundamental transformation over the past 18 months and continues to accelerate. As of January 2026, packaging manufacturers and brand owners utilizing starch-based materials face a multifaceted compliance landscape shaped by federal Food and Drug Administration requirements, rapidly proliferating state-level Extended Producer Responsibility programs, evolving compostability standards, aggressive PFAS phase-outs, and unprecedented truth-in-labeling enforcement (Code of Federal Regulations, 2024; MacKerron, 2012; Oregon Truth in Labeling Oregon Truth In Labeling Task Force, 2022) (Fig. 9.2). The convergence of these regulatory streams creates both substantial compliance obligations and significant commercial opportunities for companies that understand the nuances of each jurisdiction's requirements.

Today, it is obvious that we are witnessing a critical inflection point: the transition from voluntary sustainability initiatives to mandatory compliance frameworks that will fundamentally reshape the economics and strategic positioning of starch-based packaging suppliers and their customers.

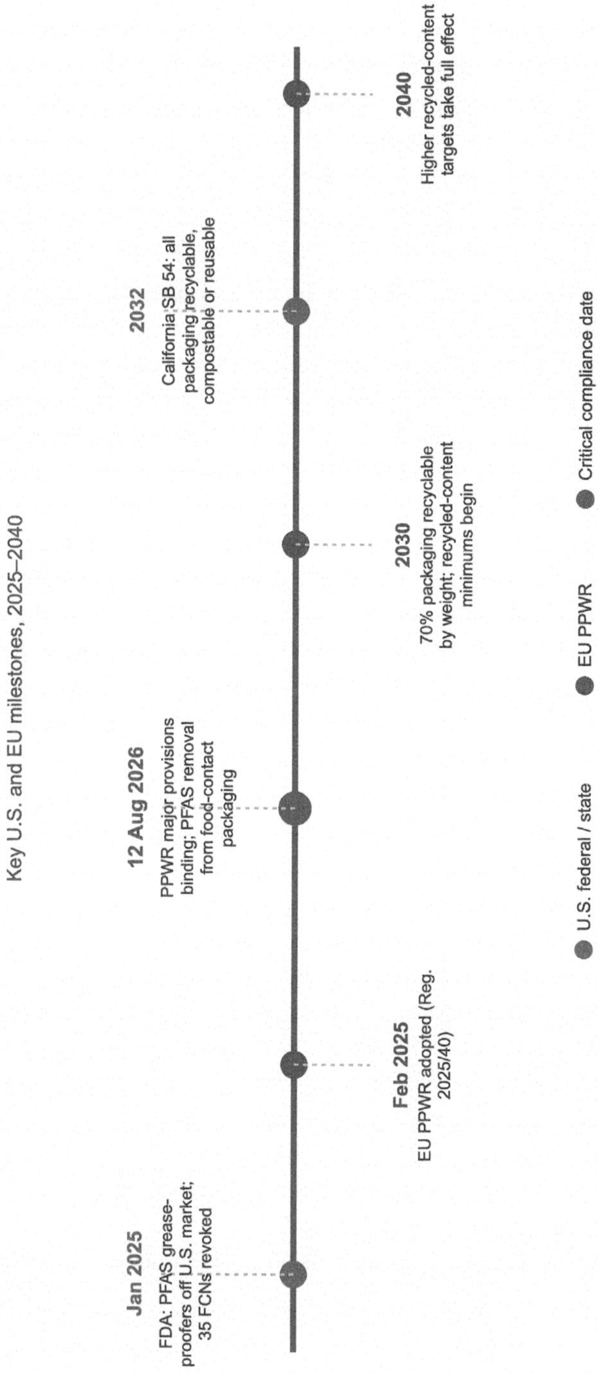

Figure 9.2 ***Regulatory timeline reshaping starch-based packaging.*** Key U.S. and EU milestones, 2025–2040.

9.4 Understanding the current federal food and drug administration (FDA) regulatory framework

9.4.1 Food contact substance approval and migration testing

The Federal Food and Drug Administration maintain rigorous authority over all materials that come into direct contact with food through Title 21 of the Code of Federal Regulations. (21 C F R. pts. 170–199). For starch-based packaging materials, including pure cornstarch products, modified starch formulations, and starch-polymer blends, manufacturers must ensure compliance with specific CFR sections that govern paper and paperboard in food contact applications. The FDA does not issue individual product certificates; rather, it publishes regulations that define allowable materials and the testing conditions under which they may be used. A critical distinction that manufacturers frequently misunderstand is that "FDA approved" is a colloquial term rather than a formal regulatory status. Instead, manufacturers must demonstrate through rigorous testing that their products comply with applicable CFR sections, most notably 21 CFR 176.170 for paper and paperboard in contact with fatty and watery foods, and 21 CFR 176.180 for paper in contact with dry foods.

For starch-based packaging to achieve FDA compliance, manufacturers must conduct migration testing under different food contact conditions, including exposure to oily, acidic, hot, and cold foods (U.S. Food and Drug Administration, 2007). The fundamental principle underlying this requirement is that any substance capable of migrating from packaging into food must not pose a safety risk at anticipated exposure levels. Independent, accredited laboratories such as SGS, Intertek, and TÜV conduct these tests and issue compliance reports that substantiate claims of FDA compliance (U.S. Food and Drug Administration, 2025). The process requires comprehensive documentation demonstrating that residual monomers, additives, and potential degradation products from the starch formulation remain below migration limits established in the Code of Federal Regulations. Companies importing or manufacturing starch-based containers destined for the United States market must obtain this documentation before market entry, as customs holds and significant fines result from non-compliance (Electronic Code of Federal Regulations (eCFR), 2026).

9.4.2 The food contact notification program and beyond

When a company develops a novel material or a new manufacturing process intended to contact food, it may seek review under the FDA's Food Contact Notification program by submitting comprehensive safety data and intended-use information; if the FDA raises no objection within 120 days, the notification becomes effective and the substance may be marketed for the specified use (U.S. Food and Drug Administration, 2018). This program has proven especially important for polylactic acid (PLA) and other bioplastics that require specific authorization for their intended uses. Importantly, Food Contact Notifications are proprietary to the specific manufacturer and supplier listed in the notification; other companies wishing to use the same substance must submit their own independent notifications.

The current state of FDA authorizations for starch-based and starch-derived polymers reflects extensive prior regulatory work. Polylactic acid, produced from fermented corn starch, is recognized as Generally Recognized as Safe (GRAS) for food contact applications (U.S. Food and Drug Administration, 2018). Similarly, polybutylene succinate and polyhydroxyalkanoates, which can be produced from starch-derived feedstocks, have effective Food Contact Notifications allowing their use in food-contact articles. However, manufacturers must remain vigilant regarding any changes in authorized uses, as the FDA maintains active oversight of food contact substances and can modify authorizations based on new safety data.

9.4.3 The PFAS phase-out: a critical compliance milestone

9.4.3.1 Recent FDA actions on PFAS

One of the most significant regulatory developments affecting starch-based packaging, particularly molded fiber containers and kraft paper-based products, has been the complete phase-out of PFAS authorized for use as grease-proofing agents on paper and paperboard food packaging. The FDA's regulatory position on PFAS evolved significantly between 2020 and 2025. In

January 2025, the FDA announced that all grease-proofing agents containing PFAS are no longer being sold in the United States market (U.S. Food and Drug Administration, 2025). This announcement formalized the conclusion of a voluntary phase-out process that began when the FDA raised safety concerns regarding certain PFAS compounds used in paper food packaging. As a conclusion, FDA had formally revoked all 35 Food Contact Notifications related to PFAS-containing grease-proofers applied to paper and paperboard, determining that these uses had been abandoned by manufacturers.

9.5 Extended producer responsibility and packaging regulations

The biggest change in shaping packaging rules worldwide is the spread of Extended Producer Responsibility (EPR). A major example is the EU's new Packaging and Packaging Waste Regulation (PPWR), which was adopted in February 2025 and starts to take effect in August 2026. The law covers packaging from creation to disposal and aims to harmonize national rules so the single market works better while cutting environmental and health harms (Bishop et al., 2026).

EU's PPWR rules are being tightened to cut waste and make packaging part of a circular economy. By 2030, 70% of packaging by weight must be recyclable, and most provisions become binding across the EU on August 12, 2026 (with some transitional arrangements running to 2040). From 2030, specific minimum recycled-content requirements kick in: PET food-contact packaging 30% (rising to 50% by 2040); other plastics in food contact 10% (25% by 2040); single-use plastic beverage bottles 30% (65% by 2040); and other plastic packaging 35% (65% by 2040). By 2030, minimum recycled content will also apply to other materials: glass 50%, paper/cardboard 70% and metal 25%. The PPWR also requires removal of PFAS from food-contact packaging by August 12, 2026 prompting development of PFAS-free alternatives such as starch-based barrier coatings and establishes substantial financial penalties for non-compliance, which will necessitate coordinated changes across manufacturers, recyclers and regulators. This requirement has catalyzed significant innovation in starch-based barrier coatings as alternatives to traditional PFAS-based treatments (Guillén, 2025).

In North America, individual states are implementing their own EPR laws creating a complex regulatory patchwork (Pierce, 2026). California's SB54, which was set to take effect in 2025, aims to reduce single-use plastic packaging by 25% by 2030. As of 2032, any package sold in California must qualify as recyclable, compostable, or reusable (Reynolds, 2026). New Jersey has two active EPR and recyclability bills in the state senate, with S208 enacting EPR and requiring producers of packaging to develop and implement stewardship plans for packaging lifecycle management (Pierce, 2026). Washington was one of the states to pass an EPR bill in 2025, and the state is moving rapidly to implement the regulation with upcoming producer registration deadlines. Over the next 5 years, Washington will be implementing their program in three phases (Washington State Department of Ecology, 2026).

9.6 Compostability standards and certifications

International standards governing compostability have become increasingly critical as starch-based packaging adoption accelerates. The most widely used compostability standards are the ASTM D6400, ISO 17,088, and EN 13,432, which are largely equivalent (Hwang et al., 2026).

The ASTM D6400 standard determines whether plastics, including polylactic acid (PLA), are capable of degrading in industrial composting environments (Plastics Industry Association, 2016). BPI certification ensures the product meets ASTM D6400 or D6868 standards for compostability in industrial facilities and is trusted across North America, especially by composting facilities and food service businesses (Jacobus, 2026).The European EN 13,432 standard requires complete biodegradation, eco-toxicity testing, and disintegration, and applies to PLA-based cups, coated paper, bioplastics, and compostable liners (European Bioplastics, 2015). The Seedling Logo issued by DIN CERTCO provides visual confirmation of compliance with EN 13,432, and clarifies where and how the product breaks down important for end-of-life disposal (EcoTopia, 2024). For home composting, French NF T51-800 standard specifies that products must achieve complete biodegradability, disintegration, plant compatibility, and earthworm safety within home composting conditions (Gioia et al., 2021).

9.7 Consumer awareness and trends

Consumer acceptance of starch-based packaging materials represents a critical factor in the commercialization of bio-derived packaging solutions. Consumer demand for environmentally conscious packaging designs like starch-based stems from multiple interconnected psychological, social, and environmental factors supported by empirical research (Thomas, 2024). About 85% of people say climate change has affected their daily lives, and because of that many are more aware of environmental issues and more likely to choose products that help the planet (Barabas, 2024). This awareness is particularly critical regarding packaging materials, as the packaging industry consumes the highest volumes of plastics produced globally and represents the main source of plastic waste entering the environment (Barbu et al., 2022; PackNode, 2023). Consumers demonstrate significant concern about plastic pollution, with over 82% of respondents in global studies acknowledging awareness of plastic waste issues and taking practical actions to address it (Seiberlich, 2024).

Younger consumer cohorts demonstrate particularly strong preferences for sustainable packaging. Gen Z and millennials exhibit approximately 90% willingness to pay premium prices for sustainable packaging, significantly higher than older demographics (Nordigården et al., 2025a, 2025b). Among consumers aged 18–34, environmental considerations influence purchasing decisions more substantially, with 67% contemplating sustainability when making purchases (Deloitte, 2024; MarketingCharts, 2024). This generational pattern extends across geographic regions, with high-income millennials showing 25% willingness to pay substantially more for sustainable packaging compared to 1% of high-income Gen X consumers (Fig. 9.3).

9.8 Perceived benefits and product attributes

Recyclability emerges as the most critical sustainability characteristic globally, with 77% of US consumers rating recyclable packaging as extremely or very important (Nordigården et al., 2025a, 2025b). While recycled content is the front-runner, compostability, reusability, and material volume represent secondary considerations of roughly equal importance for consumer. Products marketed with sustainability-related claims

Figure 9.3 ***Consumer awareness and willingness to pay.*** Selected findings shaping demand for sustainable packaging. *PwC, 2024. https://www.pwc.com/gx/en/news-room/press-releases/2024/pwc-2024-voice-of-consumer-survey.html; Mckinsey, 2025. https://www.mckinsey.com/industries/consumer-packaged-goods/our-insights/state-of-consumer-2025; Deloitte, 2024. 2024 Gen Z and Millennial Survey: Living and Working with Purpose in a Transforming World. Deloitte. https://www.deloitte.com/content/dam/assets-shared/docs/campaigns/2024/deloitte-2024-genz-millennial-survey.pdf; Trivium Packaging. (2023). Buying green report. Accessed on 10 March 2026 from https://buyinggreen.triviumpackaging.com/.*

averaged 28% cumulative growth over 5-year periods, compared to 20% for products without such claims, demonstrating tangible market advantages (Nordigården et al., 2025a, 2025b).

Carbon footprint labeling improves consumer trust, with 60% of consumers more likely to lean on products displaying verified carbon footprint labels compared to unlabeled products (Soe and Ketkaew, 2025; Herbes et al., 2020). This transparency can translate into measurable sales increases, with products featuring green claims experiencing higher sales (Lampertet al., 2022). Approximately 82% of consumers express willingness to pay premium prices for products with sustainable packaging, indicating strong market incentives for packaging innovation (Trivium Packaging, 2023).

9.9 Material-specific preferences

Consumer perceptions of packaging sustainability vary by material substrate (Berthold et al. 2024). Paper consistently ranks highest across geographies as sustainable options, while plastic formats rank lowest (Oloyede and Lignou, 2021; Lindh et al., 2016). For starch-based materials specifically, consumer acceptance is influenced by perceived quality, durability, disposal clarity, and transparency regarding environmental benefits (Findrik and Meixner, 2023). Challenges include associations between bio-based products and inferior barrier quality, though technological advances in starch blends with other biopolymers address these performance concerns (Marichelvam et al., 2019; Ghanbarzadeh et al., 2011).

9.10 Regulatory and social context

Across regions, consumers are increasingly holding brand owners and packaging manufacturers accountable for sustainability, making design and material choices the decisive levers for change (Nordigården et al., 2025a, 2025b). This creates a clear mandate for materials innovation that delivers circular outcomes: renewable feedstocks, verifiable end-of-life pathways, and compatibility with existing processing and recovery infrastructure (Peretti et al., 2024).

Starch-based systems exemplify this potential. Inherently biodegradable and derived from renewable biomass, they offer a credible route to lower-carbon packaging, but only if engineered to meet the performance, processing, and ecosystem-level criteria required for commercial scale. When formulations and supply chains are designed to deliver product protection, processing compatibility, and demonstrable end-of-life behavior, regulatory pressure becomes a catalyst for market differentiation: packaging that preserves value, reduces environmental impact, and aligns with consumer expectations (Saveth, 2023).

9.11 Conclusion and future outlook

The starch-based packaging sector is undergoing fundamental structural realignment driven by three converging mechanisms: (1) regulatory enforcement, including EPR frameworks, single-use plastic prohibitions, and material-specific compostability standards (EN 13,432, ASTM D6400); (2) demand-side dynamics, characterized by demonstrated consumer willingness to pay 20%–25% price premiums for certified biodegradable packaging, material-specific preferences (recyclability prioritized over compostability across most markets), and geographically differentiated regulatorions; and (3) technological maturation, encompassing mainly barrier property enhancements through polymer blending strategies and improved mechanical performance via reinforcement techniques. This convergence necessitates integrated operational restructuring across manufacturing infrastructure (raw material sourcing, process modification, quality assurance protocols), logistics networks (temperature and humidity-controlled storage, inventory turnover acceleration), and circular economy infrastructure (composting facility capacity, collection systems, end-market development) (Pyzyk and Rachal, 2026).

Market fundamentals remain robust, with the global starch market projected to grow from USD 65.02 billion in 2025 to USD 95.85 billion by 2032 at a compound annual growth rate of 5.7% (Maximize Market Research, 2026). Within this broader expansion, specialized segments including clean-label starch, organic starch, modified starch, and starch-based packaging are experiencing above-market growth rates, reflecting the industry's successful navigation of sustainability challenges and consumer demands. Practically, this outlook points to three priorities for stakeholders (1) invest in R&D to improve functional properties (e.g., barrier performance): and scalable modification technologies; (2) diversify feedstock sources and strengthen supply-chain resilience to manage price volatility; and (3) align innovation with evolving regulatory and certification frameworks while filling evidence gaps through lifecycle and performance comparisons with alternative biomaterials. These projections should be interpreted with caution given potential volatility in feedstock costs, technological disruption, and policy shifts, so decision-makers are advised to triangulate market intelligence and maintain adaptive strategies.

The regulatory landscape, particularly the European Union's Packaging and Packaging Waste Regulation effective in August 2026, establishes binding requirements that will fundamentally reshape packaging material choices, production processes, and supply chain structures globally. The prohibition of PFAS in food contact packaging by August 2026 represents a watershed moment for starch-based coating technologies, which are now achieving performance parity with traditional materials while offering superior environmental profiles. Extended Producer Responsibility regulations spreading across North America, Europe, and Asia create accountability structures that incentivize innovation in circular economy solutions, where starch-based and compostable packaging alternatives offer distinct advantages.

Consumer awareness has evolved substantially, with 73% of consumers demonstrating willingness to make purchasing decisions based on environmental impact and 41% ready to pay premium prices for natural and organic products. This consumer consciousness is driving market segmentation where clean-label, organic, and non-GMO starch products command premium positioning while supporting volume growth in both developed and emerging markets. The convergence of consumer demand with regulatory requirements creates a virtuous cycle where companies investing in sustainability, supply chain transparency, and product innovation gain competitive advantages while establishing market leadership positions.

Strategic collaborations, substantial capital investments, and technological innovations position the industry to meet projected growth targets while addressing supply chain resilience challenges. Regenerative agriculture initiatives, renewable energy investments, and circular economy business models demonstrate the industry's commitment to decarbonization and environmental stewardship. The transition from commodity starch production to specialty ingredient manufacturing, driven by clean-label consumer preferences and functional food market expansion, is creating higher-margin business opportunities and attracting ongoing investment.

The transition to starch-based and compostable packaging requires coordinated action by manufacturers, retailers, regulators, brand owners and consumers to establish the composting infrastructure and market mechanisms necessary for effective end-of-life management. Organizations that invest early in sustainable production practices, secure independent certification, and adopt transparent performance reporting will gain competitive advantage while advancing the United Nations Sustainable Development Goals (SDG). The recent study finds that because of strong research systems and resources from countries like the United States, United Kingdom, China, and Europe that are the leading producers of SDG research, generating valuable insights and solutions (Yumnam et al., 2024). On the other hand, research output from developing countries is rising, signaling greater recognition of SDG importance and the need for region-specific approaches, and suggesting a shift toward a more inclusive global effort. As a result, international collaborations are required to enable knowledge exchange, capacity building, and more effective, interdisciplinary responses.

References

Agrana (2026). Home compostable & bio based polymers. Accessed 10 March 2026 from https://www.agrana.com/en/products/all-productportfolios/starch-portfolio/products-for-technical-applications/bioplastics.

Barabas, D. (2024). PwC 2024 voice of consumer survey. PwC, May 15. Accessed on 8 March 2026 from https://www.pwc.com/gx/en/news-room/press-releases/2024/pwc-2024-voice-of-consumer-survey.html.

Barbu, A., Catană, Ș.A., Deselnicu, D.C., Cioca, L.I., Ioanid, A., 2022. Factors influencing consumer behavior toward green products: a systematic literature review. Int. J. Environ. Res. Publ. Health 19 (24). https://doi.org/10.3390/ijerph192416568.

Barton, T., 2025. How a cedar rapids company's expansion is helping amazon improve packaging and boost sustainability. The Gazette, August 2025. https://www.thegazette.com/business/how-a-cedar-rapids-companys-expansion-is-helping-amazon-improve-packaging-and-boost-sustainability.

Berthold, A., Guion, S., Siegrist, M., 2024. The influence of material and color of food packaging on consumers' perception and consumption willingness. Food Human. 2, 100265. https://doi.org/10.1016/j.foohum.2024.100265.

BioLeaderPack. n.d. Cornstarch tableware. bioleaderpack. Accessed on 10 March 2026 from https://www.bioleaderpack.com/product-category/cornstarch-tableware/.

Bishop, T. S., Kupfer, A. M., Leake, T. B., Nair, S. (2026). EPR packaging laws moving from concept to compliance. Mayer Brown, February 2026. Accessed on 10 March 2026 from https://www.mayerbrown.com/en/insights/publications/2026/02/epr-packaging-laws-moving-from-concept-to-compliance.

Cargill (2025). Cargill inaugurates new corn milling plant to meet growing demand for safe and high-quality food solutions. Cargill. Accessed 20 January 2026 from https://www.cargill.com/2025/cargill-inaugurates-new-corn-milling-plant.

Chen, H., He, J., Li, W., Wang, Z., Du, M., Kan, J., 2025. Structural and functional characteristics of esterified starch and microencapsulation for urate-lowering probiotic: effect of hydrophobic side-chain length. Carbohydr. Polym. 358, 123529. https://doi.org/10.1016/j.carbpol.2025.123529.

Cheng, Y., Wang, B., Li, B., Li, G., Zhong, Y., Lv, W., Xiao, H., 2025. Modified starch-based emulsion-filled gels for 3D printing: preparation, characterization, and printability. Food Bioprocess Technol. 18 (4), 3732–3746. https://doi.org/10.1007/s11947-024-03669-6.

Cruz, R.M.S., Krauter, V., Krauter, S., Agriopoulou, S., Weinrich, R., Herbes, C., Scholten, P.B.V., Uysal-Unalan, I., Sogut, E., Kopacic, S., Lahti, J., Rutkaite, R., Varzakas, T., 2022. Bioplastics for food packaging: environmental impact, trends and regulatory aspects. Foods 11 (19). https://doi.org/10.3390/foods11193087.

Dammak, M., Fourati, Y., Tarrés, Q., Delgado-Aguilar, M., Mutjé, P., Boufi, S., 2020. Blends of PBAT with plasticized starch for packaging applications: mechanical properties, rheological behaviour and biodegradability. Ind. Crop. Prod. 144, 112061. https://doi.org/10.1016/j.indcrop.2019.112061.

Deloitte, 2024. 2024 Gen Z and Millennial Survey: Living and Working with Purpose in a Transforming World. Deloitte. https://www.deloitte.com/content/dam/assets-shared/docs/campaigns/2024/deloitte-2024-genz-millennial-survey.pdf.

EcoTopia, 2024. Seedling Certification: The Gold Standard for Compostable Products. GoEcoTopia. https://www.goecotopia.com/blog/a/seedling-certification-the-gold-standard-for-compostable-products.html.

Electronic Code of Federal Regulations (eCFR), 2026. Indirect food additives: Adjuvants, production aids, and sanitizers — Industrial starch-modified, 21 C.F.R. § 178.3520. U.S. Food and Drug Administration, Department of Health and Human Services. From: https://www.ecfr.gov/current/title-21/chapter-I/subchapter-B/part-178/subpart-D/section-178.3520. (Accessed 10 March 2026).

Ellen MacArthur Foundation, 2019. Plastics and the Circular Economy – Deep Dive. Ellen MacArthur Foundation, September from. https://www.ellenmacarthurfoundation.org/plastics-and-the-circular-economy-deep-dive. (Accessed 9 January 2026). Accessed on.

Ernst, I. (2025). US group ingredion teams up with austrian agrana to boost starch production in Romania. Romania-Insider, 05 March. Accessed 21 January 2026 from https://www.romania-insider.com/ingredion-agrana-starch-romania-march-2025.

Findrik, E., Meixner, O., 2023. Drivers and barriers for consumers purchasing bioplastics – a systematic literature review. J. Clean. Prod. 410, 137311. https://doi.org/10.1016/j.jclepro.2023.137311.

Garavito, J., Peña-Venegas, C.P., Castellanos, D.A., 2024. Production of starch-based flexible food packaging in developing countries: analysis of the processes, challenges, and requirements. Foods 13 (24). https://doi.org/10.3390/foods13244096.

Ghanbarzadeh, B., Almasi, H., Entezami, A.A., 2011. Improving the barrier and mechanical properties of corn starch-based edible films: effect of citric acid and carboxymethyl cellulose. Ind. Crop. Prod. 33 (1), 229–235. https://doi.org/10.1016/j.indcrop.2010.10.016.

Gioia, C., Giacobazzi, G., Vannini, M., Totaro, G., Sisti, L., Colonna, M., Marchese, P., Celli, A., 2021. End of life of biodegradable plastics: composting versus re/upcycling. ChemSusChem 14 (19), 4167–4175. https://doi.org/10.1002/cssc.202101226.

GoneShells. n.d. The next generation of food packaging. GoneShells. Accessed on 26 February 2026 from https://goneshells.com/.

Grand View Research. (2025). Starch-based packaging market (2025 - 2030). grand view research. Accessed on 29 January 2026 from https://www.grandviewresearch.com/industry-analysis/starch-based-packaging-market-report.

Guillén, C., 2025. PPWR 2026: Essential Guide to the New European Packaging and Waste Regulation. xNova International. https://www.xnovainternational.com/post/ppwr-2026-essential-guide-to-the-new-european-packaging-and-waste-regulation.

Herbes, C., Beuthner, C., Ramme, I., 2020. How green is your packaging—a comparative international study of cues consumers use to recognize environmentally friendly packaging. Int. J. Consum. Stud. 44 (3), 258–271. https://doi.org/10.1111/ijcs.12560.

Hwang, D.K., Park, J., Oh, D.X., Jeon, H., Koo, J.M., 2026. Assessing the role of compostable plastics in circular economy transition. ChemSusChem 19 (1). https://doi.org/10.1002/cssc.202501938.

Jacobus, K. (2026). What does bpi compostable mean?. good start packaging. Accessed on 10 March 2026 from https://www.goodstartpackaging.com/what-does-bpi-compostable-mean/.

Jayarathna, S., Andersson, M., Andersson, R., 2022. Recent advances in starch-based blends and composites for bioplastics applications. Polymers 14 (21), 4557. https://doi.org/10.3390/polym14214557.

Lampert, C. (2022). Will carbon-labeled products sell more? here's what we know. Sustainable Brands. Accessed 10 March 2026 from https://sustainablebrands.com/read/will-carbon-labeled-products-sell-more-here-s-what-we-know#:~:text=Key%20takeaway%20%231:%20Carbon%2D,been%20shown%20to%20sell%.

Lindh, H., Olsson, A., Williams, H., 2016. Consumer perceptions of food packaging: contributing to or counteracting environmentally sustainable development? Packag. Technol. Sci. 29 (1), 3–23. https://doi.org/10.1002/pts.2184.

Liu, D., Zhao, P., Chen, J., Yan, Y., Wu, Z., 2022. Recent advances and applications in starch for intelligent active food packaging: a review. Foods 11 (18), 2879. https://doi.org/10.3390/foods11182879.

MacKerron, C. (2012). Unfinished business: the case for extended producer responsibility for post-consumer packaging. As You Sow. Accessed on 10 March 2026 from https://static1.squarespace.com/static/59a706d4f5e2319b70240ef9/t/5a7e426ff9619ab1a90b004d/1518223990918/REPORT-2012-UnfinishedBusiness_TheCaseforEPR.pdf.

Marichelvam, M.K., Jawaid, M., Asim, M., 2019. Corn and rice starch-based bio-plastics as alternative packaging materials. Fibers 7 (4), 32. https://doi.org/10.3390/fib7040032.

MarketingCharts (2024). Gen Zers and millennials are making climate-conscious buying decisions. MarketingCharts, June 7. Accessed on 8 March 2026 from https://www.marketingcharts.com/brand-related/csr-233222.

Market.US (2025). Global food starch market size, share, and business benefits by type (native starch, modified starch), by source (maize, wheat, potato, others), by form (powder, liquid), by application (bakery and confectionery, snacks, soup, sauces and dressings, dairy products, meat and meat products, others), by region and companies — industry segment outlook, market assessment, competition scenario, Trends, and Forecast 2025–2034. Report no. 152942. July. Accessed on 10 March 2026 from https://market.us/report/food-starch-market/.

Maximize Market Research (2026). Global starch market – starch market - industry structure evaluation, demand drivers analysis, regional growth analysis and identification, competitive positioning review & global market size forecast to 2032. Maximize Market Research, Report No. 112948, January. Accessed 10 March 2026 from https://www.maximizemarketresearch.com/market-report/global-starch-market/112948/.

Mordor Intelligence. n.d. Food Starch Market Size & Share Analysis — Growth Trends and Forecast (2026–2031). Mordor Intelligence.

Moreno-Bohorquez, E., Arias-Tapia, M.J., Jaramillo, A.F., 2026. Recent advances in thermoplastic starch (TPS) and biodegradable polyester blends: a review of compatibilization strategies and bioactive functionalities. Polymers 18 (2), 289. https://doi.org/10.3390/polym18020289.

Nordigården, D., Feber, D., Grünewald, F., Pley, M., Conrad, M., 2025a. Do US Consumers Care about Sustainable Packaging in 2025? McKinsey & Company. https://www.mckinsey.com/industries/packaging-and-paper/our-insights/do-us-consumers-care-about-sustainable-packaging-in-2025.

Nordigården, D., Feber, D., Grünewald, F., Pley, M., Seidner, M., Conrad, M., 2025b. Sustainability in Packaging 2025: Inside the Minds of Global Consumers. McKinsey & Company. https://www.mckinsey.com/industries/packaging-and-paper/our-insights/sustainability-in-packaging-2025-inside-the-minds-of-global-consumers#/.

Oloyede, O.O., Lignou, S., 2021. Sustainable paper-based packaging: a consumer's perspective. Foods 10 (5), 1035. https://doi.org/10.3390/foods10051035.

Oregon Truth In Labeling Task Force (2022). Truth In labeling final report and recommendations. Submitted to the Oregon Legislature. Oregon Department of Environmental Quality, June 1. Accessed 10 March 2026 from https://www.oregon.gov/deq/recycling/Documents/TIL-Report.pdf.

PackNode, 2023. Sustainable Packaging Trends for 2026. PackNode. https://www.packnode.org/en/sustainability/sustainable-packaging-trends-for-2026.

Park, H.J., Cho, D.H., Chung, H.J., Lim, S.T., 2024. Enhanced gelling property and freeze–thaw stability of potato, tapioca and corn starches modified by mild heating in aqueous ethanol solution. J. Sci. Food Agric. 104 (12), 7228–7237. https://doi.org/10.1002/jsfa.13544.

Peretti, K., Robinson Turner, A., Seitz, M., 2024. Circularity for Secure and Sustainable Products and Materials: A Draft Strategic Framework. U.S. Department of Energy, Washington, DC, October. https://www.energy.gov/sites/default/files/2024-10/circularity-for-secure-sustainable-products-materials-report.pdf.

Pierce, L., 2026. Packaging Policy News. Sustainable Packaging Coalition from. https://sustainablepackaging.org/2026/02/20/packaging-policy-news/.

Plastics Industry Association. (2016) Industrial compostability claims checklist. Accessed on 10 March 2026 from https://plasticsindustry.org/wp-content/uploads/2022/11/Industrial-Compostability-Claims-Checklist.pdf.

Pyzyk, K., Rachal, M., 2026. 5 Packaging Industry Trends to Watch in 2026. Packaging Dive. https://www.packagingdive.com/news/packaging-industry-trends-2026-epr-labeling-containerboard-automation/809071/.

Reynolds, M., 2026. Live at East: CA's 2032 SB 54 to Reshape Packaging in '26. Packaging World. https://www.packworld.com/sustainable-packaging/article/22961081/live-at-east-californias-2032-sb-54-mandate-already-reshaping-packaging.

Roquette (2024). Roquette and BONUMOSE GLOBAL AGREEMENT ON TAGATOSE. July. Accessed on 10 March 2026 from https://www.roquette.com/press-releases/roquette-and-bonumose-global-agreement-on-tagatose.

Royte, E., 2006. Corn Plastic to the Rescue? Smithsonian Magazine, August. https://www.smithsonianmag.com/science-nature/corn-plastic-to-the-rescue-126404720/.

Saveth, B. (2023). Embracing sustainability: the rise of eco-friendly packaging solutions. Forbes, June 23. Accessed on 10 March 2026 from https://www.forbes.com/councils/forbesbusinesscouncil/2023/06/23/embracing-sustainability-the-rise-of-eco-friendly-packaging-solutions/.

Seiberlich, H. (2024). Sustainable packaging solutions and product appeal. Meyers. November 27. Accessed on 8 March 2026 from https://meyers.com/meyers-blog/sustainable-packaging-solutions-and-product-appeal/.

Sharma, A. (2025). Food starch market size & outlook, 2025–2033: food starch market size, share & trends analysis report by raw material (maize, wheat, potatoes, others), by type (modified starch, native starch, sweeteners), by application (animal feed, bakery and confectionery, beverage, cosmetics), by end-user (business to business, business to consumer) and by region (north America, europe, APAC, middle east and Africa, LATAM) Forecasts, 2025–2033. Straits Research, report SRFB54722DR. Accessed on 10 March 2026 from https://straitsresearch.com/report/food-starch-market.

Soe, A.C., Ketkaew, C., 2025. Understanding consumer responses to carbon footprint product labels: a multi-group analysis of online and offline shopping engagement in retail contexts. Digit. Bus. 5 (2), 100158. https://doi.org/10.1016/j.digbus.2025.100158.

Sutaria, I., 2025. Starch-based packaging market: starch-based packaging market size and share forecast outlook 2025 to 2035. Future Market Insights 20. https://www.futuremarketinsights.com/reports/starch-based-packaging-market.

Thomas, D., 2024. How Sustainable Packaging Influences Consumer Behavior. C.L. Smith. https://www.clsmith.com/sustainable-packaging/how-sustainable-packaging-influences-consumer-behavior/.

Trivium Packaging. (2023). Buying green report. Accessed on 10 March 2026 from https://buyinggreen.triviumpackaging.com/.

U.S. Food and Drug Administration (2007). Guidance for industry: preparation of premarket submissions for food contact substances (chemistry recommendations). December. Accessed on 10 March 2026 from https://www.fda.gov/regulatory-information/search-fda-guidance-documents/guidance-industry-preparation-premarket-submissions-food-contact-substances-chemistry#iid1a.

U.S. Food and Drug Administration. (2018). Determining the regulatory status of components of food-contact materials (FCMs). February 21. Accessed on 10 March 2026 from www.fda.gov/food/packaging-food-contact-substances-fcs/determining-regulatory-status-components-food-contact-material#:~:text=Submit%20a%20Food%20Contact%20Formulation,the%20FCS%20in%20.

U.S. Food and Drug Administration, 2025. ASCA-accredited testing laboratories. https://www.fda.gov/medical-devices/division-standards-and-conformity-assessment/asca-accredited-testing-laboratories.

Yumnam, G., Gyanendra, Y., Singh, C.I., 2024. A systematic bibliometric review of the global research dynamics of united nations sustainable development goals 2030. Sustain. Futures 7, 100192. https://doi.org/10.1016/j.sftr.2024.100192.

Further reading

Agnihotri, J.A., 2025. Milestones in the technology of modified starch in pharmaceutical formulation. Nat. Prod. J. 15, 9. https://doi.org/10.2174/0122103155319243240927062730.

Department of Ecology State of Washington. (2025). Recycling reform act. Accessed on 10 March 2026 from https://ecology.wa.gov/waste-toxics/reducing-recycling-waste/our-recycling-programs/recycling-reform-act.

Food and Drug Administration. (2024). Components of paper and paperboard in contact with aqueous and fatty foods. 21 C.F.R. § 176.170.

Foundation for Food & Agriculture Research (2024). Sustainable packaging innovation lab. Accessed on 10 March 2026 from https://foundationfar.org/grants-funding/opportunities/sustainable-packaging-innovation-lab/.

Hillyer, M.B., Nam, S., 2024. Materials Post Processing Challenges Addressing Replacing PFAS in Food Packaging and Fabrics. Cotton Fiber Bioscience & Utilization Research. USDA-ARS, SRRC, October, New Orleans, LA. In: https://www.ars.usda.gov/ARSUserFiles/np212/015%20Hillyer%20Materials%20Post%20Processing.pdf.

Market Data Forecast. (2025). Global industrial starch market size, share, trends, & growth forecast report, segmented by source (corn, wheat, potato, cassava), type (native starch, starch derivatives & sweeteners (modified starch, other derivatives & sweeteners)), application (feed, food & beverages, paper industry), form (liquid, dry), and region (north america, europe, apac, latin america, middle east and africa), industry analysis from 2025 to 2033. market data forecast. Last modified December 2025. Accessed on 10 March 2026 from https://www.marketdataforecast.com/market-reports/industrial-starch-market.

Norton, V., Waters, C., Oloyede, O.O., Lignou, S., 2022. Exploring consumers' understanding and perception of sustainable food packaging in the UK. Foods 11 (21), 3424. https://www.mdpi.com/2304-8158/11/21/3424.

Oktay, S., Pizzi, A., Köken, N., Bengü, B., 2024. Chemical modification techniques of corn starch for synthesis wood adhesive. Int. J. Adhesion Adhes. 128, 103545. https://doi.org/10.1016/j.ijadhadh.2023.103545.

Research and Markets. (2025). modified starch global market insights 2025: analysis and forecast to 2030, by manufacturers, regions, technology, application, product type. November, Report ID 5573698. Accessed on 10 March 2026 from https://www.researchandmarkets.com/reports/5573698/modified-starch-global-market-insights-2025?srsltid=AfmBOorDenKkZIRggmAQFp3mwwI2kLi0GK4x8XSRtRgK4RsSk415RUUN.

Siracusa, V., Blanco, I., 2020. Bio-polyethylene (Bio-PE), bio-polypropylene (Bio-PP) and bio-poly(ethylene terephthalate) (Bio-PET): recent developments in bio-based polymers analogous to petroleum-derived ones for packaging and engineering applications. Polymers 12 (8), 1641. https://doi.org/10.3390/polym12081641.

Wang, F., Ma, R., Zhu, J., Ma, W., Li, J., Tian, Y., 2025. Design principles and emerging applications of starch-involved superwettable systems. Eng. Times 53, 155–175. https://doi.org/10.1016/j.eng.2025.04.022.

CHAPTER TEN

From theory to practice: Case studies in starch-based food packaging innovations

Hatice Birtane[1], Aslı Beyler Cigil[2]
[1]Department of Chemistry, Faculty of Science, Marmara University, Istanbul, Turkiye
[2]Department of Chemistry and Chemical Processing Technologies, Technical Sciences Vocational School, Gazi University, Ankara, Turkiye

10.1 Introduction

The economic and environmental impacts of food packaging using petroleum-based plastics are becoming increasingly concerning with plastic waste causing severe environmental damage, especially in developing countries where waste management and recycling facilities are rarely available (Glaser et al., 2026; Edo et al., 2025) ("Yin and Wai Woo (2024). https://doi.org/10.1039/D4FB00028E.,". With the sustainability-focused mindful consumers, the need for biobased packaging solutions is escalating. Thus, owing to their renewable and environmentally friendly characteristics, starch-based films are becoming increasingly attractive.

Because of its availability, renewability, and cost-effectiveness, starch has become the center of attention for researchers as a potential biopolymer in the preparation of bio-based edible coatings (Vatanpour et al., 2026; Jiang et al., 2020; Wang et al., 2025a). Although starch is becoming a viable sustainable substitute to petroleum-based plastic for packaging applications since it is biodegradable, nontoxic, and edible, it has poor barrier and mechanical properties when compared to polyethylene. However, natural starch cannot be processed as a thermoplastic material because of intermolecular interactions and hydrogen bonds. To solve this issue, starch is transformed into thermoplastic starch (TPS), which is moldable, generally along with low molecular weight plasticizers such as glycerol, sorbitol, urea, and water (Jiang et al., 2025; Zhang et al., 2024; Rahardiyan et al., 2023; Tutek et al., 2025). Such plasticizers enhance starch flexibility and processability by decreasing glass transition temperature (Tg) and enhancing chain mobility. Applying extrusion as a successful commercial film-forming technology,

Starch-Based Sustainable Food Packaging
ISBN: 978-0-443-36346-7
https://doi.org/10.1016/B978-0-443-36346-7.00001-1

the thermoplastic characteristics of starch can be applied to enhance structural and physicochemical aspects of films. However, the industry faces a few challenges with starch-based films, which prevent them from being more widely used. These issues mainly are that water and moisture sensitivity were high due to the hydrophilic nature of starch, poor mechanical properties, and barrier properties. To surpass the above restrictions, as well as to enhance films' strength, barrier properties, and additional features, a series of novel strategies in which films are reinforced with nanomaterials (e.g., cellulose nanofibers, montmorillonite nanoclays), biopolymers such as chitosan and PVA, and bioactive agents such as clove oil are being vigorously pursued (Plaeyao et al., 2025; Xia et al., 2024; Luo et al., 2025; Yuan et al., 2025; Tanwar et al., 2021). This makes the development of starch-based films a pivotal solution to the environmental problems and an expectation to be a prospective sustainable packaging material.

10.2 Starch, structure, sources, and properties

Since starch is renewable, abundant, and low cost, it is one of the most appropriate materials for the development of bio-based food packaging (Zhang et al., 2025). With increasing awareness of plastic pollution and damage to the environment, starch now offers a green alternative to the petrochemical-based plastics as it is renewable, biodegradable, and readily accessible.

10.2.1 Chemical composition

Starch is a nature-originated polysaccharide that is abundant, renewable, and biodegradable, and it is the second most abundant organic material on Earth (Siciliano et al., 2025) (Zhang et al., 2025; Chysirichote et al., 2025; Teseme et al., 2025; Li et al., 2026). Starch, which serves as a natural polymer that can be derived from many plant sources including corn, rice, wheat, potato, and even acorn, is nowadays gaining more and more attention for food packaging and other related applications because of its cheapness, nontoxicity, and biodegradable nature. Starch is a key source in bioplastics development, and its biodegradability potential enables it as an efficient alternative to traditional polymers based on petroleum.

10.2.2 Chemical and physical structure of starch

Starch is structurally comprised of two main polymers: amylose and amylopectin, made up of α-1,4- linked glucose with α-1,6-linked glucose along

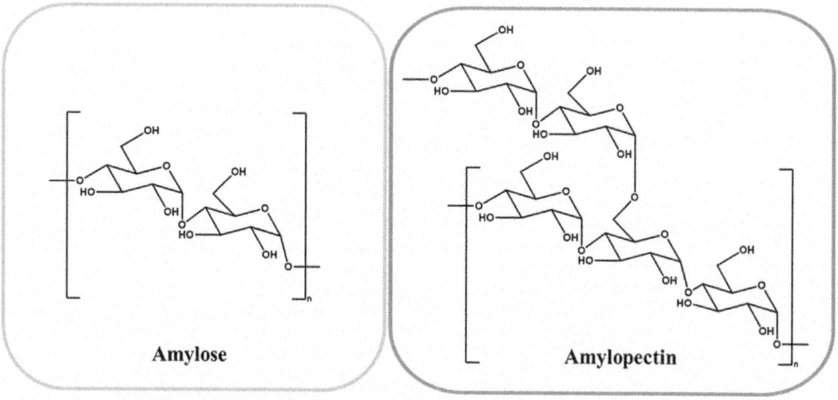

Figure 10.1 Chemical structure of amylose and amylopectin.

the chain (Jiang et al., 2020; Abdin et al., 2026). Starch types obtained by mixing these two polymers with various proportions have different physicochemical behavior (Siciliano et al., 2025). Figure 10.1 shows the chemical structure of amylose and amylopectin.

10.2.3 Structure of starch molecules

Amylose is a linear polymer consisting primarily of α-D-glucose units linked by α-1,4 bonds. Its molecular size is approximately 3000 α-glucosyl groups, and it may exhibit a low degree of branching. The long, linear structure of amylose is known to give starch its film-forming properties (Pawase et al., 2025; Song et al., 2012).

Amylopectin is a highly branched polymer composed of α-D-glucose units linked by α-1,4 bonds, similar to amylose. However, the amylopectin molecule exhibits branching with α-1,6 bonds every 20–25 glucose units. Amylopectin has a large, branched structure that can contain around 60000 glycosyl groups.

The ratios of amylose and amylopectin in starch are crucial for food packaging production, as they directly impact the mechanical, barrier, and solubility properties of the packaging film.

10.2.4 Source of starch

Starch is obtained from plants that grow in almost all temperate regions of the world. The source of the starch has a significant impact on the properties of the resulting films Table 10.1.

Table 10.1 Major starch sources worldwide (Liu et al., 2006; De Arcangelis et al., 2021; Govindaraju et al., 2022; Oliver et al., 2023).

Botanical resources	Approximate global share (%)	Amylose content (%)	Characteristic features
Corn	~80–85	22–33	It is the most common starch source worldwide. It offers good film-forming ability, stable mechanical properties, and industrial scalability.
Wheat	~5–8	18–30	It has a smaller granule size. It is common in food and packaging applications.
Potato	~4–6	20–31	It has large, polyhedral granules. It exhibits high swelling capacity and good gel-forming properties; however, its mechanical strength is limited.
Cassava, tapioca	~3–5	17–25	It offers a low gelatinization temperature and high purity. The film is advantageous in terms of transparency and flexibility.
Rice	~1–2	5–28	It has a wide amylose range. Smooth film surfaces can be obtained due to their small granule size.
Barley	<1	20–30	It has more limited industrial uses. It is generally used for special applications and modification work.
Sorghum	<1	20–30	It is obtained from drought-resistant plants.
Legumes (peas, beans, etc.)	<1	30–45	Thanks to its high amylose content, it offers good mechanical strength and barrier properties.

10.2.5 Physicochemical properties related to packaging applications

The performance of starch-based packaging depends largely on various physicochemical properties, which in turn depend on their chemical structure and the plant source from which they are derived.

Effect of Amylose/Amylopectin Ratio.

The characteristics of starch-based packaging are influenced by the proportion of amylose to amylopectin (Cheng et al., 2024; Basiak et al., 2017). Generally, higher amylose content would result in stiffer films with higher tensile strength (TS) and elastic modulus (EM), but with lower elongation at break (EAB) because of better crystallization and more ordered structure. However, the relationship is confounded by factors including the starch type and its moisture content (Jha et al., 2020).

Lower water vapor permeability (WVP) and oxygen permeability (OP) are also linked to higher amylose content in starch-based packaging. Nevertheless, films derived from potato starch (with low amylose content) display superior barrier properties to oxygen and water vapor than those from wheat and corn starch. Amylose containing α-(1, 4) glycosidic bonds can form more homogeneous stackable structures, which are less penetrable to gases than branched structures of amylopectin (Santhosh et al., 2024; Karnwal et al., 2025).

Amylose content influence on water sensitivity and solubility of the packaging. A greater solubility is associated with a higher amylose content. The hydrophilic nature of starch itself is a major limitation for high-water-sensitivity packaging applications, since water penetration negatively affects both mechanical and barrier properties of starch-based films (Shanbhag et al., 2023).

Starch-based packaging is usually opaque and thicker for products with relatively large amylose content, while they are clearer and thinner for those with relatively small amylose content (Joshi et al., 2013; Figueroa-Lopez et al., 2024). Films of potato starch-based packaging are usually colorless. Because of its reduced heterogeneity and increased density, amylopectin can be used in the production of thinner packaging films. For instance, it has been reported that the thickness of film at 112 μm for corn starch and 55 μm for potato starch (Shanbhag et al., 2023; Jansson and Thuvander, 2004).

The applicability of high Tg and the brittle nature of pure starch in packaging is limited. Because of intermolecular forces and hydrogen bonding, starch does not typically behave as a thermoplastic material. Hence, an increase in the flexibility requires a decrease in the Tg. The addition of plasticizers is one of the most widely used techniques applied to lower Tg. The addition of low molecular weight plasticizers, such as glycerol or sorbitol, or that of water breaks up hydrogen bonds and causes the chains to come farther apart. This facilitates starch degradation and allows processability of starch as a fluid thermoplastic material.

Hence, starch-based films have good potential for applications in packaging due to their biodegradability and availability, but they usually need to be modified with nanomaterials, bioactive agents, and other biopolymers to the challenges related to their high sensitivity to moisture and mechanical disadvantages. Packaging film performance is the result of the complex interaction of the amylose/amylopectin ratio, film formulation (type of plasticizer), and processing conditions.

10.3 Processing of starch-based materials

Being renewable, inexpensive, and biodegradable, natural starch is considered one of the best options to replace plastic and reduce plastic waste pollution (Tutek et al., 2025) (de Sousa et al., 2022; Lara and Salcedo, 2016). However, raw starch cannot be processed as its melting temperature is above the temperature of its thermal degradation, thus starch must be transformed into TPS through application of plasticizers, thermal and mechanical energy.

10.3.1 Gelatinization, retrogradation, and plasticization mechanisms

Starch processing is based on three basic physicochemical processes. gelatinization, retrogradation, and plasticization.

In the gelatinization mechanism, under the influence of heat and plasticizer, starch granules swell, break down, and lose their semicrystalline structure, transforming into an amorphous matrix. Lara and Salcedo stated that the ratio of water and glycerol directly affects the gelatinization temperature, and that an increase in the amount of glycerol makes it difficult for the plasticizer to penetrate the granules, thus delaying the process (Lara and Salcedo, 2016). Deep eutectic solvent (DES) based studies show that DESs can facilitate the gelatinization process of starch by weakening hydrogen bonds more effectively than water (Wan et al., 2019; Zhu, 2025).

Retrogradation (Recrystallization). As the gelatinized starch cools and is left to stand, the amylose and amylopectin chains rearrange and crystallize. This leads to the material becoming stiff and brittle. In their study, Lan et al. reported that the addition of soy polysaccharides and oil prevented retrogradation in starch-based packaging films by inhibiting the interaction between starch chains (Lan et al., 2024). In contrast, Zdanowicz et al. emphasized that glycerol allowed some crystallization, but urea and choline chloride-based DESs kept the system amorphous for a more extended period (Zdanowicz and Salasinska, 2023).

Plasticizers increase free volume and impart flexibility to the material by breaking the strong hydrogen bonds between starch chains and forming their own hydrogen bonds. Mechanical strength ensures the homogeneous distribution of the plasticizer within the matrix.

10.3.2 Thermoplastic starch (TPS) production

TPS production involves various strategies adapted to traditional plastic processing methods. While solution casting is commonly used in the literature for laboratory-scale applications, melt mixing (a melting method) is preferred for industrial efficiency. For instance, Wan et al. demonstrated that dry heat treatment improved the melt processability and mechanical properties of starches with different amylose contents (Wan et al., 2019). Another strategy reported in the literature is the single-step approach. Instead of the conventional two-step method of first plasticizing and then blending with the polylactic acid (PLA) (Jiang et al., 2025). In a simulated two-step process, first, starch is plasticized, and subsequently, it is compounded with PLA. It is energy saving and the output is high for this way of working. Another method is to chemically modify and graft. Tutek et al. studied the grafting of PLA and polycaprolactone (PCL) from the starch backbone via the hydroxyl groups of starch. This modification has been shown to enhance the mechanical properties of starch by improving its compatibility with hydrophobic polymers (Tutek et al., 2025).

Select plasticizer is one of the most important parameters influencing the final performance and storage stability of the TPS. Glycerol is the most widely employed plasticizer; however, it can increase moisture retention and crystallization with time due to its very hygroscopic properties. Murawski et al. The authors have shown that the prepared packaging films keep their stability up to 5 years (Murawski et al., 2025). Yermagambetova et al. stated that sorbitol imparted a lower degree of flexibility than glycerol but better preserved the TS of the packaging film and decreased the roughness of the surface (Yermagambetova et al., 2024). Zamanidehyaghoubi et al. similarly found that sorbitol exhibited the best plasticizing effect for pullulan starch-based packaging films, whereas fructose showed the highest glass transition temperature (Tg) (Zamanidehyaghoubi et al., 2025). DESs, referred to as "green solvent", present a better characteristic than the traditional plasticizers. De Sousa et al. demonstrated that choline chloride/oxalic acid DES is a far more efficient plasticizer than glycerol for starch films, yielding comparable mechanical properties and superior transparency at reduced concentrations. Application of DES makes the starch less crystalline than treatment with glycerol, keeping the material more in the amorphous and flexible state (de Sousa et al., 2022).

In conclusion, sorbitol is a modern alternative, promising more stable products, whereas glycerol is still the inexpensive standard for starch-based materials processing, and DESs have the advantage in processing and retrogradation retardation.

10.4 Modification of starch

Starch, being a renewable, biodegradable, and cheap biopolymer, however, has certain inherent limitations in its native form when it comes to packaging applications. The strong hydrophilic nature, weak mechanical properties, low thermal stability, and high tendency toward retrogradation of starch have adverse effects on starch-based packaging materials. Thus starch modification by physical, chemical, and enzymatic methods and starch blending with other biopolymers are essential for enhancing the mechanical strength, barrier properties, and thermal stability. In this section, starch modification approaches for packaging performance (Karnwal et al., 2025) are highlighted.

10.4.1 Physical modifications

The restructuring of the starch molecular and crystalline structure due to physical modification methods does not involve modification of the chemical structure. Advantages of these methods are that they are nonpolluting and can be applied directly to food contact materials.

Heat-moisture treatment (HMT) is a heat treatment that is carried out at a moisture level of 10%–35% and at a temperature above the gelatinization temperature. HMT enhances the binding of starch chains, thus increasing intragranular order. This results in a reduction in swelling and solubility, with the latter two exhibiting greater mechanical and thermal resistance. Modified starches by HMT are said to possess higher and stable performances particularly in packaging applications at high temperatures and high tension or compressing force. For instance, a combined effect of HMT (120°C, 25% humidity) and ultrasonic treatment on elephant foot yam starch induced molecular rearrangement by producing surface cracks on the starch granules (Medha et al., 2026). Following this modification, the swelling capacity and solubility were decreased, while the shear and heat resistance were increased, as well as the content of resistant starch was enhanced. This indicates that packaging films with greater dimensional stability and heat resistance can be obtained.

Annealing is carried out with high moisture content (40%–70%) and at a temperature below that of gelatinization. This leads to an increase in the gelatinization temperature and the thermal stability because of regularization of the crystalline areas. Annealed starches demonstrate uniform mechanical properties and are more processable as packaging films. For instance, research by Ye et al. reported that pressure annealing enhanced the crystalline perfection in corn and tapioca starches, which modulated the gelatinization behavior and reinforced mechanical properties depending on the type of starch (Ye et al., 2026).

Chemical modifications consist of reacting with other chemicals through their hydroxyl groups, which leads to improved properties (thermal, mechanical, barrier) that impacts directly packaging performance.

Acetylation increases the mobility of the chains by disrupting hydrogen bonds between starch chains. The obtained starch-based modified starches produce films that are more flexible and less moisture sensitive. Acetylated starch films show better mechanical flexibility and superior water vapor barrier performance. Acetylation of cassava starch decreased the gelatinization temperature and enthalpy, enhanced water absorption, but decreased retrogradation. Pulsed electric field (PEF)-assisted acetylation markedly enhanced the degree of acetylation and developed films that were more flexible and transparent. This has favorable implications for packaging in terms of mechanical flexibility and processability (Gagnaten et al., 2023).

During starch oxidation, carbonyl and carboxyl groups are dedicated to starch chains, which causes a reduction in molecular weight. The above processes yield films with a flat surface and good uniformity. Oxidized starches demonstrate a good adhesion profile and a modulated mechanical performance in coating applications and in mono or thin multilayer packaging. Alkali treatment followed by chemical modification and glutaraldehyde crosslinking enhanced TS (~30 MPa) and water vapor barrier properties of starch-based packaging films. Further, the inclusion of organically modified montmorillonite increased barrier properties against moisture and liquid water even more, making such systems ideal for packaging in high humidity conditions (Srinivasan et al., 2025).

Crosslinking promotes water and heat resistance through the formation of covalent bonds between starch chains. Cross-linked starches have low solubility due to their insolubility in water, high mechanical strength, and good thermal stability. Such traits are essential to packaging applications with high humidity and thermal loads (Mylkie et al., 2025).

Enzymatic hydrolysis can be controlled via enzyme modification of starch, which allows the molecular weight distribution and rheological properties of starch to be modulated. The processability and mechanical stability of the starch were improved by the above method of modification. Because of its greater environmental compatibility with respect to the chemical treatments, the enzymatic modification represents the best pathway for sustainable packaging materials. Enzymatic modification of the acetylated distarch phosphate derived from cassava starch by α-amylase led to a reduction of the stiffness of the cross-linked system, accompanied by an increase in the gel homogeneity and softness (Zhang et al., 2026). The enzymatic process enhances the flow properties and retains the functional groups of the chemical. This strategy is highly representative of a more stable mechanical performance in packaging films. However, the poor performance of starch in packaging can be improved effectively by blending starch with other biopolymers. The mixture of starch with PLA notably enhances its mechanical strength and water vapor barrier properties (Kurup et al., 2025). The combination of starch and polyhydroxyalkanoates (PHAs) leads to increased biodegradability and enhanced thermal stability (Yeo et al., 2024). Mixing chitosan with the packaging contributes to its antimicrobial characteristics as an active packaging material (Siciliano et al., 2025). It also enhances the TS of the starch matrix and gas barrier property when blended with cellulose and its derivatives (Ramakrishnan et al., 2023). These blending techniques contribute to an all-sided enhancement of the mechanical, barrier, and thermal properties of starch-based packaging materials.

Approaches to biopolymer blends, such as physical modification, chemical modification, and enzymatic modification of starch, make it possible to obtain the required mechanical strength, barrier properties, and thermal stability for packaging applications. The choice of appropriate modification techniques allows a significant improvement in the applicability of starch-based systems as green packaging materials at the commercial level.

10.5 Starch-based nanocomposites

Starch is the most common naturally occurring polymer used in different types of applications, mainly in food packaging because of its availability, low cost, biodegradability, and being a renewable source (Kumar et al., 2025a, c, 2026). But pure starch films for packaging have major drawbacks such as weak mechanical properties, high WVP, clearly hydrophilic nature, and very limited resistance against microbial attack (Dutta and Sit, 2024). These structural limitations prevent starch from being used in large-scale and long-term shelf-life applications (Adewale et al., 2022). Consequently, physical, chemical, and biocomposite modifications of starch-based materials have received considerable interest for potential enhancement of their performance (Xu et al., 2023).

Using nanotechnology in polymer science to modify and improve the functional properties of starch-based materials has been found to be a fruitful approach (Pawase et al., 2025). In particular, the incorporation of nanoparticles into the starch matrix enhances the polymer chain interaction, occupies the free space between the chains, and effectively improves mechanical properties, thermal stability, and barrier properties (Ali et al., 2025). Metal and metal oxide nanoparticles (e.g., ZnO, TiO_2, Ag, and SiO_2), nanoclays, cellulose nanocrystals, and biobased nanofillers are the most common types of reinforcement used in the formulation of starch-based nanocomposites (Panda and Mohapatra, 2025; Saleh, 2025; Jamalpoor et al., 2024). These nanoparticles do not provide only mechanical reinforcement, but also confer other functionalities such as antibacterial, antioxidant, UV barrier, and active ingredient release (Liu et al., 2025).

Nevertheless, how the addition of nanoparticles influences the starch matrix depends on the kind, size, surface characteristics, homogeneity of distribution, and concentration of the nanoparticles. Even in well-designed systems, agglomeration, enhancement of fragility, or potential toxicity may occur (Marta et al., 2022). Thus starch/nanoparticles interactions have to be assessed by means of a holistic approach in which material performance and environmental and biological (eco)toxicity are taken into consideration (Gupta et al., 2024). Thus, the literature summaries below are focused on the depiction of the trends and emerging potential for application, by reporting of the achieved structural, mechanical, barrier, and functional properties by the incorporation of different nanoparticles in starch-based (Zhu, 2025).

Metal and metal oxide nanoparticles also show a drastic enhancement in the mechanical properties of starch-based matrices. For instance, the addition of ZnO nanoparticles into starch/chitosan/vegetable mucilage composite film improved the TS from 0.28 to 1.29 MPa, EAB from 18.76% to 23.52%, and the water contact angle from 62.83 degrees to 75.17 degrees, indicating a significant enhancement of water resistance (Priyanka et al., 2024). Also, for starch-based packaging films containing dispersed Ag nanoparticles (AgNPs), the TS of the films was enhanced from 1.12 to 2.84 MPa, the EAB also increased from 21.4% to 29.6%, and the WVP was reduced by about 35% (Ortega et al., 2025). These findings suggest that nanoparticles may simultaneously improve both the strength and flexibility.

Barrier properties were greatly enhanced, to some extent, and the degree of improvement is even better depending on the type and the content of nanoparticles. Al_2O_3 nanoparticles as the composition of the films display good barrier properties with lower water vapor transmission rates (WVTR) and oxygen transmission rates at the optimal concentration (e.g., 0.4 wt%) (Jam (Ali et al., 2025). Cross-linked starch packaging films with Au nanoparticle reinforcement achieved a remarkable decrease in WVP while also boosting flexibility by about 130% as measured at EAB (Kumar et al., 2025a, b, 2026). Nevertheless, some reports indicate that because of the formation of imperfections within the film matrix, the presence of ZnO nanoparticles can enhance the permeability of water vapor at higher loadings (Priyanka et al., 2024). This underlines the critical significance of the optimum loading ratio in nanoparticles doping.

Nanoparticle reinforcement is not limited to mechanical and barrier properties; it also imparts active and intelligent functionalities to packaging films (Liu et al., 2025). ZnO nanoparticles have demonstrated vigorous antibacterial activity against both Gram-positive and Gram-negative bacteria by producing reactive oxygen species. Inhibition zones ranging from 26 to 30 mm for Bacillus sp. and *Pseudomonas* sp., and 28–31 mm for *Escherichia coli* and *Staphylococcus aureus*, have been reported (Priyanka et al., 2024). Similarly, carbon nanodots derived from ginkgo biloba leaves significantly delayed the oxidative degradation of ω-3 fatty acids due to their superior antioxidant activity (Lyu and Han, 2024). In innovative packaging applications, starch/sodium alginate packaging films containing Cu-phenylalanine metal-organic framework (Cu-Phe MOF) nanorods have been able to monitor the freshness of foods, such as shrimp and pork, in real time thanks to their ammonia gas-sensitive color change (Su et al., 2024).

The sustainability and safety profiles of nanoparticles-reinforced starch-based packaging films have also been extensively discussed in the literature. Many studies have highlighted that ZnO nanoparticles and carbon nanodots can be obtained from agricultural waste, such as mint leaves and palm seeds, through green synthesis methods, thereby ensuring both environmentally friendly and cost-effective production (Lyu and Han, 2024; Cheviron et al., 2014; Priyanka et al., 2024). Starch/chitosan/mucilage composite packaging films containing ZnO nanoparticles have been reported to be 100% biodegradable, while Al_2O_3-doped films have been shown to extend the shelf life of fruits such as bananas (Ali et al., 2025; Priyanka et al., 2024; Jam). Nevertheless, biocompatibility is dependent on the nature of the nanoparticles as well as the synthesis route. For instance, although packaging films with ZnO NPs exhibited no ecotoxicity toward Vigna angularis seeds, it was reported that AgNPs generated by laser ablation were more detrimental to the viability of Caco-2/TC7 cells than chemically produced ones (66% cell viability). The above statements demonstrate that the nanoparticles synthesis strategy is an essential element in safety (Ortega et al., 2025; Priyanka et al., 2024).

Research suggested the feasibility of employing NPs to improve the TS, barrier properties, and active functionality of starch-based packaging films. In addition, it is feasible to create green packaging systems with environmentally friendly properties by considering the suitable type of NPs, their concentration, and the synthesis technique.

10.6 Functional properties of starch-based packaging

The starch-based packaging systems films in the last years, smart packaging systems applied to starch-based packaging systems have been going further from the active packaging concept, enabling packaging materials to sense food spoilage, environmental conditions, and even to be influenced by quality changes of food (Pei et al., 2024; Lv et al., 2025). These systems are sensitized to critical parameters such as gas, pH, moisture, and oxidative degradation by taking advantage of the biodegradable and functionalized nature of the starch matrix. Literature reviewed show that smart functions are mostly conceived to control or trigger the release of active agents, to provide visual quality signals, and to modify their behavior as a function of environmental conditions. One of the most common strategies for the detection of volatile basic compounds emitted during microbial degradation of protein food in smart packaging (Jafarzadeh et al., 2024). In this context, Cu-Phe MOF nanorods integrated into starch/sodium alginate-based packaging films have been equipped with a freshness monitoring function by exhibiting a color change in response to ammonia gas. The system was tested on shrimp and pork; it was reported that the color change of the packaging showed a high correlation with the increase in total viable bacteria count and total volatile basic nitrogen (TVB-N) values (Su et al., 2024). Although the shelf life is not directly specified in days, this approach is considered one of the most prominent examples of smart packaging among starch-based packaging, as it can indicate the spoilage threshold in real time.

A more advanced example of smart packaging functionality can be seen in systems with pH-sensitive release mechanisms. Starch-based films containing tea tree oil-loaded covalent organic frameworks (CTB@TTEO) can adjust the release rate of the active substance depending on the pH of the medium. Films containing 2% (wt) CTB@TTEO exhibited the most compact structure and the highest antibacterial efficacy; this system was applied to blueberry packaging. During storage, microbial load, hardness loss, and color change were significantly delayed compared to the control group, and the retention period of quality parameters was extended. The compatibility of the emission kinetics with the Avrami model indicates that this system offers a truly intelligent and active hybrid structure capable of responding to environmental changes (Cao et al., 2025).

In controlled-release systems, which can be considered semiintelligent, starch-based packaging provides a gradual release of the active substance depending on the ambient conditions inside the packaging. The fact that starch packs containing microcellulose and β-cyclodextrin showed film adsorption capacities of 0.084 and 0.082 g/g for menthol and D-limonene, respectively, revealed that these systems are effective carriers for volatile antifungal agents. These packaging solutions were explicitly developed for blueberries, extending their shelf life from 14 to 23 days at room temperature and from 24 to 47 days in cold storage. Here, the intelligent function is defined by the automatic adjustment of the release rate in response to temperature and storage conditions (Cao et al., 2025).

Active-smart systems sensitive to oxidative degradation have been developed especially for oxidation-sensitive foods such as fish and meat products. Tea polyphenols, infused into starch/exfoliated nanofibers produced by electrospinning, exhibited potent antioxidant and antibacterial performance because of their high surface area. In preprocessed fish fillets, total bacterial count, TVB-N, and thiobarbituric acid reactive substances (TBARS) values were found to be significantly lower during an 8-day storage period at 4°C compared to the control group; the system also demonstrated thermal stability up to 302°C. In this approach, the intelligent function is based on providing adaptive protection against oxidative and microbial spoilage parameters (Wang et al., 2025b).

Smart packaging approaches that focus on moisture and leakage control also play a crucial role in starch-based systems. Cryogels produced from purple yam starch cross-linked with citric acid have exhibited water absorption capacities ranging from 846% to 1091%, making them a functional packaging component for high moisture foods. This system, used in fresh cheese packaging, completely absorbed the leakage and completely inhibited *E. coli*, while partially inhibiting *Listeria monocytogenes*. Although the shelf life was not expressed in days, it was reported that microbial safety and product stability were significantly increased (Rosas et al., 2026). Smart packaging functions are defined in some systems not through direct sensing, but through the continuity of the active function. E-Polylysine grafted starch/PLA/PBAT films have demonstrated antibacterial activity that remains unchanged over time, thanks to the covalent binding of the active agent. This packaging was tested on beef, and its shelf life was increased from 7 to 14 days at 4°C. This approach emphasizes the significance of structural strategies that prevent functional loss in smart packaging design (Liu et al., 2025).

Starch-based smart packaging systems have been shown to extend shelf life by 9–23 days for fruits, approximately double shelf life for meat and seafood, or offer real-time spoilage monitoring, while significantly delaying microbial and oxidative spoilage in fish and dairy products. These results clearly demonstrate that starch-based packaging can be designed not only as an environmentally friendly alternative but also as advanced smart systems capable of monitoring and guiding food quality.

10.7 Applications for food packaging

Rising global demand for single-use packaging materials has contributed to an increase in the use of petroleum-based plastics, especially in food packaging. Global production of plastics, which was around two million tons per year in the 1950s, has apparently reached 359 million tons in

2018, and 415 million tons in 2019. It seems that 32% of single-use plastics bypass the waste collection system, and 3% of global plastic production leaks directly into the ocean (Shen et al., 2020; Geyer et al., 2017). This situation has rendered the creation of alternative biobased polymers, particularly for short-lived packaging materials, inevitable. Accordingly, TPS has long been recognized as one of the best candidates for disposable packaging applications. TPS systems are in essence made from starch (amylose/amylopectin matrix), plasticizer, and water and can be processed by utilizing conventional thermoplastic processing methods such as casting, extrusion, blow molding, and injection molding similar to petroleum-based thermoplastics. The literature suggests that TPS can indeed be considered as a low-cost alternative for disposable plastic packaging (such as polystyrene and low-density polyethylene) (Rahardiyan et al., 2023; Orue et al., 2016).

Usually, in TPS systems, the polymer phase is the starch derived from corns, cassava, potatoes, and other tubers, and the film and packaging performance are directly influenced by the amylose/amylopectin (Vieira et al., 2021). The crystalline structure of starch granules is destroyed in the process of producing TPS under the influence of moisture, heat, and shear force, which allows it to behave as a thermoplastic. Plasticizers are an important component to improve the mechanical properties of TPS-based disposable packages. The most applied plasticizer is glycerol, whose application is favored because of its high boiling point and the fact that it is nontoxic and highly compatible with starch (Jabeen et al., 2015; Shah et al., 2021). Glycerol is typically employed in an amount of 20%–40% based on the starch content in the formulations of TPS. Glycerol addition enhances the percent elongation and flexibility of TPS films, but large plasticizer contents may lead to a reduction of TS and moisture barrier properties (Atteya et al., 2025). This means that TPS is ideal for single-use packaging, particularly for dry foods or those with low moisture. The key constraint of TPS in single-use package application is that starch is hydrophilic, and it has high WVP and a low moisture barrier. In the literature, it is pointed out that the moisture permeability of TPS films is higher than that of petroleum-based plastic films, and at the same time, it is more biodegradable and compostable (Azimi et al., 2025). The fact that starch-based packaging is biodegradable, while petroleum-based plastics take hundreds of years to decompose on Earth, has been cited in literature. Hence, to tackle these issues, TPS-based materials are mostly formulated as biopolymer mixtures or multilayer films (Suhag et al., 2020). The combination of starch with cellulose, chitosan, proteins (such as casein, gluten, and whey proteins), and synthetic biodegradable polymers enables the enhancement of the mechanical and barrier properties of starch matrices. In addition, it was also demonstrated that the TS, oxygen barriers, and thermal stability of TPS systems could be enhanced remarkably by the incorporation of nanofillers, for example, cellulose nanocrystals and cellulose nanofibers (Rahardiyan et al., 2023). In the area of single-use food packaging, applications for short-duration use are under consideration for TPS materials in the form of carrier bags, packing films, biodegradable straws, coating films, and edible films. The transparency, low density, and good film-forming ability of TPS are the factors to promote use in disposable products, and the short life issued by suitable for it. The transparency and low density of TPS, combined with the ability to form films well, render it extremely appropriate for single-use goods with a short shelf life. In addition, the transport of actives (e.g., antimicrobial or antioxidant agents) is possible through the TPS matrix, which has potential for active and smart packaging developments (Wongphan et al., 2024). To sum up, the literature suggests that starch packaging based on TPS could be a promising approach to reduce the environmental burden of petroleum-based single-use plastics. Although some aspects of mechanical and water vapor barrier properties still need to be improved, it is evident that the potential of TPS in single-use packaging material is increasing, especially with the aid of plasticizers, biopolymer blends, and nanofiller approaches.

The environmental advantages offered by starch-based thermoplastic systems, such as single-use packaging materials, have brought to the forefront the evaluation of these biopolymers not only in classic packaging forms but also in more functional systems that can come into direct contact with food (Rahardiyan et al., 2023). Especially in short-lived packaging applications, the biodegradable structure and processability of TPS provide significant advantages. At the same time, the natural, nontoxic, and edible nature of starch allows this material to be used in coating and film systems that go beyond the concept of packaging and become an integral part of the food (Kumar et al., 2025a, b, 2026). This approach directly supports the development of edible film technologies, aiming to eliminate packaging waste, and enables the evolution of starch-based materials from single-use packaging to edible protective films (Kumar et al., 2025a, b, 2026). Accordingly, starch-based edible films stand out as advanced packaging solutions that extend the shelf life of food, reduce quality losses, and facilitate the transport of active ingredients, while maintaining the sustainability advantages offered by TPS systems.

Growing environmental concerns, the waste problem caused by single-use plastics, and changing consumer demands have placed edible films and coatings at the heart of sustainable packaging technologies (Mahmud et al., 2024). Edible films are thin, biopolymer-based layers that can be applied to food surfaces and are edible, playing a significant role in sustainable packaging systems (Usman et al., 2025). Starch-based edible films have been widely studied in the literature because of their availability from renewable resources, their biodegradability, and their excellent film-forming capabilities (Pei et al., 2024). In current studies, edible film matrices are primarily based on starch/glycerol/water systems, with corn starch, arrowroot starch, and various plant-derived starches serving as the primary polymer phase. Glycerol is the most used component for plasticization, typically included in the formulation at a rate of 20%–40% of the starch content (Abdillah and Charles, 2021).

In a 2025 study, 11 different edible film formulations were prepared using a fully biobased polymer system consisting of arrowroot starch and glycerol. In this study, the rheological, mechanical, and structural properties of the films were investigated by varying the concentrations of starch and glycerol. In films produced by the dip coating method, the wet film thickness ranged from 0.78 to 1.95 mm, and the dry film thickness after drying ranged from 0.12 to 0.55 mm. The shrinkage rates during drying varied between 70.9% and 88.2%, which was attributed to the hydrophilic character of starch-based systems. In the same study, the surface adhesion properties of the films were evaluated, and the adhesion force was reported to range from 0.0195 to 3.25 N. Increasing the starch concentration enhanced structural stability and mechanical strength, while increasing the glycerol content significantly improved the flexibility and elongation ability of the films (Grazi et al., 2025).

The rheological properties of native starch-based edible films are essential for their mechanical properties. Starch/glycerol solutions have been reported to be shear-thinning and to have a high yield stress from rheological measurements. Predictions by the Herschel-Bulkley and Cross models showed a good fit ($R^2 > 0.94$) with experimental data. The formulations (high yield stress and consistency index) had the highest TS and the Young's modulus, two parameters correlated with stiffness, whereas systems showing lower yield stress lead to more flexible but mechanically weaker films. These results underline the importance of the rheological parameters in the preparation of edible films (Grazi et al., 2025), which indeed can be considered as a confirmation of the aforementioned statement.

Also, in a study in 2025, corn starch/glycerol-based edible films were developed in the form of composite matrices with active agents. The polymer matrix was 5% (w/v) gelatinized corn starch with 30% (w/w) glycerol as plasticizer, and the matrix was enriched with functional additives, olive pomace, and wine lees. Olive pomace was also added to the film at 3%–5% (w/w), and wine lees at 5%–10% (w/v). These natural additives were found to have total phenolic contents of 1.01–2.98 $mgGAEg^{-1}$ (olive pomace fractions) and 6.02 $mgGAEg^{-1}$ (wine lees). The WVP of these starch-based composite edible films was higher than that of the pure starch films. The WVP value for pure starch-glycerol film was $0.74 \times 10^{-11}\ gs^{-1}m^{-1}Pa^{-1}$, and increased in films loaded with 5% olive pomace to $2.30–2.50 \times 10^{-11}\ gs^{-1}m^{-1}Pa^{-1}$. In contrast, a drastic enhancement in optical and UV barrier properties was noticed with approximately 9.1% increased UV barrier property for packaging films, resulting in 10% wine pomace. The enhancement was attributed to the light absorption property of phenolic compounds. Mechanical tests reported starch-based edible films with 5% freeze-dried olive pomace and 10% wine pomace showed about 18% and 38% increase in TS and Young's modulus, respectively. But scanning electron microscope observations proved that maximum elongation values diminished for films with olive pulp as a consequence of particle agglomeration. Simultaneously, a more uniform microstructure was attained for the packaging films with wine sediments (Ali et al., 2025).

Analyzing these studies collectively, edible films are mainly starch/glycerol-based polymer systems and their mechanical, barrier, and surface adhesion properties can be tuned within wide ranges by adjusting the starch concentration, plasticizer ratio, and type of other functional additives. The numerical results clearly indicate that starch-based edible films have very tunable properties, allowing the films to be used for both passive and active packaging.

10.8 Environmental impact and life cycle assessment

When the environmental performance of starch-based materials is examined within the scope of life cycle assessment (LCA), it becomes apparent that these systems offer significant advantages compared to fossil-based plastics, particularly in terms of global warming potential and fossil energy consumption (Lie-Piang et al., 2021). The biogenic carbon sequestration of plant-derived starch raw materials through photosynthesis makes the carbon footprint of these materials largely dependent on energy inputs during production, processing, and disposal. Literature reports that starch-based bioplastics, when evaluated on an equivalent mass basis, provide 1.63–2.10 $kgCO_2$-eq.kg^{-1} lower greenhouse gas emissions and 29.6–76.1 MJ kg^{-1} lower fossil energy demand compared to petroleum-based plastics such as polyethylene and polypropylene (Poveda-Giraldo et al., 2025). However, the degree of starch processing is one of the key parameters determining environmental impact; it has been reported that the global warming potential of highly refined starch derivatives ranges from 1.6 to 3.3 $kgCO_2$-eq.kg^{-1} product because of the hydrolysis, purification, and intensive drying steps applied, and in some cases, even reaches levels comparable to conventional sugar and petrochemical counterparts (Suckling et al., 2025). In contrast, starch-based components with lower degrees of refining or obtained through whole-fractionation approaches show significant reductions in indicators of global warming potential, water consumption, and fossil resource scarcity when the agricultural production and processing stages are considered together (Lie-Piang et al., 2021). The starch-based material's biodegradability also influences the overall impact on the environment. It has been found that the biodegradation rates are greater than 70% in industrial composting, while these rates are around from 15% to 40% in home composting and soil biodegradation. In situ biogas utilization by energy recovery in anaerobic digestion scenarios provides environmental credits through the displacement of electricity and thus results in reducing net GHG emissions of about 2.7–2.9 $kgCO_2$-eq kg^{-1}, contributing to the life cycle environmental burden improvements. This renders the approach as one of the most environmentally favorable end uses (Poveda-Giraldo et al., 2025).

On the other hand, the potential environmental advantages of starch-based materials are greatly diminished by poor biodegradation and methane escape from landfill environments. This integrated evaluation makes it crystal clear that the superiority of starch-based plastics in environmental terms can be realized, not only when aligned with the renewable raw material origin, but also with low energy-intensive production technologies and suitable end-of-life management options.

10.9 Challenges, limitations, and future perspectives

Despite the starch-based biodegradable films having great potential as environmentally friendly substitutes to petroleum-based plastics, they have several issues that make it difficult for them to be widely utilized in industrial applications (Shah et al., 2016). The main problem is the hydrophilic nature of starch due to the abundance of hydroxyl (-OH) groups in its molecular structure. Films thus become highly sensitive to moisture, taking up water, swelling, and rapidly losing their mechanical properties when exposed to humid conditions (Othman et al., 2023). Another distressing drawback is the poor mechanical properties of natural (nonderivatized) starch film. Low TS and poor elongation values limit the applicability of starch films for numerous food packaging applications (Shah et al., 2015). Although physical, chemical, and nanofill-assisted modification methods can substantially enhance the mechanical properties, these methods commonly bring complexity in the process and cost in production. In addition, the poor water vapor and oxygen barrier of starch-based films restricts these materials mainly to dry or low moisture foods with a short shelf life (DENG et al., 2022).

Starch film from a manufacturing point of view, starch films are highly susceptible to process parameter changes. Slight variations in moisture content, temperature, drying condition, and amount of plasticizer may result in noticeable changes in film thickness, transparency, and mechanical property (Yao et al., 2022). This makes process standardization and scale-up difficult, particularly in processes involving solution casting, extrusion, and blow molding. In addition, retrogradation-based recrystallization associated with storage represents a critical issue because it results in a deterioration of the film's properties with time (Pawase et al., 2025).

This particularly complicates the standardization of process and scale-up in techniques such as solution casting, extrusion, and blow molding. In addition, retrogradation and recrystallization during storage are major concerns, resulting in undesirable variations of film properties with time. Although inorganic nanofillers (such as ZnO, TiO_2, carbon nanotubes, and graphene oxide) enhance mechanical and barrier properties to a great extent, nanoparticle migration, toxicity, and regulation are still major issues. The paucity of long-term toxicological information on nano-enabled packaging materials, combined with a nonharmonized international regulation, is a barrier to the commercialization of such systems. Thus it is clear that to overcome these issues, future work will inevitably have to be in the direction of integrated and multi-disciplinary modification methodologies. The combination of physical, chemical, and nanotechnological methods in one hybrid system can offer advantages in terms of its mechanical strength, moisture barrier, and processability. Microwave-assisted chemical modifications, two-step nonthermal processes (high-pressure/heat and water activity modification), and a controlled crosslinking process are seeing implementation as methods that can potentially improve functionalities of starch films while providing the benefits of energy saving and sustainability.

In parallel, the development of biobased nanofillers such as nanocellulose, starch nanocrystals, and protein-based nanostructures) are more advantageous in terms of safety and regulations, and the improvement of their dispersion within the starch matrix, constitute a critical research area for future studies. For industrial applications to become widespread, scalable processes such as extrusion and roll-to-roll production need to be optimized, rather than relying on laboratory-scale solution casting methods. At the same time, the widespread use of life cycle analyses and technoeconomic evaluations, which reveal energy consumption, carbon footprint, and economic feasibility, will enable a realistic comparison of starch-based films with conventional plastics.

In conclusion, the future of starch-based packaging films as high-performance and commercially viable materials depends not only on physical and mechanical improvements but also on a holistic approach encompassing regulatory compliance, food safety, consumer acceptance, and sustainable production. Studies conducted in this direction will enable starch-based films to be positioned as a strong and competitive alternative in sustainable food packaging systems.

10.10 Industrial applications of starch-based materials: From academia to commercialization

Increased environmental awareness, the scarcity of fossil resources, and the need for sustainable production have significantly increased interest in bio-based materials in the packaging, paper, and printing industries. In this context, starch stands out as one of the most important biopolymers due to its abundant availability, renewability, biodegradability, and low cost. Starch-based systems are developed in the form of films, coatings, foams, and composite structures and are widely used, especially in food packaging and paper coating applications. Styrene-butadiene (SB) and styrene-acrylic latex binders, traditionally used in paper coating systems, are environmentally disadvantageous due to their petroleum-based nature (Altay et al., 2021). These binders are known to make recycling difficult and increase the carbon footprint. Therefore, intensive academic and industrial studies have been conducted in recent years on the development of starch-based binder systems.

10.10.1 Industrial development of starch-based systems

In recent years, starch-based materials have moved from laboratory-scale studies to widespread use in industrial applications (Kumar et al., 2026). This transformation is closely related to increasing environmental concerns, regulatory restrictions on single-use plastics, and the fact that starch is a renewable and biodegradable resource (Kegl et al., 2026). Starch and its derivatives are currently used in many commercial applications, such as paper coating, biodegradable films, foams, and composite materials (Li et al., 2019). One of the most common industrial uses of starch is as a binder for paper and cardboard coatings (Altay et al., 2017). Modified starches are widely used to increase mechanical strength, improve surface smoothness, and enhance print quality (Jiang et al., 2025). Companies such as Cargill, Ingredion, Roquette, and Royal Avebe offer functional starch products specifically developed for food-contact paper packaging. These systems provide more sustainable solutions by replacing synthetic latex binders. Cargill develops paper coating binders, bio-based adhesives, and barrier coatings by offering modified starches and thermoplastic starches (TPS) systems. These systems particularly improve oxygen barrier properties while maintaining biodegradability. However, performance degradation under high humidity conditions is a significant limitation. Similarly, Ingredion develops functional and "clean-label" starch systems used in film-forming and food-contact-safe coatings. These systems are optimized for compatibility with industrial production processes. Roquette and Royal Avebe work on plant-based starch derivatives and high-performance starch systems, which are widely used, especially in paper coating and barrier applications. In these systems, starch acts as a binder, improving surface properties, print quality, and oil resistance (Mujtaba et al., 2022).

Another important application area is thermoplastic starch (TPS) systems. TPS is obtained by processing starch under heat and shear in the presence of plasticizers. TPS is widely used in the production of biodegradable film and molded packaging (Atteya et al., 2025). For example, commercial biopolymer systems such as Mater-Bi® developed by Novamont are used for flexible packaging and compostable films.

In the packaging sector, companies such as PaperFoam and Storopack offer sustainable alternatives to petroleum-based materials such as polyethylene and expanded polystyrene (EPS) by developing starch-based foam and molded packaging products. These products are widely used, especially in protective packaging applications, thanks to their properties, such as low density, shock-absorbing structure, light weight, and biodegradability (Hussain et al., 2024).

In the field of bioplastics, companies such as Novamont (Mater-Bi), NatureWorks (PLA-based systems), and BASF (PBAT-starch blends) have developed starch-based systems widely used in the production of flexible packaging and compostable films. In addition, within the scope of emerging technologies, edible films, water-soluble packaging, and polysaccharide-based coatings are being developed by combining starch with other biopolymers such as alginate, cellulose, and proteins, and these systems offer significant potential, especially in terms of food preservation and active packaging applications (Usman et al., 2025).

10.11 Starch-Based Binder Systems and Coating Technologies

Starch-based binders are widely used in paper and cardboard coating applications. On an industrial scale, modified starches offer a sustainable alternative to synthetic latex binders. These coatings are used in food-contact applications, such as fast-food packaging, by improving surface strength, print quality, and oil resistance (Du et al., 2017; Altay et al., 2021).

10.11.1 Nanoparticle starch binders

One of the most important innovations developed in recent years is the use of starch-based nanoparticle binder systems. These systems eliminate solubility, viscosity control, and gelatinization problems associated with traditional starch use. Studies have shown that starch-based nanoparticle binders can be used directly in place of synthetic latex. In studies, replacing the synthetic binder with starch nanoparticles at a 1:1 ratio resulted in performance equivalent to or better than that of the synthetic binder, with no loss in mechanical or structural properties (Altay et al., 2024). Starch-based nanoparticle binders have been shown not only to provide environmental advantages but also to be competitive in terms of performance. Evaluations in terms of print quality have reported that these systems improve color intensity, color gamut, and dot properties, and increase toner transfer efficiency (Altay et al., 2024).

Another important advantage is that these binders reduce paper weight, allowing for lighter products at the same thickness. This helps reduce logistics costs and the carbon footprint.

10.11.2 Structural and Mechanical Properties

The mechanical performance of bio-based starch binders is highly competitive with that of synthetic binders. Studies show that starch-based binders exhibit performance comparable to that of other binders in key mechanical properties, including burst strength, tensile strength, and tear resistance. Furthermore, it has been determined that starch-based systems improve wettability by increasing surface energy, thereby improving ink penetration and print quality.

10.11.3 Coating Performance and Print Quality

The type of binder used in the coating layer directly affects parameters such as porosity, surface energy, and roughness, thus determining print quality. Starch-based nanoparticle systems have been shown to improve toner transfer, increase color saturation, and yield a more homogeneous print surface. This offers significant advantages, particularly in digital printing technologies (electrophotography) (Altay et. al., 2024).

10.12 Starch-Based Packaging Systems

10.12.1 Film and Coating Systems

Starch-based films are highly effective at preventing oxygen transmission. However, due to their high moisture sensitivity, they are usually used mixed with polymers such as PLA or PBAT. Such hybrid systems improve properties such as flexibility, mechanical strength, and moisture resistance, increasing commercial viability. For example, thermoplastic starch (TPS) and starch-polyester blends are widely used in the production of compostable films. For instance, Mater-Bi systems developed by Novamont are used in flexible food packaging and offer sufficient mechanical performance along with biodegradability (Surendren et al., 2022).

10.12.2 Foam and Molded Packaging

Starch-based foams are becoming increasingly common, particularly in protective packaging applications. These materials offer an alternative to petroleum-based foams due to their low density, impact absorption, and biodegradability properties. For example, Storopack's starch-based foam and filler materials serve as an environmentally friendly alternative to petroleum-derived expanded polystyrene (EPS). These materials are biodegradable and compostable and exhibit sufficient impact absorption properties (Rabelo et al., 2025).

Starch-based materials are biodegradable and renewable resources that offer significant advantages for environmental sustainability. These materials are particularly noteworthy for food packaging applications due to their strong oxygen barrier performance. However, their high moisture sensitivity and relatively low mechanical strength, stemming from their hydrophilic structure, are among the main disadvantages that limit their areas of use. Compared to traditional plastics, starch-based systems are superior in biodegradability but lag in moisture resistance and mechanical properties. Therefore, starch-based materials stand out as a more suitable alternative, especially in dry food packaging. To achieve wider application areas, their performance needs to be improved through chemical modification, the use of plasticizers, or hybrid systems created with other polymers (Kumar et al., 2026; Karnwal et al., 2025).

The industrial-scale application of starch-based materials poses multifaceted challenges that require a comprehensive assessment of material performance, production processes, costs, and long-term stability. The most fundamental limitation of these systems is their high moisture sensitivity, due to their hydrophilic nature; this negatively impacts both barrier properties and mechanical strength, restricting their use, particularly in humid environments. Furthermore, maintaining mechanical and functional improvements achieved at the laboratory scale is often not possible in continuous production lines (Garavito et al., 2024). Factors such as compatibility with existing industrial equipment, processability, plasticizer requirements, and the need for precise control of process parameters complicate the production process and increase costs. Practical issues such as raw material standardization, batch-to-batch variability, and long-term storage stability also play a critical role in scaling up. At this point, parameters such as economic feasibility, process integration, and product lifecycle-often overlooked in academic studies but crucial in industry-constitute the main obstacles to the widespread adoption of starch-based systems. Therefore, a successful scaling-up strategy requires a holistic approach to material design, process engineering, and industrial requirements (Namjoufar et al., 2026; Requena et al., 2016).

Industrial applications of starch-based packaging systems are evident in products and technologies developed by various companies. These companies have commercialized starch-based materials for food and protective packaging by scaling them up into films, coatings, foams, and composite structures. Below, the systems developed by leading companies, their areas of use, and basic performance characteristics are presented comparatively.

Table 10.2 clearly shows that starch-based systems are not limited to laboratory scale but have been successfully commercialized in various industrial sectors in film, coating, foam, and biocomposite forms. While these systems offer environmental advantages such as biodegradability, renewability, and a low carbon footprint, limitations in moisture sensitivity, mechanical strength, and long-term stability remain significant areas of research. However, significant progress is being made towards overcoming these limitations thanks to hybrid polymer systems and nanoparticle-supported approaches. This strongly suggests that starch-based materials will play a central role in future sustainable packaging technologies.

10.12.3 Innovative Approaches

Next-generation initiatives are enhancing the functional properties of starch-based systems by combining them with other biopolymers. For example, Evoware is developing innovative approaches by combining starch-based systems with other biopolymers in edible packaging films; CuanTec is developing food preservative coatings; and Oimo is developing water-soluble packaging systems. These approaches offer next-generation sustainability solutions. In the commercialization of starch-based systems, moisture sensitivity, thermal stability issues, viscosity control during processing, and mechanical strength limitations are the main technical challenges. In traditional starch systems, the gelatinization process presents a significant production challenge. In addition to these challenges, there are economic difficulties, including cost competition with petroleum-based plastics, investment and scaling-up costs, and issues with raw material supply chains. There are some important differences between academic studies and industrial applications. Academic studies generally remain at the laboratory scale. Industrial systems, on the other hand, utilize larger-scale, multi-component, and optimized formulations. This limits collaboration between academic research and industry.

Starch-based materials have become a significant commercial reality in the packaging and paper coating sectors today. While large companies

Table 10.2 Starch-based commercial packaging systems.

Company	Product/Technology	Application area	Advantage
Cargill	Modified starch, TPS	Paper coating, barrier materials, adhesives	Biodegradable, industrial compatibility
Ingredion	Functional starch systems	Food coatings, biodegradable films	Food-grade, sustainable
Roquette	Plant-based starch derivatives	Films and coatings	Industrial scalability
Royal Avebe	Potato starch derivatives	Paper coating, barrier applications	Good film-forming and surface properties
PaperFoam	Starch-based molded materials	Food and electronic packaging	Compostable, lightweight
Storopack	Starch-based foam (RENATURE®)	Protective packaging	Biodegradable, shock-absorbing
Novamont	Mater-Bi (starch-based bioplastic blends)	Films, flexible packaging	Compostable
NatureWorks	PLA-starch blends	Films, packaging	Good mechanical properties, bio-based
BASF	PBAT–starch blends (e.g., ecovio®)	Flexible packaging	Balance of strength and biodegradability
Evoware	Seaweed- starch films	Edible packaging	Zero-waste, edible
CuanTec	Polysaccharide-based coatings	Food preservation	Shelf-life extension
Algopack	Algae-starch composites	Packaging materials	Sustainable, bio-based structure
Sea6 Energy	Seaweed-based biopolymers (with starch blends)	Packaging, biomaterials	Renewable resource
Oimo	Water-soluble starch packaging	Single-use products	Dissolvable, zero waste

optimize this technology, innovative startups are developing functional, sustainable solutions. Further integration of academic research with industrial applications will drive progress, particularly in developing highly moisture-resistant systems, optimizing multilayer and hybrid structures, and modeling scaling-up processes.

In conclusion, innovative approaches developed in recent years for starch-based packaging systems have transformed these materials from mere sustainable alternatives into high-performance engineering solutions. Nanoparticle-based binder systems, hybrid polymer blends, and functional coating technologies have significantly improved the mechanical strength, barrier properties, and processability of starch-based materials. However, to increase their applicability on an industrial scale, it has become necessary to consider parameters such as process optimization, energy efficiency, and cost-effectiveness alongside material design. Today, starch-based systems have proven their value in commercial applications, including paper coating, biodegradable film production, and foam packaging. They are poised to play a central role in developing multifunctional, intelligent, and environmentally friendly packaging solutions. In this context, strengthening academic-industry collaborations and adopting interdisciplinary approaches will be crucial in unlocking the true potential of starch-based materials.

References

Abdillah, A.A., Charles, A.L., 2021. Characterization of a natural biodegradable edible film obtained from arrowroot starch and iota-carrageenan and application in food packaging, 18790003 Int. J. Biol. Macromol. 191, 618–626. https://doi.org/10.1016/j.ijbiomac.2021.09.141.

Abdin, M., Metwally, A.A., Younis, M., Khalil, M., Arafa, S.G., 2025. Enhanced functionalities of cassava starch-based films incorporating berberine nanoparticles for sustainable food packaging. Food Chem. 498. https://doi.org/10.1016/j.foodchem.2025.147089 pp.147089–147089.

Adewale, P., Yancheshmeh, M.S., Lam, E., 2022. Starch modification for non-food, industrial applications: Market intelligence and critical review. Carbohydr. Polym. 291, 119590. https://doi.org/10.1016/j.carbpol.2022.119590, 01448617.

Ali, H., Ali, S., Ali, K., Ullah, S., Ismail, P.M., Humayun, M., Zeng, C., 2025. Impact of the nanoparticle incorporation in enhancing mechanical properties of polymers. Results Eng. 27, 106151. https://doi.org/10.1016/j.rineng.2025.106151, 25901230.

Altay, B.N., Bloembergen, S., Aydemir, C., Karademir, A., Fleming, P.D., 2017. Use of nanoparticle binders for paper coatings: a review. J. Graph. Eng. Des. 8 (1), 39–43. https://doi.org/10.24867/jged-2017-1-039.

Altay, B.N., Klass, C., Chen, T., Fleck, A., Aydemir, C., Karademir, A., et al., 2021. The effect of green biobased binder on structural, mechanical, liquid absorption and wetting properties of coated papers. Clean. Eng. Technol. 5, 100274. https://doi.org/10.1016/j.clet.2021.100274.

Altay, B.N., Carver-Kubik, A., Williams, S., Huq, A., Sugiyama, M., Yash, Dhote, et al., 2024. Starch-based nanoparticles as a replacement for synthetic latex: a comprehensive assessment of printability and colorimetric characteristics. Color Res. Appl. 49 (6). https://doi.org/10.1002/col.22941.

Atteya, Y.M., Aly, M.F., Farag, M.M., 2025. TPS and natural composites elements, techniques, challenges, and applications, 2590048X Results Mater. 28. https://doi.org/10.1016/j.rinma.2025.100816.

Azimi, N.R., Sonchaeng, U., Leelaphiwat, P., Tongdeesoontorn, W., Bumbudsanpharoke, N., 2025. Comparative study of packaging performance: thermoplastic starch blends and polyester blends for biodegradable personal care packaging. Int. J. Biol. Macromol. 320 (1), 145739.

Basiak, E., Lenart, A., Debeaufort, F., 2017. Effect of starch type on the physico-chemical properties of edible films, 18790003 Int. J. Biol. Macromol. 98, 348–356. https://doi.org/10.1016/j.ijbiomac.2017.01.122.

Cao, P., Song, Y., Jin, D., Zhang, Y., Zhang, P., Tang, K., Li, Y., Wang, R., Zhang, K., 2025. Preparation and characterization of a starch-based active packaging with pH-sensitive release via tea tree essential oil loaded in a covalent organic framework. Food Chem. 486, 144643. https://doi.org/10.1016/j.foodchem.2025.144643, 03088146.

Cheng, Y., He, Y., Xiao, L., Gao, W., Kang, X., Sui, J., Cui, B., 2024. Impact of starch amylose and amylopectin on the rheological and 3D printing properties of corn starch. Int. J. Biol. Macromol. 278 (Pt 1), 134403. https://doi.org/10.1016/j.ijbiomac.2024.134403, 01418130.

Cheviron, P., Gouanvé, F., Espuche, E., 2014. Green synthesis of colloid silver nanoparticles and resulting biodegradable starch/silver nanocomposites, 01448617 Carbohydr. Polym. 108 (1), 291–298. https://doi.org/10.1016/j.carbpol.2014.02.059.

Chysirichote, T., Khanthong, T., Khampanin, N., 2025. Sustainable active packaging: Essential oil-enhanced rice starch film for preserving fried foods. Appl. Food Res. 5 (2), 101419. https://doi.org/10.1016/j.afres.2025.101419, 27725022.

De Arcangelis, E., Trivisonno, M.C., Angelicola, M., Quiquero, M., Di Nardo, V., Falasca, L., Sestili, F., Messia, M.C., Marconi, E., 2021. Milling and rheological properties of high amylose wheat, 10959963 J. Cereal. Sci. 102. https://doi.org/10.1016/j.jcs.2021.103335.

de Sousa, A.S.B., Lima, R.P., da Silva, M.C.A, Moreira, D.N., Pintado, M.M.E., de Melo Silva, S., 2022. Natural deep eutectic solvent of choline chloride with oxalic or ascorbic acids as efficient starch-based film plasticizers. Polymer 259. https://doi.org/10.1016/j.polymer.2022.125314.

Deng, W., Yibin, L.I., Wu, L., Chen, S., 2022. Pickering emulsions stabilized by polysaccharides particles and their applications: A review. Food Sci. Technol. 42. https://doi.org/10.1590/fst.24722, 1678-457X.

Du, Y., Liu, J., Wang, J., Wang, B., Li, H., Su, Y., 2017. Starch-based bio-latex redistribution during paper coating consolidation. Prog. Org. Coat. 106, 155–162. https://doi.org/10.1016/j.porgcoat.2017.02.016.

Dutta, D., Sit, N., 2024. Preparation and characterization of potato starch-based composite films reinforced by modified banana fibers and its application in packaging of grapes. Int. J. Biol. Macromol. 254 (Pt 2), 127791. https://doi.org/10.1016/j.ijbiomac.2023.127791, 01418130.

Edo, G.I., Mafe, A.N., Ali, A.B.M., Iwanegbe, I., Yousif, E., Oghroro, E.E.A., Igbuku, U. A., Zainulabdeen, K., Owheruo, J.O., Essaghah, A.E.A., Umar, H., Ahmed, D.S., Alamiery, A.A., Aliyu, M.R., 2025. Evaluating the environmental impact of biopolymer-based packaging compared to conventional plastics in the food industry, 29503574 Sustain. Chem. One World 8. https://doi.org/10.1016/j.scowo.2025.100145.

Figueroa-Lopez, K.J., Villabona-Ortíz, Á., Ortega-Toro, R., 2024. Sustainable starch-based films from cereals and tubers: A comparative study on cherry tomato preservation. Polymers 16 (20), 2913. https://doi.org/10.3390/polym16202913, 2073-4360.

Gagnaten, M., Caceres, S.G., Osuna, I.A.R., Olaiz, N.M., Schebor, C., Leiva, G.E., 2023. Modification of cassava starch by acetylation and pulsed electric field technology: analysis of physical and functional properties. Innov. Food Sci. Emerg. Technol. 85, 103344.

Garavito, J., Peña-Venegas, C.P., Castellanos, D.A., 2024. Production of starch-based flexible food packaging in developing countries: analysis of the processes, challenges, and requirements. Foods 13 (24), 4096. https://doi.org/10.3390/foods13244096, 4096.

Geyer, R., Jambeck, J.R., Law, K.L., 2017. Production, use, and fate of all plastics ever made, 23752548 Sci. Adv. 3 (7). https://doi.org/10.1126/sciadv.1700782.

Glaser, S.J., Faisal, M., Blennow, A., Lo Leggio, L., 2026. Additives as external plasticizers in starch-based bioplastics - towards improving the properties of food packaging: A review. Int. J. Biol. Macromol. 335 (Pt 2), 148754. https://doi.org/10.1016/j.ijbiomac.2025.148754, 01418130.

Govindaraju, I., Zhuo, G.Y., Chakraborty, I., Melanthota, S.K., Mal, S.S., Sarmah, B., Baruah, V.J., Mahato, K.K., Mazumder, N., 2022. Investigation of structural and physico-chemical properties of rice starch with varied amylose content: A combined microscopy, spectroscopy, and thermal study, 0268005X Food Hydrocoll. 122. https://doi.org/10.1016/j.foodhyd.2021.107093.

Grazi, A., Nascimento, A., Almeida, R., Costa, M.E., Sousa, S., Lisboa, H.M., 2025. First-principles modeling and optimization of the dip-coating process for starch-glycerol edible films using viscosity measurements, 18734359 Colloids Surf. A Physicochem. Eng. Asp. 716. https://doi.org/10.1016/j.colsurfa.2025.136712.

Gupta, R.K., Guha, P., Prakash Srivastav, P., 2024. Investigating the toxicological effects of nanomaterials in food packaging associated with human health and the environment, 26669110 J. Hazard. Mater. Lett. 5, 100125. https://doi.org/10.1016/j.hazl.2024.100125.

Hussain, S., Akhter, R., Maktedar, S.S., 2024. Advancements in sustainable food packaging: from eco-friendly materials to innovative technologies. Sustain. Food Technol. 2 (5), 1297–1364. https://doi.org/10.1039/d4fb00084f.

Jabeen, N., Majid, I., Ahmad Nayik, G., Yildiz, F., 2015. Bioplastics and food packaging: A review. Cogent Food Agric. 1 (1), 1117749. https://doi.org/10.1080/23311932.2015.1117749, 2331-1932.

Jafarzadeh, S., Yildiz, Z., Yildiz, P., Strachowski, P., Forough, M., Esmaeili, Y., Naebe, M., Abdollahi, M., 2024. Advanced technologies in biodegradable packaging using intelligent sensing to fight food waste. Int. J. Biol. Macromol. 261 (Pt 1), 129647. https://doi.org/10.1016/j.ijbiomac.2024.129647, 01418130.

Jamalpoor, Z., Ahmadi, H., Abdous, M., Rahdar, A., 2024. CMC/Starch/SiO2/GQDs nanoemulsion for targeted delivery of 5-fluorouracil. J. Mol. Liq. 408, 125407. https://doi.org/10.1016/j.molliq.2024.125407, 01677322.

Jansson, A., Thuvander, F., 2004. Influence of thickness on the mechanical properties for starch films. Carbohydr. Polym. 56 (4), 499–503. https://doi.org/10.1016/j.carbpol.2004.03.019, 01448617.

Jha, P., Dharmalingam, K., Nishizu, T., Katsuno, N., Anandalakshmi, R., 2020. Effect of amylose–amylopectin ratios on physical, mechanical, and thermal properties of starch-based bionanocomposite films incorporated with CMC and nanoclay. Starch - Stärke 72 (1–2). https://doi.org/10.1002/star.201900121, 0038-9056.

Jiang, T., Duan, Q., Zhu, J., Liu, H., Yu, L., 2020. Starch-based biodegradable materials: Challenges and opportunities, 25425048 Adv. Ind. Eng. Polym. Res. 3 (1), 8–18. https://doi.org/10.1016/j.aiepr.2019.11.003.

Jiang, G., Zhang, W., Fan, L., Luo, C., Leng, P., Tao, D., 2025. One-step strategy for preparing renewable polylactic acid (PLA)/thermoplastic starch (TPS) composite filaments for 3D printing. Ind. Crop. Prod. 233, 121444. https://doi.org/10.1016/j.indcrop.2025.121444, 09266690.

Joshi, M., Aldred, P., McKnight, S., Panozzo, J.F., Kasapis, S., Adhikari, R., Adhikari, B., 2013. Physicochemical and functional characteristics of lentil starch. Carbohydr. Polym. 92 (2), 1484–1496. https://doi.org/10.1016/j.carbpol.2012.10.035, 01448617.

Karnwal, A., Rauf, A., Jassim, A.Y., Selvaraj, M., Al-Tawaha, A.R.M.S., Kashyap, P., Kumar, D., Malik, T., 2025. Advanced starch-based films for food packaging: Innovations in sustainability and functional properties. Food Chem. X 29, 102662. https://doi.org/10.1016/j.fochx.2025.102662, 25901575.

Kegl, T., Fan, Y.V., Čuček, L., 2025. The role of biodegradable plastics in a sustainable economy: progress and challenges. Clean. Eng. Technol. 30, 101129. https://doi.org/10.1016/j.clet.2025.101129, 101129.

Kumar, A., Kumar, A., Shuai, W., Chopra, S., Gaur Rudra, S., Rabbani, A., 2025a. Biodegradable and smart packaging films for food quality and safety: A review. Appl. Food Res. 5 (2), 101491. https://doi.org/10.1016/j.afres.2025.101491, 27725022.

Kumar, N., Ali, S.A., Kundu, S., 2025b. Highly flexible and optically active starch-based nanocomposite film: Effects of organo-capped gold nanoparticles. Mater. Today Commun. 49, 113911. https://doi.org/10.1016/j.mtcomm.2025.113911, 23524928.

Kumar, A., Saranyadevi, S., Selva Kumar, T., Shankar, N., Pawde, S.V., Liu, S., Ali, S., Wei, S., 2026. Starch based biodegradable packaging systems and their interactions with food components, shelf-life implications. A review. Carbohydr. Polym. Technol. Appl. 13, 101067. https://doi.org/10.1016/j.carpta.2025.101067.

Kurup, G., Fadzillah, M.F.F.B.M., Royan, N.R.R., Radzuan, N.A.M., Sulong, A.B., 2025. Impact of processing parameters on the compatibility and performance of PLA/tapioca starch biocomposites for short-term food packaging applications, 23524928 Mater. Today Commun. 43. https://doi.org/10.1016/j.mtcomm.2025.111651.

Lan, G., Xie, S., Duan, Q., Huang, W., Huang, W., Zhou, J., Chen, P., Xie, F., 2024. Effect of soybean polysaccharide and soybean oil on gelatinization and retrogradation properties of corn starch. Int. J. Biol. Macromol. 264 (2), 130772. https://doi.org/10.1016/j.ijbiomac.2024.130772.

Lara, S.C., Salcedo, F., 2016. Gelatinization and retrogradation phenomena in starch/montmorillonite nanocomposites plasticized with different glycerol/water ratios. Carbohydr. Polym. 151 (3), 206–212. https://doi.org/10.1016/j.carbpol.2016.05.065.

Li, H., Qi, Y., Zhao, Y., Chi, J., Cheng, S., 2019. Starch and its derivatives for paper coatings: a review. Prog. Org. Coat. 135, 213–227. https://doi.org/10.1016/j.porgcoat.2019.05.015.

Li, M., Zou, Y., Yao, Q., Zhao, Y., Zhang, J., Zhang, Q., Huang, Y., Liu, F., Zeng, X.A., 2026. Cinnamaldehyde-enhanced starch/chitosan composite films: A promising solution for eco-friendly food packaging and postharvest quality control, 18790003 Int. J. Biol. Macromol. 337. https://doi.org/10.1016/j.ijbiomac.2025.149567.

Lie-Piang, A., Braconi, N., Boom, R.M., 2021. Albert van der Padt, Less refined ingredients have lower environmental impact – A life cycle assessment of protein-rich ingredients from oil- and starch-bearing crops. J. Clean. Prod. 292, 126046. https://doi.org/10.1016/j.jclepro.2021.126046, 09596526.

Liu, H., Yu, L., Xie, F., Chen, L., 2006. Gelatinization of cornstarch with different amylose/amylopectin content. Carbohydr. Polym. 65 (3), 357–363. https://doi.org/10.1016/j.carbpol.2006.01.026, 01448617.

Liu, X., Liu, Y., Fan, M., Qian, K., Ma, W., Li, D., Li, L., 2025. Development of antibacterial starch-based PLA/PBAT active packaging films for enhanced beef preservation. Food Chem. 493 (Pt 2), 145804. https://doi.org/10.1016/j.foodchem.2025.145804, 03088146.

Luo, Q., Ouyang, J., Zhan, L., Liang, G., Wang, X., 2025. Development and characterization of capsaicin-enriched Dialdehyde starch-PVA films for antimicrobial food packaging. Int. J. Biol. Macromol. 330 (Pt 1), 147918. https://doi.org/10.1016/j.ijbiomac.2025.147918, 01418130.

Lv, H., He, D., Xu, E., Jin, Z., Zhao, H., Yuan, C., Zhao, M., Gao, W., Li, J., Liu, Y., Wu, Z., Liu, P., Cui, B., 2025. Starch-based biodegradable active intelligent packaging with color superimposition via emulsion electrospinning nanofiber for meat freshness monitoring and shelf-life extension. Carbohydr. Polym. 367, 23940. https://doi.org/10.1016/j.carbpol.2025.123940.

Lyu, J.S., Han, J., 2024. Fabrication of bio-inspired carbon nanodot-corn starch nanocomposite films via extrusion process for sustainable active food packaging applications. Carbohydr. Polym. 343, 122502. https://doi.org/10.1016/j.carbpol.2024.122502, 01448617.

Mahmud, MdZ.Al, Hosne Mobarak, Md, Hossain, N., 2024. Emerging trends in biomaterials for sustainable food packaging: A comprehensive review, 24058440 Heliyon 10 (1), e24122. https://doi.org/10.1016/j.heliyon.2024.e24122.

Marta, H., Wijaya, C., Sukri, N., Cahyana, Y., Mohammad, M., 2022. A comprehensive study on starch nanoparticle potential as a reinforcing material in bioplastic. Polymers 14 (22), 4875. https://doi.org/10.3390/polym14224875, 2073-4360.

Medha, T., Rajalechumi, K.A., Reddy, C.K., Surekha, C., Haripriya, S., Jonnalagadda, S. B., Katari, N.K., 2026. Dual physical modification of elephant foot yam starch: Effects on physicochemical, structural, functional, and digestibility properties, 27725022 Appl. Food Res. 6 (1). https://doi.org/10.1016/j.afres.2025.101616.

Mujtaba, M., Lipponen, J., Ojanen, M., Puttonen, S., Vaittinen, H., 2022. Trends and challenges in the development of bio-based barrier coating materials for paper/cardboard food packaging; a review. Sci. Total Environ. 851, 158328. https://doi.org/10.1016/j.scitotenv.2022.158328.

Murawski, T.T., Szałapak, J., Tryznowski, M., Świtała, M., Grzęda, D., Bednarczyk, E., Ryszkowska, J., Żołek-Tryznowska, Z., 2025. Long-term stability and properties of potato starch films plasticized with sugar–glycerol mixtures. Int. J. Biol. Macromol. 331, 148444.

Mylkie, K., Chełminiak-Dudkiewicz, D., Młynarczyk, D.T., Smolarkiewicz-Wyczachowski, A., Ziegler-Borowska, M., 2025. Starch films cross-linked with boronated chitosan enriched with cannabis oil as food packaging material. Polym. Test. 150, 108910. https://doi.org/10.1016/j.polymertesting.2025.108910, 01429418.

Namjoufar, M., Wei, Y., Maloney, K.J.-R., Laghaei, M., Patadiya, J., Hadigheh, A.S., et al., 2026. Overcoming technical barriers in mechanical recycling of plastic waste: the role of engineered additives. J. Environ. Chem. Eng. 14 (2), 121559.

Oliver, L., Meinders, M.B.J., 2011. Dynamic water vapour sorption in gluten and starch films. J. Cereal Sci. 54 (3), 409–416. https://doi.org/10.1016/j.jcs.2011.08.005.

Ortega, F., Minnaard, J., Arce, V.B., García, M.A., 2025. Biodegradable starch-based nanocomposite films with laser-synthesized silver nanoparticles: A materials approach for packaging, 18790003 Int. J. Biol. Macromol. 318. https://doi.org/10.1016/j.ijbiomac.2025.144904.

Orue, A., Corcuera, M.A., Peña, C., Eceiza, A., Arbelaiz, A., 2016. Bionanocomposites based on thermoplastic starch and cellulose nanofibers, 15307980 J. Thermoplast. Compos. Mater. 29 (6), 817–832. https://doi.org/10.1177/0892705714536424.

Othman, S.H., Syafiqah Kahar, N., Nordin, N., Ruzanna, A.Shapi'i, 2023. Properties and food packaging applications of solvent casting-made starch-based films incorporated with essential oils: a review. Starch - Stärke 75 (5–6). https://doi.org/10.1002/star.202200152, 0038-9056.

Panda, P., Mohapatra, R., 2025. Development of ZnO–resistant starch prebiotic matrix conjugate for metronidazole delivery for enhanced probiotic viability and antimicrobial activity. Int. J. Biol. Macromol. 329 (Pt 2), 147904. https://doi.org/10.1016/j.ijbiomac.2025.147904, 01418130.

Pawase, P.A., Pathare, A.M., Bashir, O., Saleem, F., Shrama, E., Mudgal, S., Ahmad, M., 2025. Physical, chemical, and nano-enabled modifications of starch for sustainable food packaging films: Recent trends, challenges, and prospects, 26668939 Carbohydr. Polym. Technol. Appl. 11. https://doi.org/10.1016/j.carpta.2025.100986. Available from: https://www.sciencedirect.com/science/journal/26668939.

Pei, J., Palanisamy, C.P., Srinivasan, G.P., Panagal, M., Kumar, S.S.D., Mironescu, M., 2024. A comprehensive review on starch-based sustainable edible films loaded with bioactive components for food packaging, 18790003 Int. J. Biol. Macromol. 274. https://doi.org/10.1016/j.ijbiomac.2024.133332.

Plaeyao, K., Talodthaisong, C., Yingyuen, W., Kaewbundit, R., Tun, W.S.T., Saenchoopa, A., Kayunkid, N., Wiwattananukul, R., Sakulsombat, M., Kulchat, S., 2025. Biodegradable antibacterial food packaging based on carboxymethyl cellulose from sugarcane bagasse/cassava starch/chitosan/gingerol extract stabilized silver nanoparticles (Gin-AgNPs) and vanillin as cross-linking agent, 18737072 Food Chem. 466. https://doi.org/10.1016/j.foodchem.2024.142102.

Poveda-Giraldo, J.A., Yoo, S., Yoo, S., Kim, I., Lan, K., Venditti, R., Park, S., 2025. Environmental life cycle assessment of renewable starch alternative plastic: a comparative analysis of end-of-life scenarios in the republic of Korea. J. Clean. Prod. 520, 146178. https://doi.org/10.1016/j.jclepro.2025.146178, 09596526.

Priyanka, S., Namasivayam, S.K.R., Kennedy, J.F., Moovendhan, M., 2024. Starch-chitosan-Taro mucilage nanocomposite active food packaging film doped with zinc oxide nanoparticles–Fabrication, mechanical properties, anti-bacterial activity and eco toxicity assessment. Int. J. Biol. Macromol. 277, 134319. https://doi.org/10.1016/j.ijbiomac.2024.134319.

Rabelo, L.S., Tanaka, F.C.S., Aouada, F.A., de Moura, R., 2025. The future of sustainable packaging: exploring biodegradable solutions through extrusion, thermo-expansion, 3D printing and supercritical fluid from Agro-industry waste. Foods 14 (23), 4027, 4027.

Rahardiyan, D., Moko, E.M., Tan, J.S., Lee, C.K., 2023. Thermoplastic starch (TPS) bioplastic, the green solution for single-use petroleum plastic food packaging – A review. Enzym. Microb. Technol. 168, 110260. https://doi.org/10.1016/j.enzmictec.2023.110260, 01410229.

Ramakrishnan, R., Vadivu Kulandhaivelu, S., Roy, S., Viswanathan, V.P., 2023. Characterisation of ternary blend film of alginate/carboxymethyl cellulose/starch for packaging applications. Ind. Crop. Prod. 193, 116114. https://doi.org/10.1016/j.indcrop.2022.116114, 09266690.

Requena, R., Jiménez, A., Vargas, M., Chiralt, A., 2016. Effect of plasticizers on thermal and physical properties of compression-moulded poly[(3-hydroxybutyrate)-co-(3-hydroxyvalerate)] films. Polym. Test. 56, 45–53. https://doi.org/10.1016/j.polymertesting.2016.09.022.

Rosas, A.L.G., Kleinübing, N.R., Bierhals, C.K., da Fonseca Antunes, B., Carloto, A.F., Silveira, D.R., da Silva, W.P., da Rosa Zavareze, E., Lopes, G.V., Meinhart, A.D., 2026. Cryogels based on Brazilian purple yam (Dioscorea trifida L.f.) starch: An antimicrobial active packaging approach for microbial and moisture control in fresh cheese, 18790003 Int. J. Biol. Macromol. 335. https://doi.org/10.1016/j.ijbiomac.2025.149281.

Saleh, H.M., 2025. Preparation and characterization of chitosan/starch/Ag@TiO2-NPs bio-nanocomposite. J. Indian Chem. Soc. 102 (10), 102022. https://doi.org/10.1016/j.jics.2025.102022, 00194522.

Santhosh, R., Ahmed, J., Thakur, R., Sarkar, P., 2024. Starch-based edible packaging: Rheological, thermal, mechanical, microstructural, and barrier properties - a review, 27538095 Sustain. Food Technol. 2 (2), 307–330. https://doi.org/10.1039/d3fb00211j.

Shah, U., Gani, A., Ahmad Ashwar, B., Shah, A., Ahmad, M., Gani, A., Wani, I.A., Masoodi, F.A., Yildiz, F., 2015. A review of the recent advances in starch as active and nanocomposite packaging films. Cogent Food Agric. 1 (1), 1115640. https://doi.org/10.1080/23311932.2015.1115640, 2331-1932.

Shah, U., Naqash, F., Gani, A., Masoodi, F.A., 2016. Art and science behind modified starch edible films and coatings: A review, 15414337 Compr. Rev. Food Sci. Food Saf. 15 (3), 568–580. https://doi.org/10.1111/1541-4337.12197.

Shah, M., Rajhans, S., Pandya, H.A., Mankad, A.U., 2021. Bioplastic for future: A review then and now. World J. Adv. Res. Rev. 9 (2), 056–067. https://doi.org/10.30574/wjarr.2021.9.2.0054, 25819615.

Shanbhag, C., Shenoy, R., Shetty, P., Srinivasulu, M., Nayak, R., 2023. Formulation and characterization of starch-based novel biodegradable edible films for food packaging. J. Food Sci. Technol. 60 (11), 2858–2867. https://doi.org/10.1007/s13197-023-05803-2, 0022-1155.

Shen, M., Huang, W., Chen, M., Song, B., Zeng, G., Zhang, Y., 2020. (Micro)plastic crisis: Un-ignorable contribution to global greenhouse gas emissions and climate change. J. Clean. Prod. 254, 120138. https://doi.org/10.1016/j.jclepro.2020.120138, 09596526.

Siciliano, S., Lopresto, C.G., Mileti, O., De Paola, M.G., 2025. Development of a bilayer starch/chitosan film for food packaging applications, 22124306 Food Biosci. 73. https://doi.org/10.1016/j.fbio.2025.107652.

Song, E.-H., Shang, J., Ratner, D.M., 2012. Polysaccharides, 9. Elsevier BV, pp. 137–155. https://doi.org/10.1016/b978-0-444-53349-4.00246-6.

Srinivasan, S., Ge, C., Lewis, C.L., Begum, S.S., Samui, A.B., Noyes, H.F., 2025. Effect of chemical modification of corn starch for the development of films for packaging applications: Impact of glutaraldehyde and organically modified montmorillonite incorporation, 01448617 Carbohydr. Polym. 367. https://doi.org/10.1016/j.carbpol.2025.124019.

Su, M., Qin, H., Tang, Q., Peng, D., Li, H., Zou, Z., 2024. Colorimetric ammonia-sensing nanocomposite films based on starch/sodium alginate and Cu-Phe nanorods for smart packaging application. Int. J. Biol. Macromol. 282 (Pt 6), 137470. https://doi.org/10.1016/j.ijbiomac.2024.137470, 01418130.

Suhag, R., Kumar, N., Trajkovska Petkoska, A., Upadhyay, A., 2020. Film formation and deposition methods of edible coating on food products: A review. Food Res. Int. 136, 109582. https://doi.org/10.1016/j.foodres.2020.109582, 09639969.

Surendren, A., Mohanty, A.K., Liu, Q., Misra, M., 2022. A review of biodegradable thermoplastic starches, their blends and composites: recent developments and opportunities for single-use plastic packaging alternatives. Green Chem. 24 (22), 8606–8636. https://doi.org/10.1039/d2gc02169b.

Tanwar, R., Gupta, V., Kumar, P., Kumar, A., Singh, S., Gaikwad, K.K., 2021. Development and characterization of PVA-starch incorporated with coconut shell extract and sepiolite clay as an antioxidant film for active food packaging applications, 18790003 Int. J. Biol. Macromol. 185, 451–461. https://doi.org/10.1016/j.ijbiomac.2021.06.179.

Teseme, W.B., Habtegebreil, S.A., Tolesa, G.N., 2025. Study on the selected engineering properties of anchote (Coccinia abyssinia) starch based biodegradable film for food packaging. Appl. Food Res. 5 (2), 101282. https://doi.org/10.1016/j.afres.2025.101282, 27725022.

Tutek, K., Kosmalska-Olańska, A., Masek, A., 2025. Modification of thermoplastic starch (TPS) with PLA and PCL via ring-opening polymerization (ROP): Comparative study of water solution and solvent-free approaches. Ind. Crop. Prod. 235, 121785. https://doi.org/10.1016/j.indcrop.2025.121785, 09266690.

Usman, I., Sana, S., Jaffar, H.M., Munir, M., Afzal, A., Sukhera, S., Boateng, I.D., Afzaal, M., Urugo, M.M., 2025. Recent progress in edible films and coatings: Toward green and sustainable food packaging technologies, 27725022 Appl. Food Res. 5 (2). https://doi.org/10.1016/j.afres.2025.101070.

Vatanpour, V., Bijari, M., Sadiksoz, B., Gul, B.Y., Koyuncu, I., 2026. Starch as an eco-friendly and sustainable option for separation membranes: A review of current status and future directions, 01448617 Carbohydr. Polym. 371. https://doi.org/10.1016/j.carbpol.2025.124475.

Wan, W., Xiong, D., Zhang, Q., Wang, J., 2019. Miscible–Immiscible Transition of Deep Eutectic Solvents in Water Switched by CO2. ACS Sustain. Chem. Eng. 7, 17882–17887. https://doi.org/10.1021/acssuschemeng.9b04430.

Wang, H., Yang, L., Yang, Y., Zhang, D., Hao, G., 2025a. Multifunctional natural starch-based hydrogels: Critical characteristics, formation mechanisms, various applications, future perspectives. Carbohydr. Polym. 357, 123458. https://doi.org/10.1016/j.carbpol.2025.123458, 01448617.

Wang, J., Si, Y., Liu, Y., Zhang, X., Xu, J., Xu, H., Kang, Z., Kong, L., Li, S., 2025b. Loading of tea polyphenols into electrospun starch-based nanofibrous films for active package of pre-processed fish fillets. Int. J. Biol. Macromol. 319 (Pt 2), 145370. https://doi.org/10.1016/j.ijbiomac.2025.145370, 01418130.

Wongphan, P., Paiva, R., Nerín, C., Harnkarnsujarit, N., 2024. Sodium erythorbate and hexametaphosphate loaded TPS/PBAT films: A multifunctional packaging for extending beef shelf-life. Food Packag. Shelf Life 46, 101396. https://doi.org/10.1016/j.fpsl.2024.101396, 22142894.

Xia, Y., Wang, S., Meng, F., Xu, Z., Qi, F., Gu, Z., Zhang, C., Peng, L., Kong, F., 2024. Eco-friendly food packaging based on paper coated with a bio-based antibacterial coating composed of carbamate starch, calcium lignosulfonate, cellulose nanofibrils, and silver nanoparticles. Int. J. Biol. Macromol. 254 (Pt 1), 127659. https://doi.org/10.1016/j.ijbiomac.2023.127659, 01418130.

Xu, H., Chen, L., McClements, D.J., Cheng, H., Long, J., Jin, Z., 2023. Development and characterization of active starch-based films incorporating graphene/polydopamine/Cu2+ nanocomposite fillers, 01448617 Carbohydr. Polym. 305. https://doi.org/10.1016/j.carbpol.2022.120498.

Yao, X., Liu, J., Hu, H., Yun, D., Liu, J., 2022. Development and comparison of different polysaccharide/PVA-based active/intelligent packaging films containing red pitaya betacyanins. Food Hydrocoll. 124, 107305. https://doi.org/10.1016/j.foodhyd.2021.107305, 0268005X.

Ye, S.J., Choi, H.W., Baik, M.Y., 2026. A new physical starch modification method: Pressure annealing treatment (PAT), 18790003 Int. J. Biol. Macromol. 338. https://doi.org/10.1016/j.ijbiomac.2025.149757.

Yeo, J.C.C., Kinyanjui Muiruri, J., Fei, X., Wang, T., Zhang, X., Xiao, Y., Thitsartarn, W., Tanoto, H., He, C., Li, Z., 2024. Innovative biomaterials for food packaging: Unlocking the potential of polyhydroxyalkanoate (PHA) biopolymers. Biomater. Adv. 163, 213929. https://doi.org/10.1016/j.bioadv.2024.213929, 27729508.

Yermagambetova, A.D., Tazhibayeva, S.M., Tyussyupova, B.B., Musabekov, K.B., Pastorino, L., 2024. Effect of plasticizers on the rheological properties of xanthan gum--starch biodegradable films. Heliyon 10, e34550. https://doi.org/10.1016/j.heliyon.2024.e34550.

Yin, Y., Wai Woo, M., 2024. Transitioning of Petroleum-based plastics food packaging to sustainable bio-based alternatives. Sustain. Food Technol. 2. https://doi.org/10.1039/D4FB00028E.

Yuan, X., Wei, F., Wang, Z., Li, X., Deng, Y., Shi, X., Lin, B., 2025. Transparent, mechanically robust and antibacterial food packaging film based on starch/PVA/citric acid/carboxycellulose nanofibers by one-pot green process. Int. J. Biol. Macromol. 302, 140573. https://doi.org/10.1016/j.ijbiomac.2025.140573, 01418130.

Zamanidehyaghoubi, G., Shahidi, F., Edalatian Dovom, M.R., Mohebbi, R., Roshanak, S., 2025. Investigating the effect of three different types and concentrations of plasticizers on physico-mechanical properties of pullulan food-packaging films. Food Packag. Shelf Life 51, 101590.

Zdanowicz, M., Salasinska, K., 2023. Characterization of thermoplastic starch plasticized with ternary urea-polyols deep eutectic solvent with two selected fillers: microcrystalline cellulose and montmorillonite. Polymers 15 (4), 972. https://doi.org/10.3390/polym15040972.

Zhang, S., Li, X., Gao, B., Zhang, S., 2024. Reactive extrusion fabrication of thermoplastic starch with Ca2+ heterodentate coordination structure for harvesting multiple-reusable PBAT/TPS films. Carbohydr. Polym. 339, 122240. https://doi.org/10.1016/j.carbpol.2024.122240, 01448617.

Zhang, H., Wu, H., Thian, E.S., 2025. Design of microcapsule-enhanced composite starch films for antibacterial inner food packaging, 22142894 Food Packag. Shelf Life 52. https://doi.org/10.1016/j.fpsl.2025.101661.

Zhang, Q., Li, X., Jin, Z., Svensson, B., Bai, Y., 2026. Enzymatic-chemical modification of cassava starch as a sauce thickener to regulate the balance between processing and edible quality. Carbohydr. Polym. 375, 124768. https://doi.org/10.1016/j.carbpol.2025.124768, 01448617.

Zhu, F., 2025. Starch based films and coatings for food packaging: Interactions with phenolic compounds. Food Res. Int. 204, 115758. https://doi.org/10.1016/j.foodres.2025.115758, 09639969.

Further reading

Abe, M.M., Martins, J.R., Sanvezzo, P.B., Macedo, J.V., Branciforti, M.C., Halley, P., Botaro, V.R., Brienzo, M., 2021. Advantages and disadvantages of bioplastics production from starch and lignocellulosic components. Polymers 13 (15), 2484. https://doi.org/10.3390/polym13152484.

Jamali, A.R., El-Reash, Y.G.A., Chandio, A.D., Khan, W., Abutalib, M., Goda, M.N., Ferjani, H., Hosseini-Bandegharaei, A., Koheil, H., Keshta, B.E., 2025. Starch-based Al2O3 nanocomposite films for enhanced shelf-life and sustainable food packaging. Results Eng. 28, 107532. https://doi.org/10.1016/j.rineng.2025.107532.

Liu, Y., Wang, H., Deng, H., Min, T., Wang, H., Yi, Y., Song, Z., Zhou, M., 2026. Multifunctional nano-enabled films and coatings for food packaging. A review of advances in active protection and intelligent monitoring. Food Res. Int. 226, 118138. https://doi.org/10.1016/j.foodres.2025.118138.

Maite, G., Sara, G.C., Rodríguez, I.A., Olaiz, N.M., Schebor, C., Leiva, G.E., 2023. Modification of cassava starch by acetylation and pulsed electric field technology. Analysis of physical and functional properties. Innov. Food Sci. Emerg. Technol. 85, 03344–103344. https://doi.org/10.1016/j.ifset.2023.103344.

Natalia, G., Melián, E., Acebal, C.C., Domini, C.E., Faccio, R., López, O.V., 2025. Harnessing antioxidant-rich byproducts from olive and wine industries to produce corn starch films as edible coatings for minimally processed foods. Food Biosci. 69, 106802. https://doi.org/10.1016/j.fbio.2025.106802.

Vilpoux, O.F., Francisco, J., 2023. Global production and use of starch. Elsevier eBooks 1, 43–66. https://doi.org/10.1016/b978-0-323-90058-4.00014-1.

Von Mühlen, M., Mörschbächer, A., Vanessa, L., Ademir, J.Z., Borsoi, C., Catto, A.L., 2025. Evaluating the properties of starch/chitosan films with the incorporation of various nanoclays for use in food packaging. Int. J. Biol. Macromol. 298. https://doi.org/10.1016/j.ijbiomac.2025.140054, 140054–140054.

Zia, ud D., Chen, L., Xiong, H., Wang, Z., Ullah, I., Lei, W., Shi, D., Alam, M., Ullah, H., Khan, S., 2020. Starch. An undisputed potential candidate and sustainable resource for the development of wood adhesive. Starch 72 (3–4). https://doi.org/10.1002/star.201900276, 1900276–1900276.

Index

Note: 'Page numbers followed by *f* indicate figures, *t* indicate tables.'

P

T

Printed and bound by CPI Group (UK) Ltd, Croydon, CR0 4YY
17/09/2026
02218538-0003